BIODIVERSITY

AND THE

ECOSYSTEM APPROACH

IN AGRICULTURE,

FORESTRY

AND FISHERIES

FAO INTER-DEPARTMENTAL
WORKING GROUP ON
**BIOLOGICAL
DIVERSITY**
FOR FOOD AND AGRICULTURE

ISBN 92-5-104917-3

F O R E W O R D

The Satellite Event held during the weekend of 12-13 October 2002 and reported in these Proceedings examined agriculture, fisheries and forestry using ecosystems approaches. The event was organized by FAO's Inter-Departmental Working Group on Biological Diversity for Food and Agriculture.

FAO's Biodiversity Programme turns a common argument on its head. It is often claimed that agriculture is the world's greatest threat to biodiversity. But at the same time over 40 percent of the land surface of the world is covered by agriculture (including pastures, rangelands, inland fisheries and managed forests). Ecological studies of agro-ecosystems reveal the same functional groups of species and essential ecosystem processes found in natural ecosystems. Agriculture depends on ecosystem services delivered through agro-biodiversity and simultaneously delivers ecosystem services (mostly through non-market channels) to wider environments. Careless management or bad policy incentives can threaten biodiversity and those services, but agriculture nonetheless represents the world's largest opportunity for ecological learning by the practical managers of ecosystems.

There are well over 500 million farm management units in the world, overwhelmingly found in developing countries. FAO's ongoing field work in over 100 000 rural communities has found that all those farm managers can understand their farms, fields, forests, range-lands and fisheries as ecosystems and thus be able to manage them better. Farmers, even in the poorest and most food-insecure regions of the world, manage genes by their decisions on crop varieties, manage species by their decisions on farm animals and manage ecosystems by their decisions on soils or pollinators. FAO's biodiversity programme includes many applications of the Malawi Principles on the Ecosystem Approach of the Convention on Biodiversity, some of which are highlighted in this volume.

Farmers, fisherfolk, and forest dwellers not only understand and can apply ecosystem approaches in their decision making, but also understand the potential impact of large-scale environmental threats to their livelihoods. Illiterate farmers are still aware of the harmful effects of insecticides on domestic animals, fish and the health of people in their communities and downstream. These rural people are eager to do their part to reduce and mitigate these threats. By valuing and understanding their own experiences using ecological concepts they can connect and help solve environmental problems at the same time that they solve production problems.

FAO's biodiversity programmes apply ecosystem approaches to stimulate community level education and experiential learning by rural people. The same approaches educate national policy makers wishing to fulfill commitments made to environmental treaties while still meeting agricultural production demands. If policy makers like these, during the few days spent each year visiting the countryside, meet farmers in their fields, then those policy makers sharply increase their confidence in agro-ecologically literate farmers. That confidence goes on to strengthen national commitment to deeper policy reforms, to reduce and eliminate perverse subsidies that threaten biodiversity and to support public investment in rural human capital through education, transport and communications.

The complementary processes of ecological education by and of farmers, as well as policy makers, drive FAO's strategy for mainstreaming biodiversity in national agricultural policies and programmes. This strategy grew from the FAO/CBD/Netherlands Technical Workshop on Sustaining Agricultural Biodiversity and Ecosystem Functions (1998) and the Programme of Work on Agricultural Biodiversity, which the Convention on Biological Diversity asked FAO to lead through its COP V/5 decision.

We would like to thank FAO's Priority Area for Interdisciplinary Action (PAIA) on Biological Diversity for Food and Agriculture and the FAO/Netherlands Partnership Programme (FNPP) Agro-Biodiversity Theme for financially supporting this Satellite Event on Biodiversity and the Ecosystem Approach in Agriculture, Forestry and Fisheries. We would also like to acknowledge the endorsements by the PAIA, the FNPP and the Commission on Genetic Resources for Food and Agriculture of FAO's programme on work on biodiversity and ecosystem approaches in food and agriculture. Finally, the intellectual contributions of these bodies, the authors of these papers, the participants in the discussions during the Satellite Event and the rural peoples whose work is documented here continue to inspire and guide the work of FAO's biodiversity programme and are gratefully acknowledged.

Peter Kenmore

Chair

Inter-Departmental Working Group on

Biological Diversity in Food and Agriculture

CONTENTS

CONTENTS

SUMMARY REPORT

BIODIVERSITY AND THE ECOSYSTEM APPROACH IN AGRICULTURE, FORESTRY AND FISHERIES

FAO, ROME, 12-13 OCTOBER 2002

PURPOSE OF THE MEETING

A Satellite Event on Biodiversity and the Ecosystem Approach in Agriculture, Forestry and Fisheries was held on the occasion of the Ninth Regular Session of the Commission on Genetic Resources for Food and Agriculture, in FAO, Rome, 12-13 October 2002. An informal presentation of the Satellite Event's discussions and results was given on 16 October 2002.

The event was organized by the FAO Inter-Departmental Working Group on Biological Diversity for Food and Agriculture[1] in support of the recommendation of the Seventh Session of the Commission on Genetic Resources for Food and Agriculture which mentioned that "countries were encouraged to develop strategies, programmes and plans for agro-biodiversity in conformity with an ecosystem approach". The meeting was supported by FAO's Priority Area for Interdisciplinary Action on Biological Diversity and by the FAO/Netherlands Partnership Programme.

More specifically, the meeting aimed to improve awareness of the importance of biological diversity, through concrete cases from around the world, and to support the integration of the ecosystem approach applied to all types of production systems and policies and within biodiversity programmes and plans.

About 100 participants attended the day and a half event, including country delegations and observers to the Interim Committee of the International Treaty and the Ninth Regular Session of the Commission. The agenda and list of participants are attached.

DISCUSSIONS

Coverage

Presentations and discussions covered grasslands in South Africa, agro-pastoral systems in Nigeria, rice-fish ecosystems in Cambodia, organic agriculture systems in 16 locations and 10 countries (Bangladesh, Brazil, Cuba, Germany, Indonesia, Italy, Mexico, Peru, Spain, South Africa), mahogany forests in Mexico, medicinal plants in India, soil systems in six locations and five countries (Australia, Brazil, India, Mexico and Sahel), apple pollination in the Himalayas, rice ecology in Asia and ingenious agricultural systems in five countries (Asia, French Guyana, Slovakia, Tanzania, Tunisia).

The case studies depicted innovations from over 20 countries and production systems which included some 70 plants (mainly crops) and a number of animals (ranging from honey bees, through fish to livestock) used for food, fibres and medicine. The studies involved universities, research institutions, inter-governmental institutions, government institutions, civil society institutions and farmers' associations. A number of these institutions are FAO partners in the field.

---

[1] Information on the Inter-Departmental Working Group on Biological Diversity and FAO's work in this area is available at: www.fao.org/biodiversity

SUMMARY

Food production ecology

The ecological relationships and intrinsic mechanisms involved in the food web were illustrated for both aquatic and terrestrial environments. More specifically, natural population regulation of pests in rice fields and nutrient cycling in soil ecosystems were illustrated.

Agro-ecological management entails a paradigm shift where agro-ecosystems are designed and managed in a manner that optimizes nutrient and energy flows below as well as above the ground.

Knowledge of ecological processes was shown in these cases to replace either excessive dependence on, or the lack of, external inputs. The understanding of ecological relations and functional groups (rather than of individual species) can unlock new potentials.

Ecological services are productive inputs

The importance of functional bio-diversity in providing goods and services to both the natural and human environments was highlighted. Habitats rich in native flowering plants and free of pesticide use increased pollinators and pollination services in the Himalayas and hence, food security of rural communities.

The conservation of wild habitats is key to the continued harvesting of medicinal and aromatic plants, which enjoy growing consumer demand in both developing and developed countries. Although the bulk of quantities of medicinal herbs come from cultivation, 60-90 percent of medicinal species traded internationally are harvested from the wild.

Ecosystem services (e.g. pollination, predation, soil nutrient cycling) are as valuable as agricultural products in providing public goods. Conserving these ecological processes allows managing agro-ecosystems for improved production and resilience. An ecosystem approach to food production implies both species and area based approaches which analyze and act upon the intrinsic relationships between species and their environment.

Coping with change through diversification and heterogeneity

The survival strategies of farmers and pastoralists include creating the conditions to harvest food, fibre and medicinal products throughout the year. Continued production relies on harnessing biodiversity goods and services within prevailing conditions of both high and low external inputs, in order to optimize the flow of energy and nutrients throughout the food chain. This is achieved by optimizing heterogeneous conditions in order to meet different needs. For example, corralling and manure practices in Nigeria's agro-pastoral systems increased soil heterogeneity in order to create conditions that allowed production under different rainfall regimes. Networking, conflict resolution and contracts between farmers and herders showed the diversity and adaptability of farmers' strategies and institutions to cope with harsh conditions.

In organic agriculture, the need to compensate for the restriction on (or lack of) synthetic inputs result in the establishment of polycultures that use local varieties and breeds resilient to local climatic conditions and resistant to pest and diseases. Conservation through the utilization of adapted genetic resources is the main productive strategy.

Even where managed biodiversity is rare, such as in the case of Mahogany tree species in Mexico and medicinal, aromatic and dye plants in India, local communities' ownership, management and marketing of a species of high ecological and socio-economic value was sustainable. In fact, the presence of 1-2 mahogany trees in a hectare of forest represents a capital that can be nurtured and sustainably exploited, especially when agro-forestry activities were integrated.

Farmers' and pastoralists' decision-making in managing biodiversity for food production, livelihoods and resilience in the face of changing and unforeseen biophysical and socio-economic conditions is key to adaptive management strategies.

Managed ecosystems increase biodiversity

About 40 percent of the world's lands are occupied by agriculture as compared to no more than 12 percent of lands occupied by reserves and protected areas. To be successful, wildlife conservation needs well connected areas and agricultural practices with positive externalities affecting both natural and semi-natural ecosystems.

Well managed agro-ecosystems have demonstrated the potential of agriculture to restore and maintain biodiversity at gene, species and ecosystem levels, through sustainable management of existing systems and restoration of degraded systems.

Markets valorize managed biodiversity

The market demand for organic and speciality products allows farmers and processors to improve household income by using biodiversity and associated ecological services. Certification of organic and other products add value to biodiversity and local economies.

The cases of quinoa and naturally pigmented cotton in Peru or Maya organic chocolate in Mexico added value to threatened genetic resources and traditional systems through processing. Consumer demand for Saraceno grain or Garfagnaga spelt in Italy for their specific medicinal or gastronomic properties allowed the rescue of these threatened species.

Good connections within the food production chain (from production to processing and marketing) ensured economic viability of marginal areas and restoration of under-utilized biodiversity and authoctonous races (e.g. Maremmana cattle in Italy).

Celebrating biodiversity knowledge systems

The main question is how agro-ecosystem complexities could be maintained in a transforming society. Complex and sound agro-ecosystems are the result of co-adaptation and co-evolution between nature and man. Key determinants for biodiversity improvements at all levels are household and community factors, which are location-specific.

Cultural heritage and knowledge systems determine biodiversity and inter-species dynamics. The potential of joint learning processes between farmers, who have a historical perspective of their ecosystems and scientific knowledge of ecology was illustrated in examples such as the use of termites for rehabilitating crusted soils and many organic agriculture projects.

Numerous studies showed that traditional knowledge and new ecological knowledge establish positive pressures, which deserve attention. Rural communities have the capacity to adapt to change. This capacity is amplified when good governance is ensured.

Ownership and participation maintain biodiversity

Access to land and water resources and changes in property boundaries, such as cropland fencing and pastoralism in Nigeria or flooding of fishponds in Cambodia create common property challenges to bio-diversity management.

Community-based management and participatory selection and evaluation of open-pollinated varieties (e.g. Association for Biodynamic Vegetable Plant Breeding in Germany) and old breeds (e.g. of native chicken in South Africa) create common ownership and benefit sharing of biodiversity.

Ecological, economic and social resilience can only be managed by building on local heterogeneity through decentralized decision-making and resource appropriation. It appears that political functions are as important as ecosystem functions.

CONNECTEDNESS

The resilience and stability of agro-ecosystems depend on the quality of connectedness within and among natural and social systems. Too few or too weak connections may marginalize communities or populations and increase their risk of collapse. Too many or too strong connections may amplify local systems shocks (like forest fires or national economic depression in regional or global markets) and cause wider disruption. Recommendations across case studies include increased focus on: ecological connections, economic connections and socio-political connections.

The quality of connectedness should be improved through ecological literacy and good governance in order to equip rural communities with the necessary tools and mechanisms to maintain biodiversity and act upon change.

AGENDA

Saturday 12 October

8:30	Registration
9:30	**Plant and animal diversity and sustainable production**

- Biodiversity and performance of grassland ecosystems in communal and commercial farming systems in South Africa (David Hoare, South Africa)
- Biodiversity management in West African pastoral and agro-pastoral systems: a case study from North-west Nigeria (Irene Hoffmann, FAO)
- Traditional use and availability of aquatic biodiversity in rice-based ecosystems: I. Kompong Thom Province, Cambodia (Peter Balzer, Cambodia)
- Organic agriculture and genetic resources for food and agriculture (Cristina Grandi, Italy and Christina Henatsch, Germany)

11:30	Coffee break
11:45	Discussion
13:00	Lunch break

14:30	**Biological diversity and wild and semi-wild ecosystems**

- Effectiveness of biodiversity conservation (Doug Williamson, FAO)
- Conservation and use of mahogany in forest ecosystems in Mexico (Patino Fenando, FAO)
- Impact of cultivation and gathering on biodiversity of medicinal plants: global trends and issues and the case of India (Kampalappa Ramakrishnappa, India and Rainer Krell, FAO)

16:00	Coffee break
16:15	Discussion
17:30	End of first day

Sunday 13 October

9:00	**Biological resources that support production systems**

- Soil biodiversity management for sustainable and productive agriculture: lessons from case studies (Sally Bunning, FAO)
- Cash crop farming in the Himalayas: the importance of pollinator management and managed pollination (Uma Partap, India)

10:00	Coffee break
10:15	**Ecological and social services provided by agro-ecosystems**

- Ecosystem management in agriculture: principles and application of the ecosystem approach (William Settle, FAO)
- Globally important ingenious agricultural heritage systems: (David Boerma, FAO)

11:30	Discussion
12:00	Wrap-up and conclusions (Peter Kenmore, FAO)
14:00	End of meeting

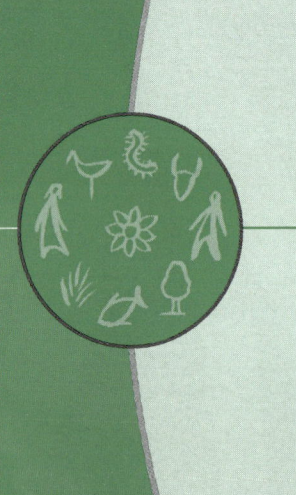

RESPONSIBLE
TECHNICAL DIVISION

**Plant Production and Protection Division
Grassland and Pasture Crops Group**

Stephen Reynolds

BIODIVERSITY AND PERFORMANCE OF GRASSLAND ECOSYSTEMS IN COMMUNAL AND COMMERCIAL FARMING SYSTEMS IN SOUTH AFRICA

1

BIODIVERSITY AND PERFORMANCE OF GRASSLAND ECOSYSTEMS
IN COMMUNAL AND COMMERCIAL FARMING SYSTEMS IN SOUTH AFRICA

AUTHOR

David Hoare

Agricultural Research Council,
Range and Forage Institute,
South Africa
dhoare@lantic.net

CONTENTS

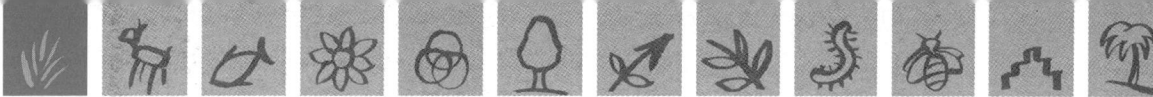

ABSTRACT

The Eastern Cape province of South Africa is home to a large human population that is rural, poor and has experienced little development. The major pressures on these communal rangelands are from intense herbivory, wood fuel harvesting and shifting cultivation. Eight study sites in communal and commercial farming systems were selected and a number of diversity indices were performed on floristic data from these sites. It was found that mean species richness, total number of species and species evenness were higher in the commercial than communal grasslands although the variation in species composition (internal heterogeneity) was similar. Analysis using plant functional traits indicated that there was a shift in the communal areas towards pioneer woody species that had little browsing, fuel or timber value, grass species with lower palatability and forb species that were tolerant of intense herbivory and disturbance to the soil surface. There were more exotic species per sample in the communal than commercial grasslands and these contributed a higher proportion of the total vegetation cover. Landcover data indicates a loss of woody vegetation cover and transformation by shifting cultivation. Production data from livestock numbers at a district level indicate no reduction in long-term stocking rates, but stock composition has changed. Data from stocking rates is problematic as a measure of production and must be considered to be inconclusive since other factors could maintain high stock numbers. There is evidence of improving rangeland condition in the area around towns, where animal biomass is also lower than was historically recorded.

INTRODUCTION

South Africa has a unique natural environment and biological diversity, for which it has been recognized at a national level that good management is essential for sustainable development. The wise use of resources requires a good understanding of the ecological processes that maintain the resource base and it is essential that the complex relationships between the social order and natural environment are well understood. The nature and intensity of resource use in South Africa has not been spatially uniform and different social structures have been imposed on the environment in different areas and at different times.

The study area is situated in the Eastern Cape Province of South Africa (Figure 1). The Eastern Cape comprises an area of approximately 170 500 km². It is an area of extraordinary complexity and diversity, encompassing three regional biodiversity "hotspots". Although literature on vegetation studies has been accumulating recently, it has in the past been understudied and to date, diversity patterns have never been adequately

FIGURE 1

Southern Africa with the study area in South Africa shaded

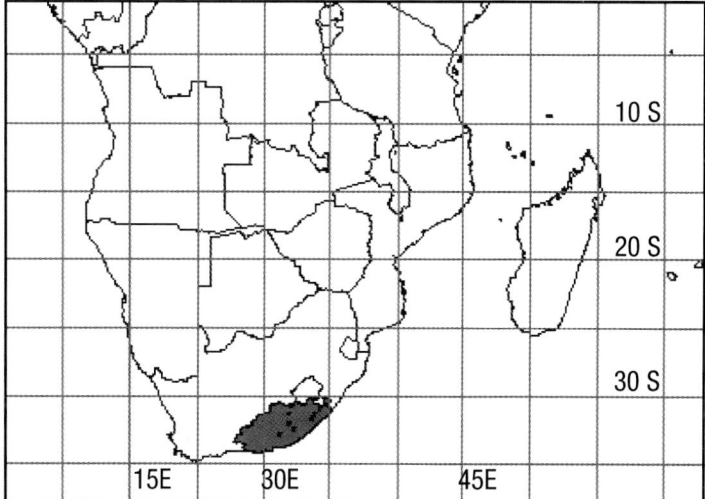

described. Environmental gradients in this area are very steep and complex leading to a rich mixture of floristic elements. This provides an ideal natural laboratory which should be taken advantage of for understanding patterns across a variety of scales. An understanding of diversity patterns in the region can lead to an understanding of ecosystem structure and function, resulting in better management of the area, especially with respect to understanding patterns across a variety of scales. An understanding of diversity patterns in the region can lead to an understanding of ecosystem structure and function, resulting in better management of the area, especially with respect to rangeland conservation planning. The region has vast economic potential from the perspectives of eco-tourism and agriculture (commercial and subsistence). Different management practices have an effect on species composition and thus diversity and it is important to understand this relationship in order to predict their effects on diversity patterns.

The Eastern Cape is home to 15.5 percent (6.3 million) of South Africa's total human population. Forty-nine percent of the province's population is unemployed compared to the national figure of 25 percent (South African Institute for Race Relations, 1991). The majority of people of the Eastern Cape are more rural, significantly poorer and less developed than in other parts of South Africa with a large proportion of the population being reliant to some degree on natural resources for direct subsistence use or indirectly as a form of income generation. A rapidly growing population coupled with increasing poverty and urbanization have a compounding impact on the resource base.

The land use of the region is divided into communal and freehold tenure systems: communal tenure areas are heavily populated (56 people per km^2) whereas in regions characterized by freehold tenure, where commercial farming generally takes place, population density is lower (3–6 people per km^2) (Palmer *et al.*, 1999). In communal areas livestock represents wealth and is a form of currency, and these areas are heavily stocked. In contrast, commercial farming systems in the freehold areas are characterized by land stocked at economically sustainable levels.

The north-eastern part of the province has been under communal land ownership for in excess of 100 years, whereas the westerly regions have been commercially farmed as stock ranches for close to 100 years (Figure 2). In the central regions there has been a recent shift from commercial to communal land ownership (Figure 2), thus providing an opportunity to compare communal impacts of different time periods. In the communal area the major pressures on the landscape and on biodiversity are from intense herbivory, fuel wood harvesting and shifting cultivation. Herbivory by domestic stock is expected to cause a change in species abundance and composition, wood harvesting changes vegetation structure and can lead to the loss of certain woody species and shifting cultivation disturbs the soil surface so that opportunities arise for invasion by exotic plant species and changes species composition by resetting the landscape to an earlier successional stage. The objectives of the study were to determine whether there are any significant differences in natural plant species and functional type diversity in communal and commercial grasslands of the Eastern Cape, South Africa.

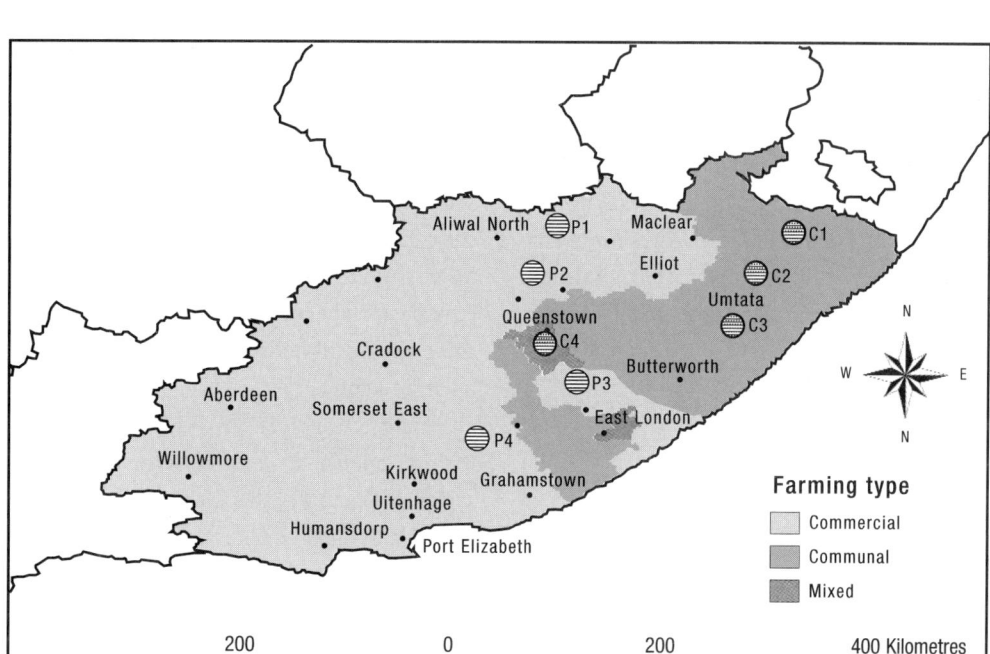

FIGURE 2

Study area showing predominant farming types in the Eastern Cape, South Africa. Sites for biodiversity analysis are indicated by circles (stripes: commercial, dots: communal. P1: Aliwal North. P2: Stormberg. P3: Amatola. P4: Smaldeel. C1: Mount Ayliff. C2: Tsolo. C3: Umtata. C4: Sada).

Physiography

The mountain regions of the study area are dominated by grassland, with numerous small patches of Afromontane forest on southern aspects, whereas lower-lying areas are covered by grassland, savanna and dwarf-shrub (Nama-karoo) vegetation. The incised river valleys of the study area are characterized by subtropical thicket, a vegetation type endemic to the Eastern Cape and containing a number of succulent species. This study concentrates on the grasslands in the mesic parts of the Eastern Cape for which detailed recent floristic survey data are available.

The dominant geological group in the study area is the Karoo Supergroup, comprising alternating bands of fine-grained sandstone, shale and mudstone (Maud, 1996). The soils in the study area may be divided into mountain and plain types. The soils of the mountain areas are generally shallow and weakly developed lithosols whereas on the plains soils may be shallow and poorly drained with high-clay subsoils or, in the dryer areas, sandy loams containing many boulders and gravel (Hartmann, 1988; Werger, 1980). Dolerite outcrops may yield more fertile, clay-rich soils. This study was located in formations of the Karoo Supergroup.

The climate according to the Köppen classification is Cfa (C = warm temperate climate with the coldest months 18°C to +3°C, f = sufficient precipitation during all months, a = warmest month is over +22°C) (Schultze, 1947). The lowland regions (< 800 m) have hot summers and frost prone, cold winters. The average daily minima for the coldest months are below freezing for the whole study area. Winter frost is common and especially severe at high altitudes. The whole study area experiences maximum rainfall, in summer although a weak bimodal (spring and autumn) pattern exists in the central regions. Data from weather stations indicate a gradient of

increasing annual rainfall from west to east ranging from 419 mm at Venterstad to 664 mm at Lady Grey. Surface response models (Dent *et al.*, 1989) suggest that median annual rainfall in excess of 1 000 mm occurs at high altitudes in the Witteberg and Drakensberg ranges and that rainfall increases with altitude and from west to east in the study area.

METHODOLOGY

Grasslands of the commercial farming regions of the study area have been floristically described in detail (Hoare & Bredenkamp, 1999, 2001), but no published descriptions are available for the communal rangelands. Floristic data from these areas has been accumulated from a number of unpublished surveys, including the author's own unpublished data. These have already been collated into a single database (Mucina *et al.*, 2000; McDonald, 1997) as an initial phase of a recent national vegetation mapping project (Mucina & Rutherford, 2002). The criteria for selecting from this database were that the data must contain full-floristic information, be geo-referenced to within 100 m accuracy (the basic expected accuracy using GPS with selective availability), and the plot-size must be known.

Eight study sites were chosen, four each within communal and commercial grazing systems (Figure 2). The sites were selected to be in grassland vegetation and, as far as possible, similar with respect to environmental conditions (rainfall, topography, elevation, geology, soils). Study sites could not be exactly matched with respect to these environmental conditions so it was necessary to measure species' richness along gradients to determine whether there were any directional changes with respect to environmental gradients. The two major environmental gradients in the study area are elevation and rainfall (Hoare, 1997; Hoare & Bredenkamp, 1999, 2001). Species richness from a total of over 500 floristic samples in mixed communal and commercial grasslands covering a range of environmental conditions was compared to rainfall and elevation gradients.

The selected sites were a circular region with a radius of 10 km (area: 314 km²) within which 10 random samples of 10 x 10 m were selected in open grassland vegetation. A complete checklist of species was available for each sample, as well as a cover estimate for each species using a modified Braun-Blanquet cover-abundance scale (Table 1). The cover-abundance classes were converted to a central class value for the purposes of further analysis (Table 1).

From the 10 samples for each site the following indices were calculated:

TABLE 1

Braun-Blanquet cover-abundance classes used for estimating aerial cover of individual plant species.

SYMBOL	CLASS LIMITS (% COVER)	CENTRAL VALUE (% COVER)
r	< 0.25	0.3
+	< 2	1
1	2 – 4	3
2a	5 – 11	8
2b	12 – 24	18
3	25 – 49	38
4	50 – 74	68
5	75 – 100	88

1. Total number of species per site,
2. Mean number of species per sample,
3. Jackknife estimate of species richness ($S' = S+(n-1/n)k$) per site,
4. Whittaker's beta diversity index ($\beta\omega = (S/\alpha)-1$) as a measure of the variation in species composition among localities within each site, which distinguishes it from species turnover along gradients (Vellend 2001),
5. Simpson's evenness index ($(1-\Sigma\ p_i^2) / (1-(1/s))$), a Type II index sensitive to changes in abundant species (Krebs, 1989),
6. The proportional cover of the five most dominant species per sample.

A simple functional type analysis was performed on the species by grouping them into major life-form groups (Table 2). Life form has been used as an easily recognizable plant functional trait (PFT) (e.g. Tilman *et al.*, 1997), but most previous studies have found that life form alone is too broad a classification by which to determine the relative sensitivity to environmental variability so the life-form categories were further sub-divided using life-cycle, morphological and utilization traits (Table 2). Traits were selected that are easy to measure (Box, 1996), but were linked to ecological function, e.g. response to

disturbance (Kindsher & Wells, 1995; Lavorel *et al.,* 1997; Campbell *et al.,* 1999; McIntyre *et al.,* 1999). At a global scale it has been possible to identify only a few traits that are consistently associated with disturbance (e.g. Lavorel *et al.,* 1997; Diaz & Cabido, 1992). A multivariate procedure was used for deriving plant functional types from functional traits (Landsberg *et al.,* 1999; McIntyre *et al.* ,1999).

The invasibility of ecosystems in the study area was evaluated on the basis of the presence and relative dominance of exotic species within floristically sampled vegetation. The causes and mechanisms of invasion were not investigated here and no data is available on the spatial extent of exotic invasions in the study area.

Land cover information for the evaluation of cultivation impacts was obtained from a national land cover database (Fairbanks *et al.,* 2000). However, although shifting agriculture was observed on the ground, it is difficult to quantify from remotely-sensed data as fallow lands develop secondary grassland-type vegetation cover that within a single or two seasons is indistinguishable spectrally from adjacent natural grasslands. These data were therefore supplemented using field observations of land cover.

LIFE FORM	LIFE HISTORY	MORPHOLOGICAL / PHYSIOLOGICAL ATTRIBUTES	
Tree/shrub		Spines, prickles, thorns Wood hard/soft Edible/inedible Single/multi-stemmed	Secondary or toxic compounds Succulent Resprouter
Grass	Annual / perennial	Habit (tussock / stoloniferous)	Palatability
Forb	Annual / perennial	Habit (prostrate or erect) Palatable / unpalatable Geophyte	N-fixing Spines, prickles, hairs Succulent

TABLE 2

Hierarchical system of plant functional traits used in determining plant functional types for an analysis of impacts of grazing, browsing, trampling, soil disturbance and wood harvesting in a comparison between communal and commercial farming areas of the Eastern Cape.

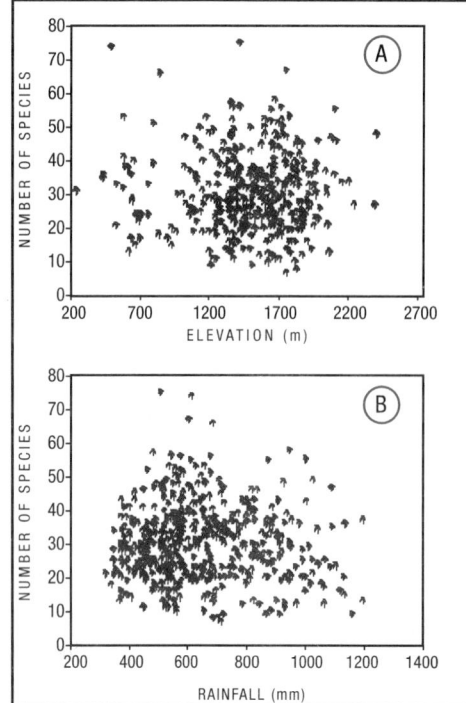

FIGURE 3

Relationship between species' richness and environment (A: elevation and B: median annual rainfall) for 679 plots of 100 m^2 in grasslands of the Eastern Cape.

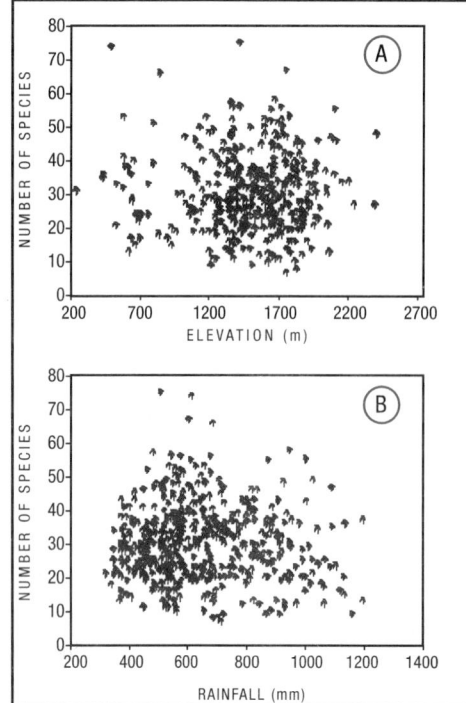

FIGURE 4

Relationship between environmental heterogeneity, measured as standard deviation in elevation, and species' richness for communal and commercial grasslands of the Eastern Cape.

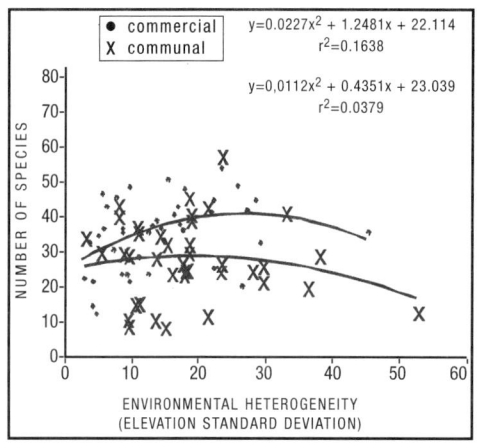

FIGURE 5

Relationship between the Normalized Difference Vegetation Index from the NOAA AVHRR sensor and species' richness for communal and commercial grasslands of the Eastern Cape.

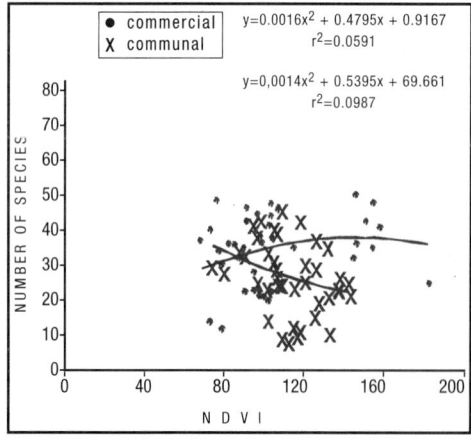

RESULTS

Diversity/environment relationships

There was no relationship between species' richness per 100 m^2 sample and elevation (Figure 3a, $r^2 = 0.004$) or species' richness per 100 m^2 sample and rainfall (Figure 3b, $r^2 = 0.015$) . However, there was a very weak positive relationship between species' richness and environmental heterogeneity for commercial grasslands ($r^2 = 0.164$), measured as the standard deviation in elevation around each sample (Figure 4). The magnitude of this relationship with environmental heterogeneity is different for communal and commercial grasslands, with communal grasslands tending to have lower species' richness for the same amount of environmental heterogeneity as the commercial grasslands. Using normalized difference vegetation index (NDVI) as a surrogate for production (Figure 5) there appeared to be similar species richness at low production levels for communal and commercial grasslands and an apparent differential in species richness at higher production levels with commercial grasslands having more species than communal grasslands. This difference requires further study at other locations.

Species diversity

There were an average of 33.8 species (32.6 without exotics) in the commercial grasslands and 26.3 (23.5 without exotics) in the communal grasslands per 100 m^2 sample (Table 3). There was a total of 145 species per site in the commercial and 126 species per site in the communal grasslands. Using a jackknife estimate of species'

SITE	MEAN SPECIES RICHNESS PER SAMPLE	MEAN EXOTIC SPECIES PER SAMPLE	TOTAL SPECIES	JACKKNIFE TOTAL SPECIES	UNIQUE SPECIES	ALIEN SPECIES	INTERNAL HETERO-GENEITY ($\beta\omega$)
Communal	26.3	2.8	126	189.5	70	11.3	3.84
Mount Ayliff	17.3	2.8	86	134.6 (30.7)	54	8	3.97
Tsolo	33.4	3.9	157	238 (27.3)	90	17	3.70
Umtata	28.2	3.2	128	188.3 (26.8)	67	17	3.54
Sada	26.1	1.4	137	197.1 (41.5)	69	7	4.25
Commercial	33.8	1.2	145	209.1	72	7	3.30
Aliwal	34.4	0.6	113	154.4 (29.9)	46	2	2.29
Stormberg	36.4	0.9	148	208.3 (18.5)	67	6	3.07
Amatola	39.1	2.2	198	298.8 (30.0)	112	13	4.06
Smaldeel	25.1	1.1	120	174.9 (26.9)	61	7	3.78

TABLE 3

Diversity indices for sites in communal and commercial grasslands of the Eastern Cape. The values given here are mean values from 10 samples of 10 x 10 m per site.

richness this can be extrapolated to 209 and 190 species in the commercial and communal grasslands respectively (Table 3). There was a similar number of unique species per sample in communal and commercial grasslands suggesting that rates of compositional turnover amongst rare species were similar in both grassland systems.

This is supported by the measure for variation in species composition among localities within each site (Vellend, 2001), which is similar for communal (mean $\beta\omega$ = 3.84) and commercial grasslands (mean $\beta\omega$ = 3.30) (Table 3). Species' dominance, measured as the proportional cover of the top five species, was on average 88 percent in the communal areas and 66 percent in the commercial areas (Table 4), indicating that fewer species are contributing to the proportional cover in communal than commercial grasslands. Species evenness, using Simpson's index (Table 4), was 0.685

in the communal grasslands and 0.843 in the commercial grasslands indicating that proportional cover was more evenly distributed amongst species in the commercial grasslands than in the communal grasslands.

SITE	SPECIES EVENNESS (SIMPSON)	DOMINANCE OF TOP 5 SPP	PROPORTIONAL COVER OF EXOTICS
Communal	0.685	0.88	0.20
Mount Ayliff	0.443	0.97	0.05
Tsolo	0.733	0.89	0.29
Umtata	0.752	0.91	0.42
Sada	0.812	0.75	0.03
Commercial	0.843	0.66	0.01
Aliwal	0.868	0.63	0.01
Stormberg	0.842	0.64	0.01
Amatola	0.855	0.59	0.02
Smaldeel	0.807	0.79	0.01

TABLE 4

Measures of species' evenness and species' dominance in communal and commecial grasslands of the Eastern Cape

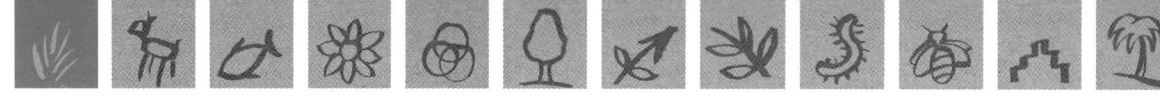

There was a strong negative relationship between species' dominance and species richness (Figure 6) in commercial grasslands ($r^2 = 0.66$), but this relationship was very weak in communal grasslands ($r^2 = 0.07$). The weakness of this relationship in the communal grasslands is probably due to the high levels of dominance in most samples thus providing fewer points lower down in the dominance scale against which to run a regression.

F I G U R E   6

The effect of species' dominance on species richness in communal and commercial grasslands of the Eastern Cape.

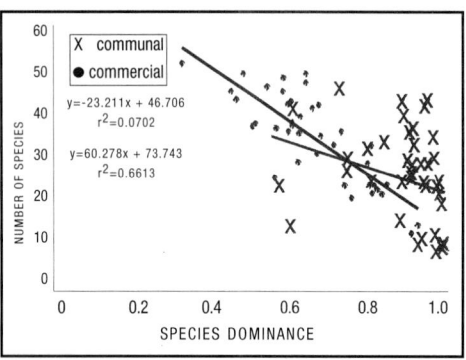

Functional types

Multivariate analysis of functional traits based on ecological responses to disturbance and utilization resulted in the definition of 20 plant functional types (Table 5). There were significant differences in the proportional representation of functional types amongst woody species, grasses and forbs. Communal areas had higher proportional cover of woody species in the functional groups representing low-growing, spreading trees or shrubs, often pioneers with thorns (PFT 1), and the group representing shrubs that are inedible for browsing (PFT 5; Figure 7). The commercial areas had higher proportional cover of single-stemmed upright trees with hard wood (PFT 2), trees with soft wood or that are succulent (PFT 3) and woody species favoured for browsing (PFT 4; Figure 7). There were six grass functional types of increasing palatability and grazing value (Table 5). Communal and

F I G U R E   7

Proportional aerial cover of plant functional types in communal and commercial grasslands of the Eastern Cape divided into trees/shrubs, grasses and forbs. Plant functional type numbers are according to Table 5.

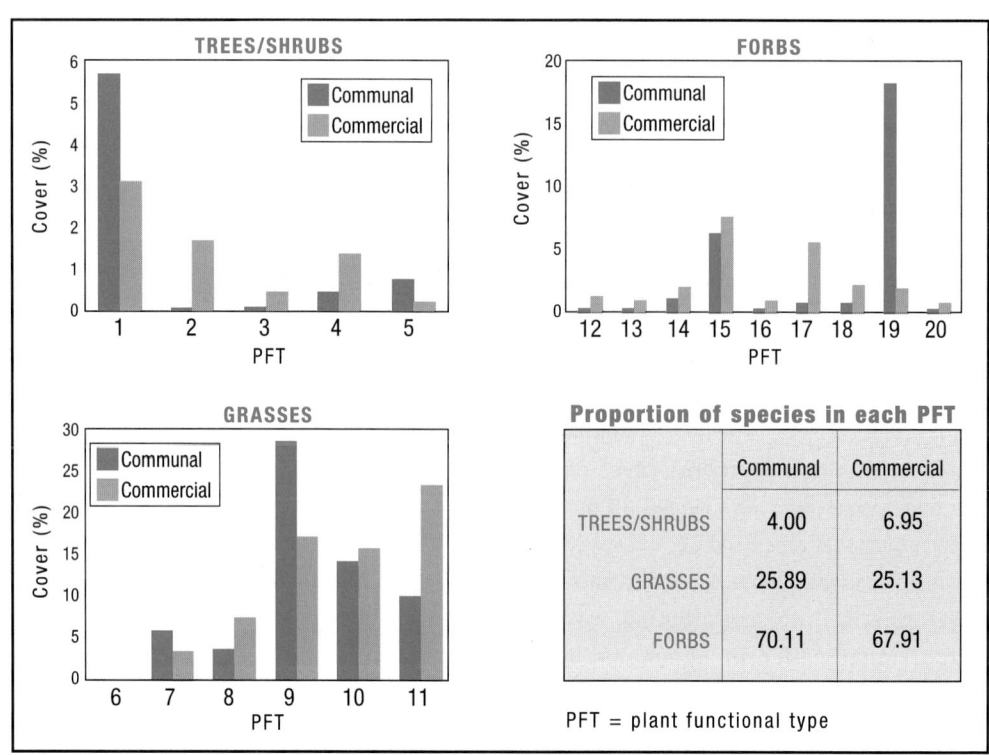

WOODY TREES AND SHRUBS	PFT
Single-stemmed trees	
Hard wood	
With thorns/prickles, often low growing, spreading	1
Without thorns/prickles, usually upright tree	2
Soft wood, often succulent or fleshy	3
Multi-stemmed shrubs (with or without prickles/thorns)	
Edible for browsing	4
Inedible, often with hairy leaves, secondary compounds, etc.	5

GRASSES/SEDGES/RESTIOS	PFT
Annual (tufted)	6
Weak perennial (tufted)	7
Perennial	
Rhizomatous/stoloniferous	8
Tufted	
Low grazing value (= low production and/or palatability)	9
Moderate grazing value	10
High grazing value (= high production and/or palatability)	11

FORBS	PFT
Annual	
Unpalatable (secondary compounds, milky latex, etc.)	12
No secondary compounds, etc.	13
Perennial	
Erect herbs	
Below-ground storage organs (tap-roots, bulbs, etc.)	14
No below-ground storage organs	
Secondary compounds	15
No secondary compounds	
Strong prickles	16
No strong prickles	17
Prostrate herbs, most with below-ground storage organs (tap-roots, bulbs, etc.)	
Bulbous geophytes	18
Non-bulbous geophytes	
Non nitrogen-fixing	19
Nitrogen-fixing	20

TABLE 5

Plant functional types (PFTs) derived from a list of functional traits occurring in species of communal and freehold grasslands of the Eastern Cape.

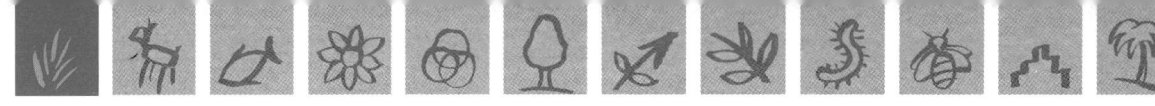

commercial grasslands showed opposite trends with the proportional cover of grasses with lower grazing value increasing in the communal areas and decreasing in the commercial areas (Figure 7). The exception was perennial stoloniferous grasses (PFT 8), which were, unexpectedly, better represented in the commercial areas. They were replaced in the communal areas by prostrate forbs. All forbs were better represented in commercial than communal areas, except prostrate, non nitrogen-fixing forbs with underground storage organs (PFT 19; Figure 7), a group which includes the typical rosette-shaped, prostrate herbs with tap roots that are often found in areas with high grazing impact (Lavorel *et al.*, 1997). Of note was the much higher representation of forbs in the commercial grasslands that belong to the class that lacks any apparent defence against herbivores (secondary compounds, prickles, etc. – PFT 17). The number of species in each of the major functional classes (trees/shrubs, grasses and forbs) was consistent between communal and commercial areas with approximately 70 being forbs, 25 percent percent grasses and the remainder woody species (Figure 7).

Species' composition

There were 12 species, mostly grasses, that were identified as having a high frequency and percentage cover in either the communal or commercial grasslands or both. The composition and grazing value of these are provided in Table 6. Only two decreaser species appear on the list, namely *Themeda triandra,* one of the most important grazing grasses in southern Africa, and *Hypparhenia hirta,* which, even though it decreases under grazing pressure, is an indicator of disturbance and is only of moderate grazing value when the plants are still young. *Themeda triandra* was moderately better represented in the commercial grasslands, but was absent from a number of communal and commercial samples. The species that have a high importance value in the communal areas all have poor grazing value or are indicators of disturbance (Table 6). In contrast the important grass species in the commercial grasslands have moderate grazing value and there is a higher diversity of different grass species at the different sites that are able to be grazed relative to the communal areas.

TABLE 6

Species with a high importance (cover and frequency) in commercial and communal grasslands of the Eastern Cape

SPECIES	COMMERCIAL	COMMUNAL	GRAZING RESPONSE
Richardia humistrata	0.02	31.06	Increaser
Eragrostis plana	0.27	13.07	Increaser
Sporobolus africanus	0.10	9.36	Increaser
Hyparrhenia hirta	0.05	7.03	Decreaser
Aristida congesta	0.81	2.69	Increaser
Cynodon dactylon	0.19	2.22	Increaser
Themeda triandra	7.23	6.64	Decreaser
Eragrostis curvula	4.97	4.53	Increaser
Cymbopogon plurinodis	3.40	2.12	Increaser
Eragrostis chloromelas	4.80	1.09	Increaser
Elionurus muticus	2.79	0.69	Increaser
Aristida diffusa	1.77	0.14	Increaser

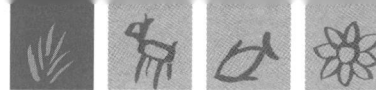
Exotic species

There were an average of 11.3 exotic species per site and 2.8 species per sample in the communal grasslands versus 7.0 and 1.2 species per site and sample for the commercial grasslands (Table 3). This is 8.9 and 4.8 percent of the relative floras of these two regions. Exotic species contributed 20 percent to the proportional cover of communal grasslands and only 1 percent to the cover of commercial grasslands (Table 4).

There was a weak positive relationship between total species richness and number of exotic species for communal grasslands ($r^2 = 0.26$), but not in commercial grasslands ($r^2 = 0.09$). There was a weak positive relationship between exotic species' cover and species' richness in communal grasslands ($r^2 = 0.22$).

Land cover

Land cover data derived from Landsat TM satellite data for the communal portions of the study area indicated approximately 21 percent cover by cultivation and 69 percent natural vegetation (National Land cover Database), but observations in the field suggest that transformation is much greater than this (Table 7). A random sample of 250 sites visited in the field in the communal areas during 2002 indicated that only 33 percent are still natural vegetation, the remainder being transformed by cultivation, urbanization, plantations and other land cover categories.

The major discrepancy between satellite and ground-based statistics are the fact that the satellite imagery appears unable to detect secondary grassland on previously cultivated lands as a separate class to natural grassland. This indicates that large

parts of the cultivated areas are left fallow in any one year. Evidence of landscape transformation can be observed from satellite imagery (Figure 8) as well as from observations in the field, where woodland remnants containing species not able to be utilized for timber or fuel may often be seen in areas where woodland previously occurred (Figure 9).

	NATURAL	TRANSFORMED	CULTIVATED
Satellite data	69%	31%	21%
Ground data (250 sites)	33%	67%	55%

TABLE 7
Land cover statistics (percentage cover) for communal areas from two different sources.

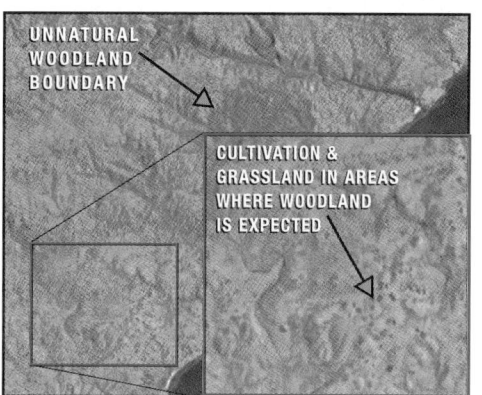

FIGURE 8

Landsat TM image of portion of study area showing extent to which woodland loss has occurred. The expectation is that the majority of this image be covered by woodland, but much of this has been transformed to secondary grassland.

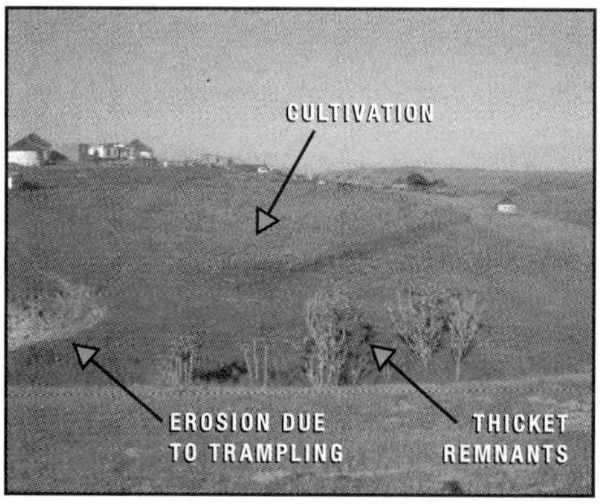

FIGURE 9

Photograph of small settlement within the boundaries of the Landsat TM image shown in Fig. 8 showing remnants of woodland. The species remaining are stem-succulent tree species that have poor wood quality.

Production

The only available data to evaluate whether system rundown is occurring comes from livestock numbers on a district level. Total livestock biomass reveals a degree of variability between years, but no indication that carrying capacity is declining (Figure 10). Although there appears to be more variability in later years and some downward trend for the Mount Ayliff district after 1990, this trend may be attributed to inter-annual rainfall variability or social factors. One indicator of potential degradation is the fact that the livestock composition has changed from predominantly cattle to predominantly goats in the last decade, goats being able to utilize components of the vegetation that are unavailable to cattle. However, the control of stock numbers and the human factors explaining stocking rates are relatively complex and are not investigated further here. Whether vegetation transformation can be invoked as an explanation for stocking rates and composition is not known, at this stage.

DISCUSSION

The communal grasslands of the Eastern Cape have come under severe utilization pressure from a number of factors. It is clear from land cover information that shifting cultivation and betterment planning has modified a large area of the communal landscape and that much of what is considered natural grassland is, in fact, secondary. In addition, expectations in terms of woody vegetation cover are much lower than would be expected from vegetation distribution models and pictorial evidence indicates that there are woodland remnants far away from any existing woodlands. Add to these two factors the high stocking rates measured for almost a century and a picture emerges of the severe pressure on all components of biodiversity in communal grasslands of the Eastern Cape.

This case study has been able to demonstrate that the particular land-use pressures in the communal parts of the study

FIGURE 10

Total numbers of livestock units (LSUs) for the districts of Tsolo and Mount Ayliff in the communal grasslands of the Eastern Cape.

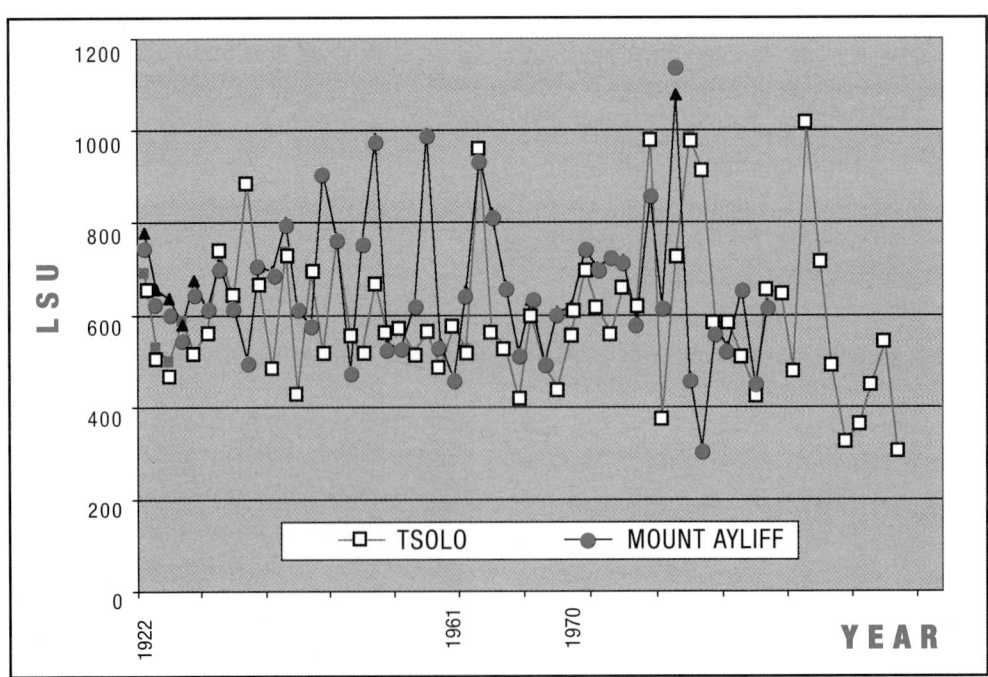

area have led to a reduction in indigenous species richness and increasing dominance by fewer species, although internal rates of species' turnover are unchanged. Functional type analysis demonstrates that those woody species most susceptible to over-harvesting or intense browsing have been reduced as a proportion of the flora in communal areas. Similarly, those grasses most susceptible to overgrazing are found in lower proportions in the communal than commercial grasslands and those forbs most tolerant of soil disturbance and high grazing pressure dominate communal grasslands.

The different grazing systems has a severe impact on the forb component of the flora, with only 12 percent of forb species shared between the Amatola commercial and Umtata communal sites. A greater degree of floristic congruency had been expected because the two sites are geographically close to one another, environmentally more similar than any other two sites and are within identical grassland vegetation types (Mucina & Rutherford, 2002). Almost 75 percent of the species are found in the forb functional groups and it is in this group that the most species change that may affect overall diversity appears to be occurring.

One of the major effects of the grazing pressure in the communal grasslands has been a reduction in species evenness (Table 4). Few studies have evaluated the effect of reduced biodiversity on energy flow and nutrient cycling, but Wilsey and Potvin (2000) determined that higher levels of evenness resulted in higher total biomass irrespective of the identity of the dominant species. Evenness may change with little effect on richness, as is the case here, but it is still regarded as a negative response to environmental stress.

There are more exotic species per sample in the communal areas and exotic species contribute a large proportion of the aerial cover of the vegetation in communal grasslands. Tilman (1999) states that invasibility is equally dependent on species composition, disturbance and other factors as on species' richness. However, this study was unable to support this hypothesis as there was no positive effect of reduced species richness on species invasibility. Although there were more exotic species in the communal than the commercial grazing areas, no causal link can be established between reduced species richness and ecosystem invasibility in this type of correlative study (Rosenzweig, 1995). It is probably the higher disturbance regime that is contributing to the increased occurrence of exotic species in communal grasslands. High exotic cover is known to affect indigenous species cover and species richness also declines, but it is impossible to determine whether it was reduced species richness (relative to potential species richness) that led to invasion in the first place in this study (as per Tilman, 1997). Exotic species appeared to make a positive contribution to species richness and, under the intense grazing pressure, to be behaving biologically as indigenous species, i.e. have become naturalized.

There was no relationship between rainfall or elevation and richness across the broad study area so it is assumed that these factors had no influence on the results presented here. However, a comparison between communal and commercial areas for elevation heterogeneity and vegetation activity (NDVI) as determinants of species richness produced positive results, although the relationship was very poor. Environmental heterogeneity measured as the standard deviation in elevation showed a partially hump-shaped curve, but this curve was lower in the communal areas

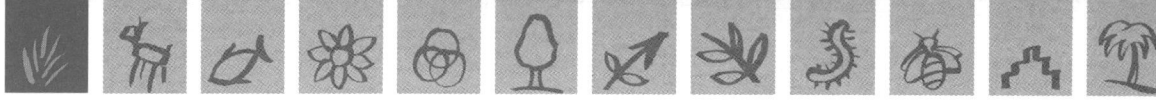

relative to the commercial areas, i.e. species richness was lower in communal areas than commercial areas for the same levels of environmental heterogeneity. It may be that the intense utilization pressure is homogenizing the species distribution patterns to some degree.

There was no apparent reduction in ecosystem production despite intense grazing pressure across many years. This is based on stocking rates (Figure 10), which appear not to have decreased despite many years of high grazing pressure. However, this is a problematic measure to use and does not take into account a host of human factors that could play a role in determining stocking rates including other inputs that could support animal numbers. There is some evidence from satellite data that production in the communal areas is less than expected for the same amount of rainfall, but the results are not conclusive and further research is currently being undertaken. The results for grassland production, as presented here, are, therefore, problematic and inconclusive. It appears that in these highly productive communal grasslands that production levels (in terms of stock numbers) are able to be maintained at high levels because functional diversity has not changed (see Tilman *et al.*, 1997) and that the only risk is that there is no standing biomass to buffer environmental uncertainty. The social response to drought is usually a reduction in stock numbers by translocation, harvesting or mortality. Of more concern from the type of information emanating from the current study is creeping environmental degradation - loss of environmental quality in imperceptible amounts that eventually emerges as a biodiversity issue.

The grasslands at the Umtata site provided some evidence that recovery of these grasslands may be possible. The grass component of this site has a number of species of high grazing value, although the forb component of the grasslands is still severely transformed. This site has undergone a reduction in stocking rates as urbanization levels have increased. The response is that overgrazed grass species have recovered and those of poor grazing value have become proportionately lower. Further reduction in grazing pressure could result in the prostrate forbs becoming overgrown, which would initially lead to a reduction in biodiversity, but indigenous forbs of other functional classes would eventually become re-established depending on their availability in the landscape. This trend in improving rangeland condition has been observed around other towns, e.g. Butterworth and Lusikisiki, where NDVI values are higher and animal biomass is lower (Palmer, pers. comm.).

Currently there is a weak national policy framework for dealing with the kinds of biodiversity/production management issues outlined here. Land redistribution occurs at a national level to address some land tenureship issues, but implementation has been slow. There are a number of research projects managed by the Agricultural Research Council and funded by the National Department of Agriculture that are attempting to quantify some of the land degradation problems in communal areas. The intention is that outcomes from this type of research will guide the National Department in developing a new legislative framework, but this process is still in its very early stages and has not yet developed into a structured programme although there are dedicated personnel at a national level working on such a programme. The current research falls partially within the sphere of this type of policy development research framework, although it is sometimes not clear to researchers what the influence of the particular research programme

may be. A third aspect of management policy development is initiatives managed by the provincial government, such as the erection of fencing, development of skills and further research projects centred on management issues. There are also extension officers in the local areas, but they have poor training and little infrastructural support so they are largely ineffectual in modifying management behaviour. Another national level intervention is the so-called Landcare project, which is an attempt to develop partnerships between the private sector and local communities with the aim of improving land management. These projects often have an economic focus and attempt to instil some entrepreneurial spirit in local communities, i.e. show local communities that they can benefit financially by managing their land in a more sustainable way. It appears that the major stumbling block for many of these projects is social inertia and customs.

The methods applied here have not been widely used in research of this nature in southern Africa and also form only a small part of the range of methodologies employed for investigating the stability of production systems. However, the particular suite of methodologies used here specifically addresses biodiversity patterns and can be used in any other natural forage production system with equal chance of successfully extracting underlying ecological processes. The intention is to attempt

comparisons across different natural vegetation boundaries using this set of methods to see if consistent results can be obtained. A number of organizations made an input, directly or indirectly, into this project, namely three universities, two ARC institutes, the Eastern Cape provincial government, the national government via the Department of Agriculture, and the FAO. There are strengths and weaknesses in this grouping since it brings together a variety of different research and management approaches, but appears to lack a single purpose, which maybe it does. A lesson learnt from fieldwork exercises is that there are conflicting views and constraints on what data to collect due to time and financial constraints and differing research objectives, but the collection of basic ecological and floristic data appears to be able to yield some very powerful results.

Sustainable agriculture relies on ecosystems with integrity and this ecological integrity is dependent on the composition and interaction of its constituent organisms. Natural biodiversity is one component of this integrity and, based on the type of work that ecologists such as Tilman have performed, it appears to play an important role in the stability of natural systems. The challenge is to determine how important biodiversity is as a factor on its own and whether it is an indicator of ecosystem health at a higher level.

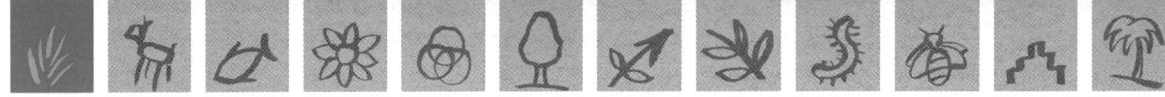

REFERENCES

Box, E.O. 1996. Plant functional types and climate at the global scale. *Journal of Vegetation Science* 7: 309–320.

Campbell, B.D., Stafford Smith, D.M. and Ash, A.J. 1999. A rule-based model for the functional analysis of vegetation change in Australasian grasslands. *Journal of Vegetation Science* 10: 723–730.

Dent, M.C., Lynch, S.D. & Schulze, R.E. 1989. Mapping mean annual and other rainfall statistics in southern Africa. Department of Agricultural Engineering, University of Natal. ACRU Report No. 27. Massachusetts: Clark University.

Diaz, S. and Cabido, M. 1992. Morphological analysis of herbaceous communities under different grazing regimes. *Journal of Vegetation Science* 3: 689–696.

Fairbanks, D.H.K., Thompson, M.W., Vink, D.E., Newby, T.S., van den Berg, H.M and Everard, D.A. 2000. The South African Land-Cover Characteristics Database: a synopsis of the landscape. *South African Journal of Science* 96: 69–82.

Hoare, D.B. 1997. Syntaxonomy and synecology of the grasslands of the southern part of the Eastern Cape. M.Sc. thesis, University of Pretoria, Pretoria.

Hoare, D.B. and Bredenkamp, G.J. 1999. Grassland communities of the Amatola/Winterberg mountain region of the Eastern Cape, South Africa. *South African Journal of Botany* 64: 44–61.

Hoare, D.B. and Bredenkamp, G.J. 2001. Syntaxonomy and environmental gradients of the grasslands of the Stormberg / Drakensberg mountain region of the Eastern Cape, South Africa. *South African Journal of Botany* 67: 595–608.

Hartmann, M.O. 1988. The soils of the Eastern Cape. In: M.N. Bruton & F.W. Gess. (eds.) *Towards an environmental plan for the Eastern Cape*. Rhodes University, Grahamstown.

Kindsher, K. and Wells, P.V. 1995. Prairie plant guilds: a multivariate analysis of prairie species based on ecological and morphological traits. Vegetatio 117: 29–50.

Krebs, C.J. 1989. Ecological methodology. University of British Columbia.

Landsberg, J., Lavorel, S. and Stol, J. 1999. Grazing response groups among understorey plants in arid rangelands. *Journal of Vegetation Science* 10: 683–696.

Lavorel, S., McIntyre, S., Landsberg, J. and Forbes, T.D.A. 1997. Plant functional classifications: from general groups to specific groups based on response to disturbance. TREE 12: 474–478.

Low, A.B. & Rebelo, A.G. (eds.) 1996. *Vegetation of South Africa, Lesotho and Swaziland. Companion to the vegetation map of South Africa, Lesotho and Swaziland*. Dept. of Environmental Affairs and Tourism, Pretoria.

Maud, R. 1996. The macro-geomorphology of the Eastern Cape. In: C. Lewis (ed.) *The geomorphology of the Eastern Cape, South Africa*. Grocott & Sherry, Grahamstown.

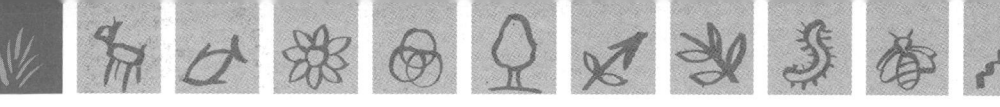
McDonald, D.J. 1997. VEGMAP: a collaborative project for a new vegetation map of southern Africa. *South African Journal of Science* 93: 424–426.

McIntyre, S. Lavorel, S., Landsberg, J. and Forbes, T.D.A. 1999. Disturbance response in vegetation—towards a global perspective on functional traits. *Journal of Vegetation Science* 10: 621–630.

Mucina, L, Bredenkamp, G.J., Hoare, D.B & McDonald, D.J. 2000. A National Vegetation Database for South Africa *South African Journal of Science* 96: 1–2.

Mucina, L. and Rutherford, M.C. (eds) 2002. Vegetation map of South Africa, Lesotho and Swaziland: an illustrated guide. *Strelitzia xx,* National Botanical Institute, Pretoria.

Palmer, A.R., Ainslie, A.M. and Hoffmann, M.T. 1999. Sustainability of commercial and communal rangeland systems in southern Africa. Proceedings of the VIth International Rangelands Congress, Townsville, Australia.

Rosenzweig, M.L. 1995. *Species diversity in space and time.* Cambridge University Press, Cambridge.

Schultze, B.R. 1947. The climate of South Africa according to the classification of Köppen and Thornwaite. South African Geographic Journal 29: 32 – 42. South African Institute of Race Relations. 1991. *Fast Facts* 1, 2. Johannesburg.

Tilman, D., Knops, J., Wedin, D., Reich, P., Ritchie and Siemann, E. 1997. The influence of functional diversity and composition on ecosystem processes. *Science* 277: 1300–1305.

Tilman, D. 1997. Community invasibility, recruitment limitation, and grassland biodiversity. *Ecology* 78: 81–92.

Tilman, D. 1999. The ecological consequences of changes in biodiversity: a search for general principles. *Ecology* 80: 1455–1474.

Velland, M. 2001. Do commonly used indices of β-diversity measure species turnover? *Journal of vegetation Science* 12: 545–552.

Werger, M.J.A. 1980. A phytosociological study of the upper Orange River valley. *Memoirs of the Botanical Survey of South Africa* No. 46. Botanical Research Institute, Pretoria.

Wilsey, B.J. and Potvin, C. 2000. Biodiversity and ecosystem functioning: importance of species evenness in an old field. *Ecology* 81: 887–892.

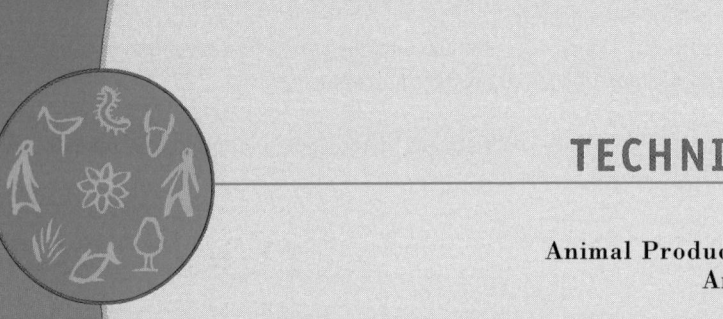

RESPONSIBLE
TECHNICAL DIVISION

Animal Production and Health Division
Animal Production Service

Irene Hoffmann

BIODIVERSITY MANAGEMENT IN WEST AFRICAN PASTORAL AND AGRO-PASTORAL SYSTEMS

2

BIODIVERSITY MANAGEMENT IN
WEST AFRICAN PASTORAL AND
AGRO-PASTORAL SYSTEMS

AUTHOR

Irene Hoffmann
Chief,
FAO Animal Production Service
irene.hoffman@fao.org

ACKNOWLEDGEMENTS

I am grateful to the DG VIII and
DG XII of the European Union
for financing the research under
the projects "Range development
and camel studies",
"Range development in the
endangered Sudan Savanna in
Sokoto State, Nigeria", and
"Development of pastoral and
agro-pastoral livelihood systems
in West Africa"
I also thank my colleagues from
Justus-Liebig-University Giessen,
Usman Danfodiyo University and
the people in the Zamfara Reserve
for their kind co-operation.

CONTENTS

DIVERSITY IN NORTHERN NIGERIAN RISK-PRONE SYSTEMS

Northern Nigeria is a risk-prone environment, with high inter-annual variations in rainfall. Farmers and herders have developed strategies, such as mobility or flexible access to natural resources, to cope with climatic risks (Scoones *et at.,* 1995). Traditionally, livestock and cropland agriculture were segregated and practised by different ethnic groups. However, these groups always interacted and exchanged their products (Van Raay, 1975; Mc Intire *et al.,* 1992; Mohammed, 2000; Hoffmann *et al.,* 2001). Today, the systems tend to be more integrated and traditional pastoral groups such as the Fulani are engaged in cropping, whereas traditional farmers such as the Hausa are increasingly keeping livestock.

THE SITE

Studies were conducted in the Zamfara Forest Reserve in northwest Nigeria between 1990 and 2000. The reserve is located between 6°30' and 7°15'E, and 12°10' and 13°05'N in the north of Zamfara State, sharing a border with the Niger Republic to the north, Sokoto State to the west and Runka Reserve of Katsina State to the east. Annual rainfall ranges from 500 mm in the north to 850 mm in the south with considerable inter-annual variations. The vegetation is of a northern Sudan savanna type.

Documents from the pre- and post-colonial eras, as well as recent research reports document the historical development of the area. From historical sources, Krieger (1954) concluded that Zamfara existed already in the

14th century. The area was virgin bush, probably inhabited by hunters from Katsina. It can be assumed that the region, which is now the two Forest Reserves, or at least a large part of it, was abandoned during the political and social disruptions following the *jihad* (holy war in 1804). Subsequently, the depopulated area became covered by a dense secondary savanna vegetation (Keay, 1949; Kueppers, 1998). The fact that dye basins, ruins of old settlements and fortification walls can be found in the reserve underline this assumption. Probably, the transhumant Fulani also avoided the area during this time. Due to its low population density and the formerly dense vegetation, an area of 2 394 km² in Zamfara became a forest reserve in 1916. It was used by the British colonists to secure their supply of tropical timber and to establish a game hunting area. In 1932, the vegetation was described as an open *Combretum* savanna of varying quality. Near villages and on farmlands most economically valuable tree species were absent.

The villages which today encompass the four enclaves in the reserve were founded

TABLE 1

Some characteristics of livestock and cropping systems in northern Nigeria

	SECTOR		
	LIVESTOCK	**AGRICULTURE**	**OFF-FARM**
Natural resources base	Rangeland	Cropland	Rangeland, cropland and others
Mobility	Pastoral agro-pastoral and sedentary	Shifting and permanent cultivation	Labour migration
Ethnic groups	Tuareg, Fulani > Hausa	Hausa > Fulani	All
Property rights	Communal/ open access	Communal private	Private

*Map of the
Zamfara Reserve*

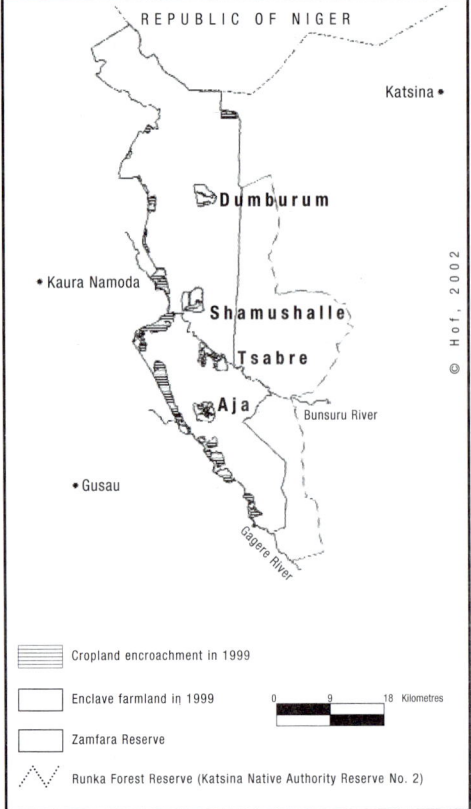

REPUBLIC OF NIGER

Katsina •

© Hof, 2002

Dumburum

• Kaura Namoda

Shamushalle

Tsabre

Aja

Bunsuru River

• Gusau

Gagere River

▤ Cropland encroachment in 1999

☐ Enclave farmland in 1999

0   9   18 Kilometres

☐ Zamfara Reserve

⟋⟍ Runka Forest Reserve (Katsina Native Authority Reserve No. 2)

The reserve, established in 1918, covers today a total area of 2 300 km^2 including four enclave villages (Dumburum, Shamashalle, Tsabre, Aja). About 50 villages are lined up on its western fringe. About 130 000 people live within or around the reserve and utilize its natural resources (ARCA, 1995). The area is populated by Hausa farmers and by Fulani pastoralists. The clear distinction between the livestock and cropping systems is becoming blurred. About 80 percent of the Fulani, sedentary and transhumant herders, are engaged in crop production (Schaefer, 1998).

In this paper, the use and management of biodiversity, both in the livestock and the cropping systems, will each be analyzed.

DIVERSITY IN THE RANGELAND-BASED LIVESTOCK SYSTEM

between the end of the last century and the 1940s, after the establishment of the Forest Reserve. The first settlers were hunters of Hausa origin who later cleared the bush for cropland. Both hunting and bush clearing resulted in a decline in the amount of wildlife and opened up the vegetation. As a result, an influx of pastoral herds took place and grazing became more frequent in the 1930s (Hoffmann, 1998). Initially, the villages had a low population of less than 100 persons each. Taking the figures given in the Forestry Ordinance 1916, there were 19 households in Aja, 40 in Tsabre and 63 in Dumburum in 1916. Population estimates indicate that 27 250 people lived in the enclave villages in 1994 (ARCA, 1995). The upsurge was due to natural population growth and to immigration after the droughts and famines during the twentieth century.

Most large ruminant livestock is kept by Fulani pastoralists. However, sedentary Hausa farmers are increasing their livestock holdings with small ruminants and some cattle. The reserve is an important rainy season grazing area for transhumant pastoralists as well as for the herds of the sedentary farmers living in the enclaves or the bordering villages. After the grain harvest, most livestock are fed on stubble. Later in the dry season, transhumant pastoral herds leave the region in search of pasture and water. However, about one third of the Fulani are sedentary and stay in the region with their herds throughout the year (Schaefer, 1998). Therefore, the livestock system encompasses transhumant and sedentary animals.

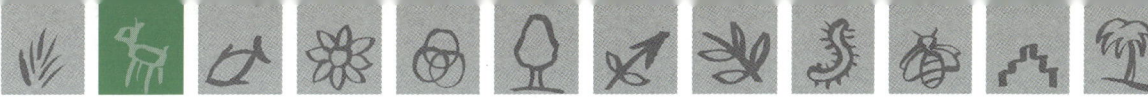

Livestock management on rangeland and cropland

Cattle are the basis of livelihood for pastoralists and agro-pastoralists in the northern parts of West Africa. They are used for milk, manure, meat and draught power, and they serve as savings and insurance. Composition of pastoral herds, particularly cattle, has been analyzed repeatedly. Usually, the sex and age groups of individual herds, directly counted or resulting from interviews by the owner or herder, are recorded (Sutter, 1987; FDLPCS, 1992; Vabi, 1993; Amanor, 1995). Data on livestock numbers are generally hard to obtain in northern Nigeria. Collecting data on herd size by counting or interviews poses some problems. Firstly, pastoralists are reluctant to give information about their livestock numbers due to cultural reasons and fear of taxation (Gefu, 1992; Mohammed and Bello, 1994). Secondly, risk management practices in an ecologically and economically highly variable environment augment the problems of getting accurate data on livestock ownership. Such management practices lead to an overlay of ownership and management patterns, hence, the term 'herd' applies to the unit of management rather than to the ownership. This is specifically due to: the widespread exchange of animals in a social network, the division of herds into management units which might be herded far away from the owner's homestead, the herding of animals of different (absentee) owners in one herd, and the waged herder's reluctance to admit being a "poor" employee who has to herd for others.

The stocking density information for this paper was obtained from transect counts. Livestock density in the Zamfara Reserve was estimated by monthly repeated animal counts (June 1993 to May 1994) along five lines transects from North to South (Schaefer, 1998). Each transect crossed the reserve in east-west direction, thereby crossing transhumance routes at a right angle. Animal counts were then converted into tropical livestock units (TLU) of 250 kg liveweight.

An average of 1.5, 0.6 and 0.3 TLU per hectare (ha^{-1}) was found on rangeland during the rainy, early and late dry seasons. Over the whole year, an average of 0.84 head of cattle, 0.55 sheep and 0.38 goats was found per ha of range, resulting in a stocking rate of 0.81 TLU ha^{-1}. This stocking rate exceeds the recommended rate (Boudet, 1991). Cattle represent about 80 percent of livestock biomass. The highest density of cattle was observed in August with 2.3 head ha^{-1}. This coincides with the peak of rainfall, growth rate of the vegetation, and the feed supply of pastures in quantitative and qualitative terms. The continually decreasing animal density on rangeland during the dry season reflects the decrease in feed and water availability and the subsequent migration of pastoral herders out of the Zamfara Reserve (Figure 1).

FIGURE 1

Monthly stocking density on rangeland and cropland in the Zamfara Reserve (data: Schaefer)

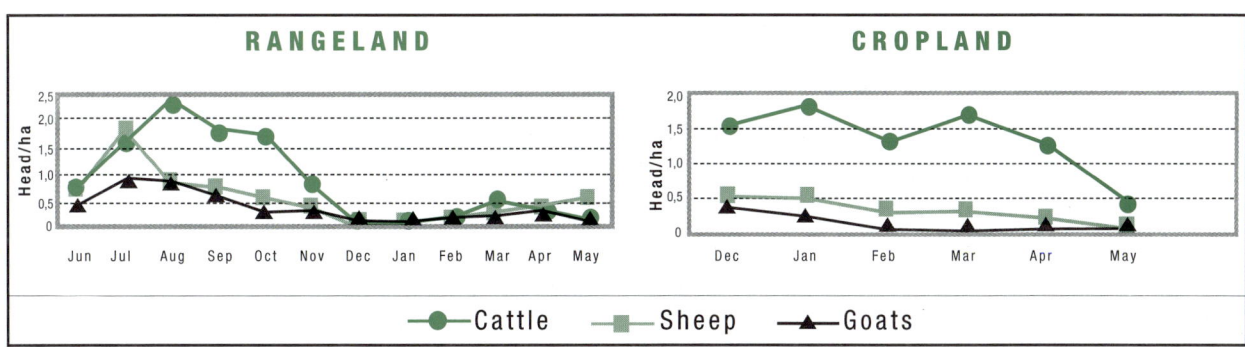

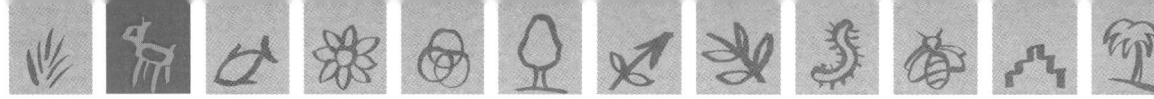

No livestock was found in the cropland areas during the rainy season, when access is forbidden by the village authorities. Cattle numbers are kept at a rate of 1.6 head per ha of cropland from December to March, rapidly declining thereafter (Figure 1). In contrast, small ruminant stocking rate declined gradually from 0.3 to 0.1 animals ha^{-1} during the dry season. The high livestock density on cropland leads to considerable nutrient input through manure (Hoffmann et al., 2001). After planting in May/June, no livestock is allowed in the cropland areas.

Stocking density of cattle during the rainy season varies with the distance to the village, although not significantly (Figure 2). At a distance of four km 2.3 heads of cattle were counted per ha. This value decreased by about 30 percent at six and eight km. The small ruminant counts did not show an effect of settlement distance. Possibly, six km represents the limit of village herds' grazing orbit, as was found in farmer interviews by Malami et al. (1998) and Hassan (2000). The slight but insignificant increase in to more than eight km distance from the village might be due to transhumant Fulani herds, who settle at a certain distance from the village and have a five km grazing orbit during the rainy season (Mohammed and Bello, 1994).

Grazing orbits are enlarged during the dry season due to search for feed and water. Fifty-four percent of interviewed pastoralists cover more than five km grazing distance from their settlements, including those transhuming southwards (Mohammed and Bello, 1994). In contrast to the rainy season, when livestock are concentrated on the rangeland, they are more evenly distributed in the landscape during the dry season. After the harvest, village authorities declare the opening of the fields. Therefore, livestock use rangeland and cropland. An average of 1.2 and 0.3 TLU ha^{-1} was found on cropland and rangeland during the six months of dry season.

Composition of herding units

When asked about their livestock holding in different surveys, the Fulani in the reserve gave figures ranging from 69 to 75 head of cattle, 33 to 43 sheep and 34 to 36 goats (Kyiogwom et al., 1994). Similar cattle, but lower small ruminant holdings were observed in settled Fulani in northern Cameroon (Vabi, 1993). These figures should be considered cautiously, because the Fulani are reluctant to give information about their livestock numbers. However, another approach of indirect estimates of pastoral herd size reveals similar results. The pastoralists told in interviews that three to four young men or children per household are engaged in livestock herding (Kyiogwom et al., 1994). The average herding unit in the Zamfara Reserve consisted of 20.2

FIGURE 2

Mean stocking density of livestock in two seasons by distance to the villages (data: Schaefer)

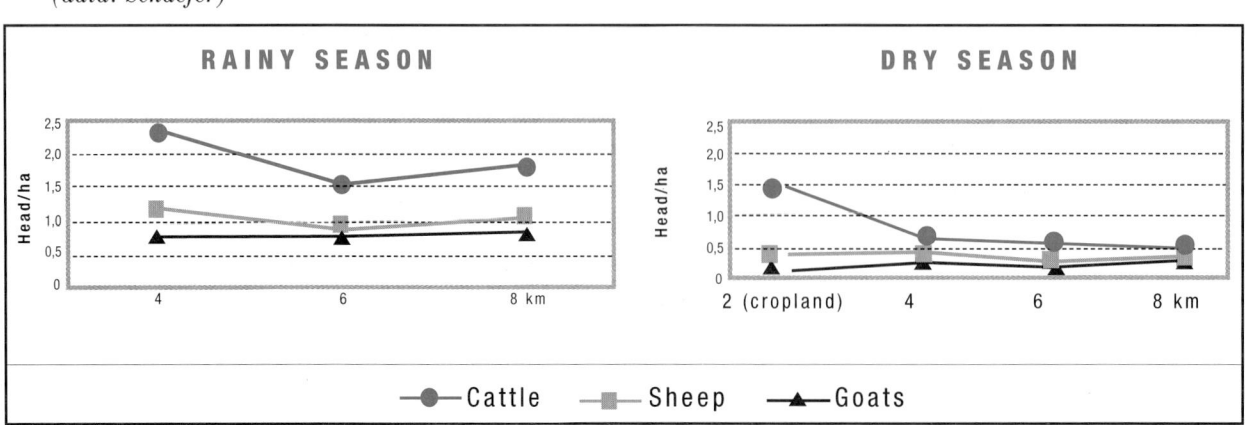

cattle. Hence, 70 head of cattle per pastoral household seems to be a realistic estimate. These cattle holdings are above the figures given for agro-pastoral Fulani which vary between 32 in the Udubo Grazing Reserve (Gefu, 1992) and 50 in central Nigeria (Powell and Taylor-Powell, 1984; Powell, 1986). Evidently, sedentary pastoralists keep smaller herds than transhumant or nomadic ones (Gefu, 1992). Ownership of larger herds in the Zamfara Reserve might be due to the presence of nomadic and transhumant pastoralists (70 percent of the Fulani) (Mohammed and Bello, 1994; Schaefer, 1998) and the relative abundance of rangeland resources. Herd size of the Fulani is larger than that of Hausa. Seventy-seven percent of the Hausa farmers keep an average of 13 sheep, and 75 percent of the farmers keep 11 goats. They also own a few cattle, mainly adult males for draught purposes, but only seven percent of the farmers keep more than 10 cattle.

The number of cattle or small ruminants observed per herding unit did not differ between cropland and rangeland. The average cattle number in 1 264 herds was 20.2 (median 13; range 1 to 183). Two-thirds of the herds were comprised of less than 20 head of cattle. Although the observation of herds does not allow direct conclusions on the ownership pattern, this figure can be explained by a high percentage of smaller herds owned by farmers, who keep only few cattle with a high proportion of work bulls (Hassan, 2000; Hoffmann *et al.*, 2001). The pastoralists stated in the interviews that large herds are split and the cattle grouped by category. Bulls, cows, young stock and calves are then herded separately to better suit their feed requirements and walking ability (Schaefer, 1998).

The condition of cattle differed significantly by season. The cattle observed

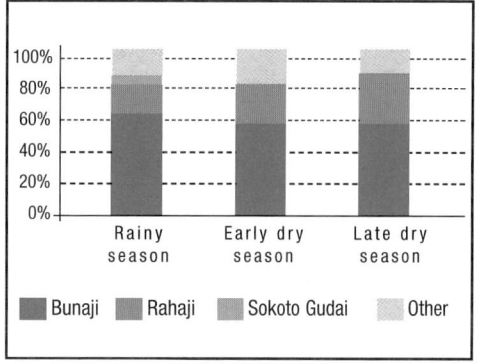

FIGURE 3

Cattle breeds on rangeland in the Zamfara Reserve, by season

(data: Schaefer)

in the rainy and early dry seasons were in good condition, but in the late dry season, the majority was only in medium condition.

The major ecotypes or breeds in the area are the Bunaji (White Fulani), the Rahaji (Red Bororo), and the Sokoto Gudali. Rahaji is a dual-purpose breed producing milk and beef. Bunaji and Sokoto Gudali supply additional draught power, Bunaji is utilized principally for milk and the well-muscled Sokoto Gudali for meat and labour (FDLPCS, 1992).

Seven hundred thirty-four herds (58 percent) consisted of one breed. If herds are fairly pure, reproductive animals of a certain type are selected. This implies that the livestock owners have a good idea about adaptation of cattle for particular environments and purposes and select their bulls accordingly. Season and region within the reserve significantly influenced the distribution of these breeds.

The Rahaji, better adapted to the harsher arid environment, are more frequently found in the northern parts of the reserve. Rahaji is the most prestigious pastoral breed and best adapted to arid environments. They are sensitive to humidity-related diseases (Blench, 1999).

The Bunaji, which is the most important breed in 42 percent of the herds in the north, is more frequently found in the central and southern parts of the reserve, where it clearly dominates (62 and 90

percent, respectively). Docility of the breeds increases from Rahaji over Bunaji to Sokoto Gudali (FDLPCS, 1992). Therefore, Rahaji are solely kept by pastoralists - regardless of whether they farm or not. Bunaji are kept by pastoralists and agro-pastoralists, and Sokoto Gudali are mainly kept by farmers (Hausa and Fulani).

Replacement of Rahaji and Bunaji by Gudali cattle in the herds of settled Fulani was also observed by Vabi (1993) and Blench (1994). Hence, the distribution of cattle breeds reveals a deliberate choice of particular breeds for particular purposes, both for use and in cognizance of their adaptation to ecological conditions.

In 374 herds (30 percent) a second breed or more were observed: These herds consisted mainly of a combination of other breeds together with Bunaji (168 herds) or Rahaji (92 herds). Rahaji were counted together with Bunaji in 74 herds. A possible reason for finding herds with different breeds might be contract herding, i.e. herding of animals belonging to other families. A family herd is more likely to consist of a single breed because animals are inherited from family members and related to each other. Herds of consigned animals are more likely be composed of different breeds.

Model calculation of mobility

Figure 4 shows an estimate of the cumulated number of cattle in the reserve during the year, derived from extrapolation of transect counts to the whole area. From the lowest figure in December it can be assumed that sedentary Fulani and Hausa farmers keep about 50 000 estimated 16 250 households in the enclaves and the bordering villages (ARCA, 1995; Hassan, 2000). Sixty-nine percent of these households own about four cattle, resulting in 45 000 head of cattle. Assuming an average herd size of pastoral herds of 70 cattle, a maximum of 6 400 migratory pastoral households with 450 000 heads of cattle can be found in the reserve at the peak of the rainy season in August. The real figures might be lower because the arid northern part of the reserve was not covered by the transects, and because the extrapolation of line transects to the whole area is necessarily very rough.

However, they clearly show that the predominant mobile grazing management involves mobility of a large number of livestock and humans (Figure 5). The absolute number of humans and animals involved, and their spatial and temporal movements, have social and ecological implications. On the social side, this implies functioning social and information networks, logistics, communication and the solution of conflicts over resources between the

Cattle breeds on rangeland and cropland in the Zamfara Reserve, by region (data: Schaefer)

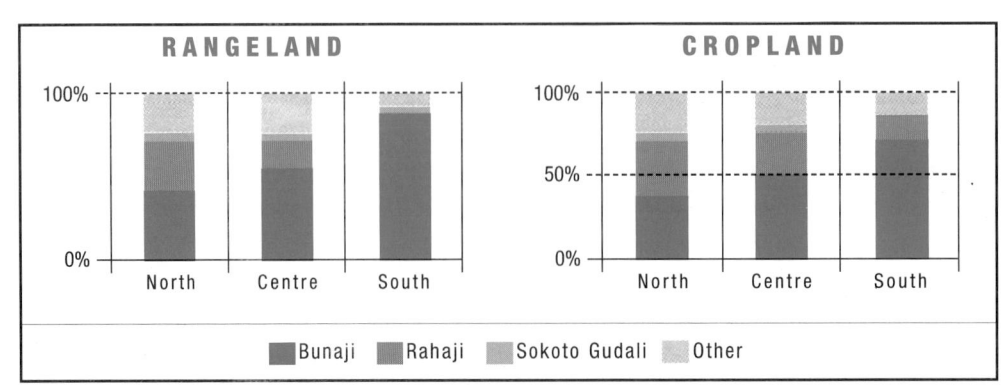

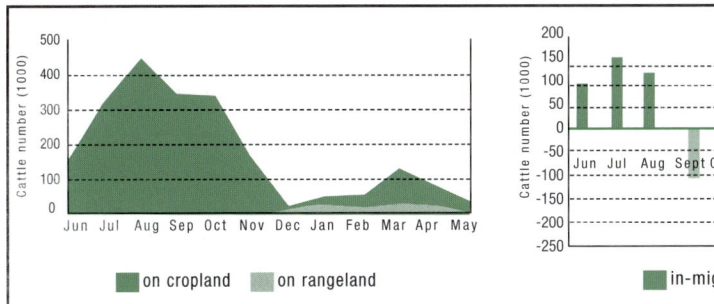
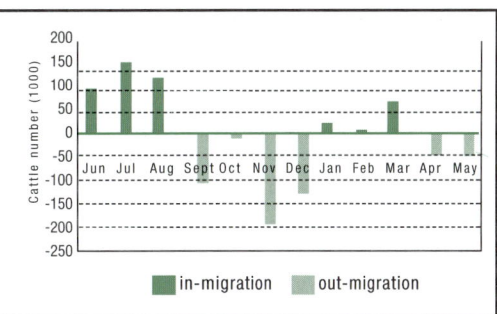

FIGURE 5

Total number of cattle, and their in- and out-migration in the Zamfara Reserve, by month

different user groups. Mohammed and Bello (1994), when asking the Fulani about their perception of problems related to transhumance, found that the incidence of livestock disease (32 percent), theft and conflicts with settled farmers (12 percent each), family split and social discrimination (two percent each) were mentioned as problems.

Rangeland biodiversity

Vegetation types and species' number

Animals of pastoralists and farmers feed on the rangelands, particularly during the rainy season. Therefore, pasture quantity and quality have to be looked at. Ground cover of the herbaceous layer increased from the arid northern (29 percent) to the central and southern parts of the reserve (54 and 49 percent). On the same gradient, ground cover of the woody vegetation increased from about three percent tree and 37 percent shrub cover to about 40 percent tree and 46 percent shrub cover (Hoffmann *et al.*, 1998), (Table 2).

Although the number of species identified is only slightly lower than that found in the 1940s (Keay, 1949), there seems to be a change in the vegetation composition with an increase of unpalatable species. Herders and farmers are conscious observers of their environments. Answers from both interviewed Fulani herders and Hausa farmers, suggest evidence of a change in the plant community as a result of increased grazing pressure in the long run. According to them, vegetation cover was much denser and palatable forage species more frequent in earlier times when Zamfara Reserve was only

	HILLS	RIPARIAN FORESTS	SUDANIAN SAVANNA		
Species number (n)			**SOUTHERN TREE TYPE**	**NORTHERN TREE TYPE**	**NORTHERN SHRUB TYPE**
Trees	13	18	12	27	4
Shrubs	15	24	15	29	9
Herbs and grasses			22	18	13
Ground cover (%)					
Herbaceous layer			49	54	29
Woody species			46	15-30	3-15

TABLE 2

Species' number and ground cover on natural rangeland in the Zamfara Reserve (Elsholz, 1996; Kueppers, 1998)

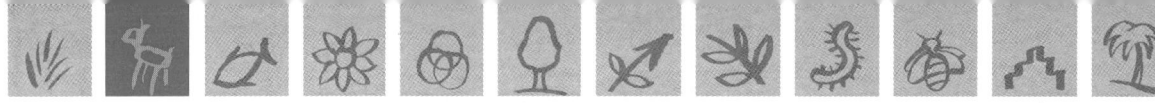

lightly used for grazing livestock. Perennial grasses like *Andropogon gayanus,* which were abundant in earlier times according to the Fulani, have more or less disappeared (Eckert and Hoffmann, 1998). Their assessment is supported by Keay (1949). In the 1940s he still found perennial grasses in the herb stratum, but noticed that the woody vegetation was influenced by cattle grazing and fire. Today, rapidly maturing annual plants such as *Zornia glochidiata* or *Dactyloctenium aegypticum* are increasing (Kueppers, 1998; Schaefer, 1998). These species have good feed quality but little leaf biomass and are less susceptible to defoliation by grazing. Their ground cover reaches 25 percent. *Cassia tora,* the most important increaser which locally may reach 100 percent ground cover, is mostly unpalatable for grazing animals (Kueppers, 1998; Kreimer and Steinbach, 1998).

Characteristics and use of woody vegetation

Size and crown diameter tend to increase from north to south, following the rainfall gradient. Trees and shrubs found in the rangeland are lower and have a smaller crown than those found on cropland, although trees on cropland are more intensively used (Figure 6). This might indicate a higher age of trees in cropland.

The tree composition in the fields reflects the former open savanna woodland which covered most of the area before settlement took place (Keay, 1949). Trees found on the fields were in most cases indigenous species left at random while the land was cleared and undesirable vegetation was removed. Other species were explicitly left over and used in a multi-purpose way except *Prosopis africana,* the characteristic species of the former dry deciduous forests (Aubreville, 1950), which is today mostly encountered as dead stumps since they are hardly removable by farmers. The majority of dead tree stumps is found on cropland.

The most common tree species found on farmland are *Adansonia digitata, Anogeissus leiocarpus, Diospyros mespiliformis, Lannea microcarpa, Parkia biglobosa, Piliostigma reticulatum, Sclerocarya birrea, Tamarindus indica* and several species of *Ficus.* They were left on the farmland because of their usefulness, but no effort was made to replace old trees or to increase their number. Only *Adansonia digitata, Azadirachta indica* and several fruit trees were planted *(Moringa oleifera, Mangifera indica).*

FIGURE 6

Characteristics of woody species on rangeland and cropland in the Zamfara Reserve (CBH = circumference at breast height; data: Hoffmann and Malami, unpublished)

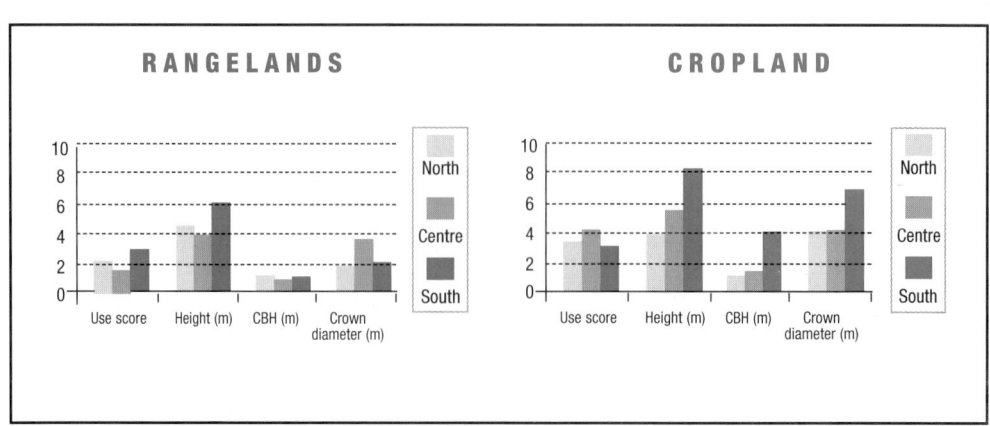

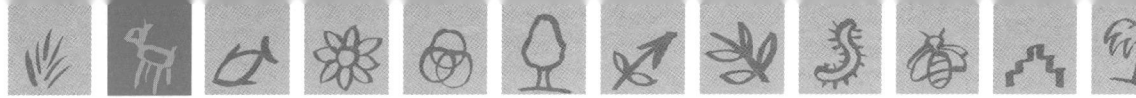
Traditionally, lopping of trees provides important feed resources during the dry season, especially for pastoralists with limited access to crop residues or stubble. Important browse species include *Butyrospermum parkii, Vitex doniana, Ziziphus mauritania, Pterocarpus erinaceus, Khaya senegalensis, Adansonia digitata, Acacia seyal* and *Celtis integrifolia.*

Due to intensive utilization (fodder, medicinal purposes, human nutrition and firewood), nearly all tree species are heavily lopped near the villages (Figure 7). However, at greater distances from the villages the number of lopped species and the degree of utilization were generally found to decrease. RIM (1991) reported that the proportion of trees utilized for browse alone was very small and most trees were lopped for multiple use. In Zamfara, no rejuvenation was found in 13 tree species (including *Pterocarpus erinaceus, Khaya senegalensis* and *Celtis integrifolia*). Adult specimens are utilized very intensively so that recruiting for these species is seriously threatened. The intensity of use, estimated by the use score, is closely linked to the usefulness of the trees as animal fodder. Only the unpalatable species were mainly untouched. Species which are valued because of their fruits in other West African countries like *Lannea microcarpa, Sclerocarya birrea, Vitex doniana* or *Butyrospermum parkii* were only lopped for browse (Kueppers, 1998). Since the Hausa know about the usefulness of those trees, they preserve the tree while using it to feed their animals to a certain extent (Eckert *at al.,* 1998).

THE FARMING SYSTEM

T here is a widespread belief that increasing population pressure will lead to further natural resources degradation (Drechsel *et al.,* 2001). However, the "Boserupian" development paradigm that population growth induces changes in factor scarcities which in turn alter the relative factor prices, stresses the role of people as producers and innovators.

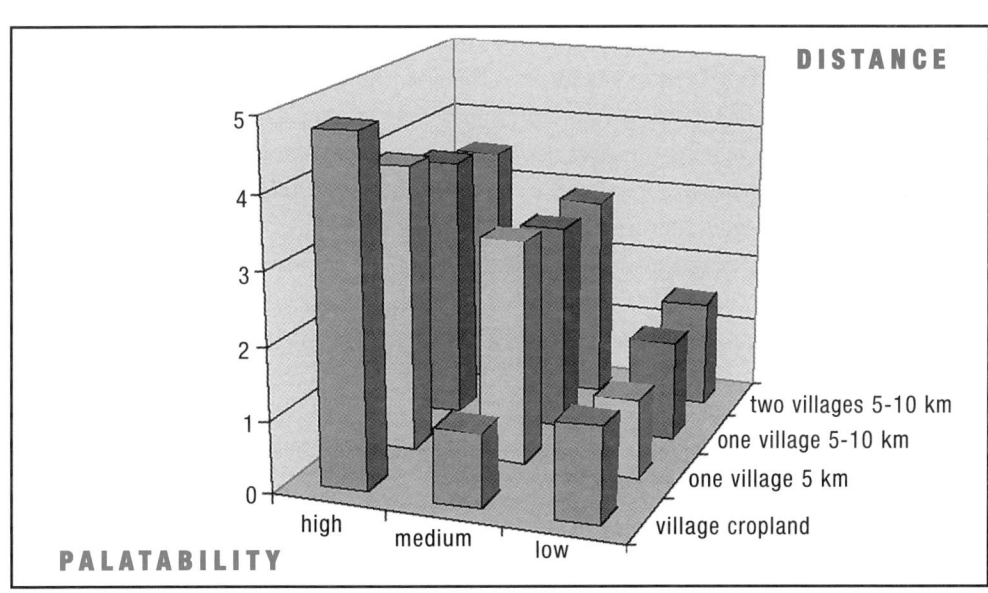

FIGURE 7

Use score of woody vegetation, by palatability and settlement density (Use score low = 1, high = 5, all branches removed from the tree) (data: Hoffmann and Malami, unpublished)

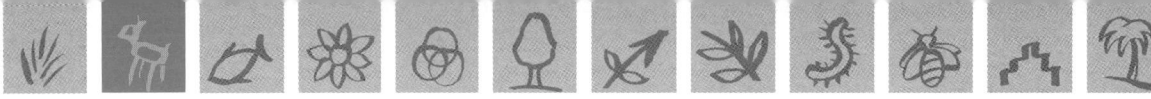

The improvement of yield-enhancing and land-conserving methods serves to increase the productivity of the available resources (Boserup, 1965). Moreover, not only technological but also institutional changes are induced in response to changing resource endowments (Hayami and Ruttan, 1985).

In closely settled zones of northern Nigeria, increased population density has been accompanied by more productive and diversified farming systems (Mortimore *et al.*, 1990; Harris, 1998; Mortimore and Adams, 1998), based on close tree-crop-livestock integration. Densely populated zones in semi-arid northern Nigeria, such as the Kano Close Settled Zone (KCSZ), have supported population densities of more than 300 people per km^2 for centuries. There, livestock and crop systems become more integrated and cropping intensity as well as labour input increase with population density. Small ruminants increase whereas cattle decline. Livestock feeding becomes more labour-intensive because of the scarcity or complete loss of natural range. Crop residue, cut grass and browse are gathered to feed livestock kept in confinement. Manure production is an important component of livestock production (Harris, 1998).

Although the Zamfara Reserve is a large and remote area where population pressure is not expected to play a big role, this does not hold true due to its particular forest reserve status. Except for the land allocated to the enclave villages at the establishment of the reserve in 1919 and under the Forestry Ordinance of 1957, bush clearing to enlarge the cropping area is not permitted under the Forestry Laws, but has always been practised to a limited extent. As pressure on cropland is high, shifting cultivation and fallow are no longer possible. An interpretation of air photographs and satellite images shows that about 2 200 ha around the enclaves and about 8 000 ha on the western fringe of the reserve had been illegally cleared (Hof, 2000). About two-thirds of the total encroachment took place between 1990 and 1994 (ARCA, 1995). Farm size estimates range between 1.5 and 1.8 ha for the enclaves Dumburum, Aja and Shamashalle, and 5.4 ha for Tsabre, with an average of 3.2 ha. Farms in the bordering villages, without legal restriction, are larger (4.3 ha) (Hassan, 2000). The average field size within the reserve is 1.1 ha (median 0.7 ha).

Soil diversity management

In the villages in the Zamfara Reserve, intensive application of farmyard manure was found to be the most important source of nutrients, followed by dung voided by livestock directly on the field (Hoffmann *et al.*, 2001). No manuring gradient depending on the proximity to the village could be found, indicating intensive soil fertility management in all fields. Nutrient transfer from the rangeland to the cropland through farmyard manure application or corralled animals is an important strategy for maintaining soil fertility. This is usually based on exchange relations between pastoralists and farmers, where the pastoralists provide herd manure through corralling on fields and farmers remunerate them in grains. This manure contract fulfils ecological functions in terms of cropland soil fertility maintenance, and institutional functions as a response to pressure on natural resources. Livestock was corralled on 49 percent of the fields in the Zamfara Reserve later in the dry season. The majority of fields was corralled by Fulani livestock, that provided 1 306 kg of faeces ha^{-1}.

Any manuring practice resulted in local concentrations of the manure due to the

rotation of manure application between different fields and within a field. Farmyard manure is rotated in a one to two year cycle, but it is widely spread on the field. A more pronounced local manure concentration was observed for the corrals, which covers only about 10 percent of the area of a field. On instruction of the field owner, the night-corral is regularly shifted to achieve an adequate spread of the manure, depending on the livestock species and the nutrient status and former yield of the soil. They result in a two to four year rotation, where one part of the field has been intensively manured during the preceding dry season, and the other parts in previous years (Hoffmann *et al.*, 2001; Figure 8).

Although even spreading of farmyard manure and faeces is the farmers' goal, the rotation of both types of manure results in micro-variability on the field, and residual effects of nutrients, which have to be taken into account for the calculation of nutrient balances. The variable pattern of manuring and planting might also influence N-fixation of legumes. The farmers explained that manure scarcity is the main reason for the rotation. Since they aim at maintaining and equilibrating the soil fertility of their different fields, soils considered as fertile receive less manure than those considered to be poor. To better deal with the scarce manure, farmers take into account the slope of the field to minimize run-off and the risk of "crop burning" on recent and intensively manured areas in case of low rainfall in the early growing period. Crop burning has also been mentioned by Sandford (1989). In years of good rainfall, spots with high soil nutrient content will result in high yields, whereas in years of little or poorly distributed rainfall, spots with low nutrient content will still produce minimal yields. According to the farmers, the second year of

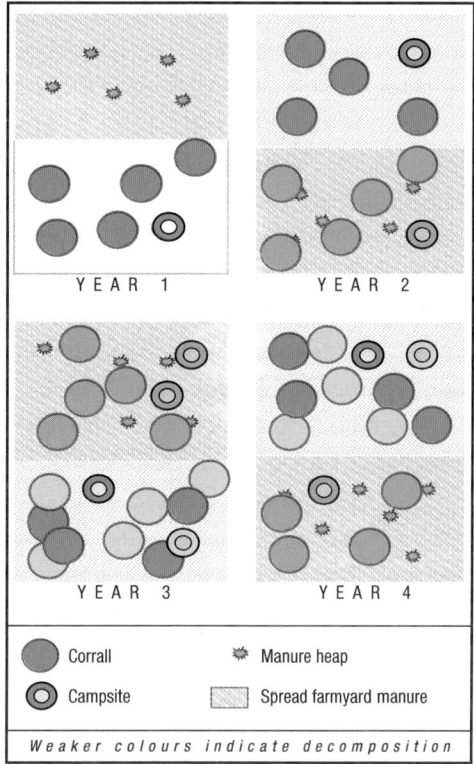

FIGURE 8

Model of the rotation of corrals, campsites and the spread of farmyard manure on a field in north-western Nigeria, resulting in spatial heterogeneity of soil fertility (Hoffmann et al., 2001)

intensively manured spots results in good crop yields without risk of burning. Hence, such differences in soil fertility on one single field would help to reduce risk of crop losses and equalize yield variations. Brouwer *et al.* (1993) reported similar use of micro-variability for millet planting as a risk aversion strategy in Niger and argued that farmers aim for reliable but not maximum yields even in years of bad rainfall.

Biodiversity management on cropland: species, varieties and planting patterns

The farmers try to increase species and spatial heterogeneity through their choice of crop species, varieties and planting patterns. Most common features are a multi-species mix, a large number of local varieties for different targets and various crop species

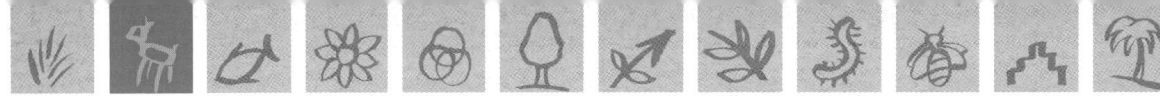

combinations and planting patterns. They will be explained in the following paragraphs.

The land in the reserve has been continuously cropped for the last 40 years. Fifty different crop species combinations were found, mostly in the relay intercropping system (Mané-Bielfeldt *et al.,* 1998). In the vast majority of fields, subsistence gramineous staple crops were intercropped with more cash generating and nitrogen-fixing legume crops.

Vegetables, spices and some economic trees were grown to a lesser extent.

The most common combinations were millet *(Pennisetum ssp.)* and cowpea *(Vigna unguiculata)* with 35 percent of investigated fields, millet, cowpea and groundnut *(Arachis hypogaea)* (11 percent) and millet, sorghum *(Sorghum bicolor)* and cowpea in nine percent of the fields (Table 3). Crop rotations are uncommon. Pure millet cropping was rare (3.4 percent), while various two-crop combinations, millet plus another crop, occurred most frequently (47.5 percent). One-quarter of the fields was planted with three crops. Cotton *(Gossypium hirsutum) is* sometimes part of the system. In *fadama* land, low-lying areas that are seasonally

flooded or have a high water table and can be cultivated during the dry season as well, rice *(Oryza sativa)* is grown, sometimes together with maize *(Zea mays)*. However, these fields make up only a small percentage of the arable land in the enclaves.

The farmers were asked about the sources of their crop seeds and the factors influencing their choice of varieties. Results showed that farmers stored millet, sorghum and cowpea seeds from their previous harvest or bought seeds on the local markets. Groundnut and cotton seeds are more commonly bought on the market than those of the other crops. Seven varieties each were identified for millet and cowpea, eight for sorghum and three for groundnut. The varieties of sorghum and cowpea are chosen for their yield of grain as well as for by-product depending on the specific needs of the farmer. Livestock keeping farmers will chose a variety yielding high amounts of by-product, thereby integrating fodder production with grain production. Other criteria used in the choice are availability of seeds or the length of the growing period. Sometimes a mixture of several millet or sorghum varieties was planted in the same field to extend the

TABLE 3

Main combinations of species on the crop fields in the three villages Dumburum, Shamashalle and Tsabre in the Zamfara Reserve in 1996 (figures are percentages of all investigated fields)

SPECIES COMBINATION	DUMBURUM (n = 100)	SHAMASHALLE (n = 109)	TSABRE (n = 113)	TOTAL OF VILLAGES (n = 322)
mi+cp	11.0	49.5	42.5	35.1
mi+gn+cp	15.0	7.3	10.6	10.9
mi+so+cp	15.0	8.3	5.3	9.3
mi+so	11.0	2.8	8.0	7.1
mi+so+gn+cp	11.0	1.8	0.9	4.3
mi+gn	6.0	4.6	1.8	4.0
Mi	2.0	6.4	-	2.8
mi+so+gn	7.0	-	-	2.2
so+gn+cp	2.0	4.6	-	2.2
mi+co	2.0	1.8	1.8	1.9
ri+ma	-	-	5.3	1.9
So	3.0	1.8	0.9	1.9
Gn	-	-	0.9	0.3
Others	22.0	11.1	22.0	18.3
mi = millet; so = sorghum; cp = cowpea; gn = groundnut; ri = rice; co = cotton; ma = maize				

harvesting period and minimize risk of failure. The species, varieties and mixtures of crops grown on particular soils depended on how farmers rate the crop production potentials of the soils (Mané-Bielfeldt and Schaefer, 1996; Kyiogwom *et al.,* 1998).

The main crop combinations with millet and cowpea were found to be planted in a great variety of patterns, with a wide range of different row and plant distances (Tables 4 and 5). Sorghum rows were more widely spaced than millet (2.2-2.3 m vs. 1.0-1.2 m). In both combinations with groundnut, the average row distance for millet was wider.

The high spatial heterogeneity of planting patterns impedes the calculation of yield averages per single crop. The following model calculations of grain and crop residue yields are derived from the interviews and relate to a multitude of crop mixtures (Figure 9).

The estimated grain harvest of 1 856 kg ha^{-1} is sufficient to support of about two persons per ha on a cereal-based diet throughout the year (Oltersdorf and Weingärtner, 1996) and even leaves a surplus for sale. This finding is in agreement with the interview results that the people never

CROP COMBINATION	ROW 1	ROW 2	ROW 3	ROW 4	ROW 5
mi+cp (n=20)[1]	cp	mi	(mi)	(mi)	(mi)
	mi/cp	mi	(mi)	(mi)	
	cp	mi			
mi+cp+so (n=17)[2]	cp	so	mi	mi	
	so/cp	mi	mi		
	so	cp	Cp/mi		
mi+cp+gn (n=16)[3]	mi/gn	cp	(cp)	(cp)	(cp)
	mi	gn	cp	(cp)	(cp)
	mi/gn	mi/cp			
	mi/gn	mi	cp		
[1] 3 different patterns	[2] 8 patterns		[3] 8 patterns observed	In brackets: optional	

T A B L E   4

Most frequent planting patterns of the main crop combinations in the Zamfara Reserve enclave villages

	n	Average row distance (m)			Plants per 100 m row	
		range	most often	mean	range	mean
Millet	63	0.7-6.6	1.0-1.2	1.7±1.1	50-145	82±22
Sorghum	23	1.4-5.0	2.2-2.3	2.9±1.0	50-145	85±27
Cowpea	61	1.0-5.0	2.3	2.4±0.7	35-145	82±25
Groundnut	15	0.7-2.9	1.1	1.3±0.7	95-265	167±50

T A B L E   5

Average inter- and intra-row distances for the main crop species in the Zamfara Reserve enclave villages

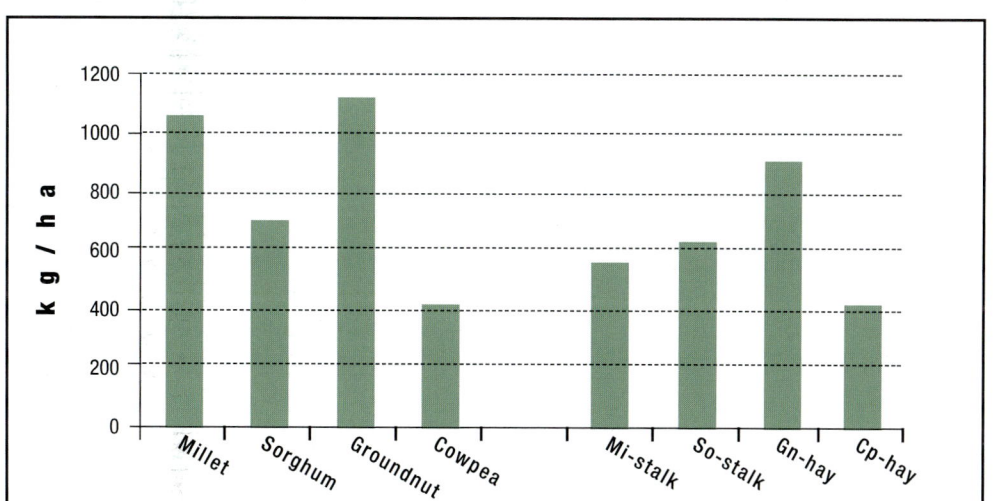

F I G U R E   9

Grain and crop residue yield estimates in villages in the Zamfara Reserve and bordering villages

experienced hunger, and that about 10 percent of the farmers indicated cereal sales (Mane-Bielfeldt *et al.,* 1998).

CONCLUSIONS AND RECOMMENDATIONS

The case study of the Zamfara Reserve shows that farmers and pastoralists have developed similar strategies to use and manage biodiversity. In general, they are based on local knowledge of soils, wild and domestic plants, and livestock, on the temporal and spatial variations of access to natural resources, including mobility and flexible property rights, on the exchange of goods and services within and between systems, and a mix of income generating activities. This is typical for West African drylands, where both livestock and farming production systems have been maintained facing variable rainfall, demographic expansion and changing market conditions. All the strategies are based on high diversity, flexibility and adaptability in order to better deal with incommensurables, as was defined by Mortimore and Adams (1999):

- flexibility – the day-to-day decisions allocating resources of land, labour, capital, knowledge or other scarce factors in response to opportunities and constraints;
- adaptability – the year-by-year modifications or transformation of their production systems in response to economic or environmental realities; and
- diversity – of available choices, for example, of animals, farm technologies, fodder sources, farm crops, markets, or off-farm incomes.

The biodiversity management of the livestock system can be summarized as **Use of rangeland biodiversity through the management of livestock mobility and deliberate choice of livestock breeds.** The experiences of the rangeland-based livestock system in the Zamfara Reserve show that the predominant mobile grazing management involves mobility of a large number of livestock and humans. This has social implications: The mobile system implies functioning social and information networks, logistics, communication and the solution of conflicts over resources between the different user groups. It also has ecological implications: In the present set-up, rangeland is an open access resource. Stocking density in the Zamfara Reserve exceeds the recommended rate. The related high grazing pressure seems to have ecological implications, affecting the herbaceous and woody layers of the vegetation. As a result, rangeland biodiversity is being reduced. Long-term ecological studies are needed to show whether this trend is reversible. Most long-term studies done so far in West Africa indicate that a certain vegetation resting or patchiness of grazing favour vegetation recovery and species' richness (Miehe, 1990; Hiernaux, 1998). The needed management practices, such as rotational grazing, however, cannot be achieved in an open-access situation. Ongoing studies investigate whether the rangeland in the reserve is common property resource with somehow regulated access or an open-access resource without any binding rules.

In order to generalize the experience of the cropping system of the Zamfara Reserve, it was found that soil fertility maintenance is a key issue for the farmers. Farming is based on relatively secure property rights to land. The farmers' strategy is the active **Management of soil and crop biodiversity,** through maintaining diversity on various spatial and temporal scales, within and between species of crops and animals, and they do so to reduce risk. Farmers rely not only on close crop-livestock integration on the farms (crop-residue feeding, nutrient cycling), but also on exchange and interaction with the mobile livestock system. Animal genetic resources are exchanged between the two systems.

Policy recommendations include:

- Support farmers and pastoralists to maintain and enhance their diversity management.
- Support livestock mobility through flexible property rights and secure access of mobile pastoralists to pasture and water.
- Enable choice of breeds and varieties for various purposes and agro-ecological conditions.
- Support adaptation to future climatic and economic challenges.

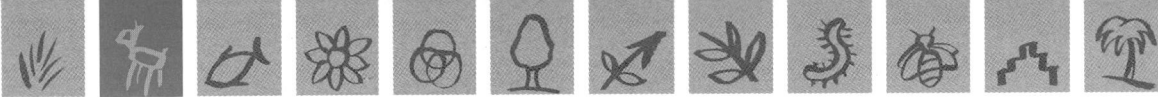

REFERENCES

Amanor, K.S. 1995. Dynamics of herd structures and herding strategies in West Africa: A study of market integration and ecological adaptation, *Africa* 65(3): 351-393.

ARCA, 1995. Forest Reserve study: Ruma Kukar Janjarai and Zamfara Forests. Final report Vol. 1. The environment: present conditions and degradation factors. Katsina Arid Zone Programme, Federal Republic of Nigeria, National Planning Commission.

Aubreville. A., 1950. La flore forestière soudano-guinéene. Paris.

Blench, R. 1994. The expansion and adaptation of Fulbe pastoralism to subhumid and humid conditions in Nigeria, *Cahiers d'études Africaines* 133-135: 197-212.

Blench, R. 1999. Traditional livestock breeds: Geographical distribution and dynamics in relation to the ecology of West Africa, ODI *working paper* 122, 67 pp.

Boserup, E. 1965. The conditions of agricultural growth. Allen and Unwin, London.

Boudet, G. 1991. Manuel sur les pàturages tropicaux et les cultures fourragères. Manuels et précis d'élevage No.4, Ministère de la Coopération, IEMVT.

Brouwer, J. Fussell, L.K., Herrmann, L., 1993. Soil and crop growth micro-variability in the West African semi-arid tropics: a possible risk-reducing factor for subsistence farmers. Agric. Ecosyst. Environ., 45, 229-238.

Drechsel, P. Gyiele, L., Kunze, D., Cofie, O., 2001. Population density, soil nutrient depletion, and economic growth in sub-Saharan Africa. Ecological Economics 38: 251-258.

Eckert, B. Hoffmann, I. 1998. Statutory versus customary land allocation: Implications on land tenure and resource management in the Zamfara Reserve. In: Hoffmann. I. (ed.). Prospects of pastoralism in West Africa. Giessener Beiträge zur Entwicklungsforschung. Reihe I. Vol. 25. 169-178.

Eckert, B., Kueppers, K., Hoffmann I., 1998. Indigenous trees as an important resource in rural life - the case of Zamfara Reserve. In: Hoffmann. I. (ed.). Prospects of pastoralism in West Africa. Giessener Beiträge zur Entwicklungsforschung. Reihe I. Vol. 25. 228-237.

Elsholz. C. 1996. Studies on structure and botanical composition of the vegetation in Zamfara Reserve. Sokoto State. Nigeria. UDU/JLU EEC Linkage Programme. unpublished research report. Gießen.

FDLPCS (Federal Department of Livestock and Pest Control Services) 1992a. Livestock in Sokoto Sate, in *Nigerian livestock resources*. Vol II: National synthesis. Vol III: State reports. RIM, UK, 26pp.

FDLPCS (Federal Department of Livestock and Pest Control Services) 1992b. *Nigerian livestock resources*. Vol I: Executive summary and atlas. RIM, UK, 53pp.

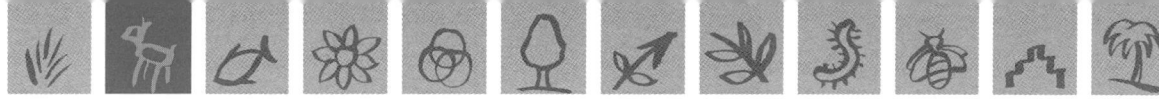

Gefu, J.O. 1992. Pastoralist perspectives in Nigeria. The Fulbe of Udubo Grazing Reserve. Research Report No. 89. The Scandinavian Institute of African Studies, 106 pp.

Harris, F.M.A. 1998. Farm-level assessment of the nutrient balance in northern Nigeria. *Agric. Ecosyst. Environ*. 71, 201-214.

Hassan, W.A. 2000. Biological productivity of sheep and goats under agro-sylvo-pastoral systems in the Zamfara Reserve in north-western Nigeria. Cuvillier, Goettingen, 260pp.

Hayami, Y., Ruttan, V.W. 1985. Agricultural Development: An International Perspective, Baltimore. Johns Hopkins University Press.

Hof, A. 2000. Developing a GIS methodology for the Analysis of agro-pastoral livelihood and ecosystems in the Zamfara Forest reserve, NW Nigeria. MSc dissertation, Dept. of Geography, University of Durham, September 2000.

Hoffmann, I. 1998. Zamfara Reserve - Past and present. In: Hoffmann. I. (ed.). Prospects of pastoralism in West Africa. Giessener Beiträge zur Entwicklungsforschung. Reihe I. Vol. 25, 1-9.

Miehe, S. (1990). Inventur und Überwachung der Vegetation in Weideparzellen. Ergebnisse der Untersuchungen 1988-1990 und vorläufige Gesamtauswertung des Standweideversuchs nach 10jähriger Laufzeit. GTZ Projekt Desertifikationsbekämpfung durch agro-pastorale Landnutzungsmodelle. Eschborn (unpublished).

Hiernaux, P. 1998. Effects of grazing on plant species composition and spatial distribution in rangelands of the Sahel, *Plant Ecology* 138, 191-202.

Hoffmann, I., Gerling, D., Kyiogwom U.B., Mané-Bielfeldt, A. 2001. Farmers' management strategies to maintain soil fertility in a remote area in Northwest Nigeria. *Agric. Ecosyst. Environ*. 86 (3): 263-275.

Hoffmann. I., Willeke-Wetstein. C., Schäfer. C. 1998. Desription of a grazing ecosystem in Northwestern Nigeria. using environmental indicators. *Animal Research and Development* 48: 69-83.

Keay, R.W.J. 1949. An example of Sudan Zone Vegetation in Nigeria. J. Ecol. 37. 335-364.

Kreimer. D., Steinbach. J. 1998. Control of *Cassia tora* Linn. in three pasture ecotypes in the Sudan Savanna in north-western Nigeria. In: Hoffmann. I. (ed.) Prospects of pastoralism in West Africa. Giessener Beiträge zur Entwicklungsforschung. Reihe I. Bd. 25. 87-95.

Krieger, K. 1954. Die Zamfarawa - Ein Stamm der Hausa in Nord-Nigeria. Geogr. *Rundschau* 6, 387-393.

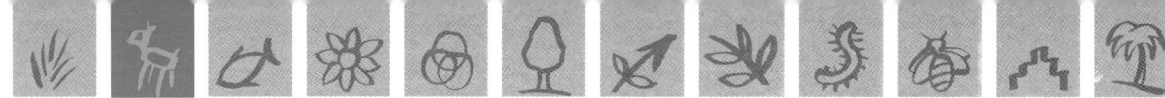

Kueppers, K. 1998. Evaluation of the ligneous strata of the vegetation of the Zamfara Reserve. In Hoffmann. I. (ed.) Prospects of pastoralism in West Africa. Giessener Beiträge zur Entwicklungsforschung. Reihe I. Vol. 25, 41-47.

Kyiogwom U.B., Umaru, B.F., Bello, H.M. 1998. The use of indigenous knowledge in land classification and management among farmers in the Zamfara Reserve. In: Hoffmann, I. (Ed.). Prospects of pastoralism in West Africa. Giessener Beiträge zur Entwicklungsforschung, Reihe I, Vol. 25, pp. 220-227.

Kyiogwom, U.B., Mohammed, I., Bello, H.M., Maigandi, S.A., Schaefer, C. 1994. The economic situation of the livestock farmer in Zamfara, in: Range development in the endangered Sudan savanna in Sokoto State. Unpublished report, Giessen, 63-70.

Malami. B.S.. Hassan. W.A.. Tukur. H.M.. Maigandi. S.A. 1998. Communal grazing of sheep and goats in the Zamfara Reserve. In: Hoffmann. I. (ed.). Prospects of pastoralism in West Africa. Giessener Beiträge zur Entwicklungsforschung. Reihe I. Vol. 25. 191-195.

Mané-Bielfeldt, A., Schaefer, C. 1996. Evaluation of crop production in the enclaves of Zamfara reserve, NW Nigeria, Part 2. Unpublished research report, Giessen.

Mané-Bielfeldt, A., Schaefer, C., Gefu, J., Hoffmann, I., Mohammed, I., Steinbach, J. 1998. Crop production and utilisation in the Zamfara Reserve. In: Hoffmann, I. (Ed.). Prospects of pastoralism in West Africa. Giessener Beiträge zur Entwicklungsforschung, Reihe I, Vol. 25, pp. 238-246.

McIntire, J., Bourzat, D., Pingali, P. 1992. Crop-Livestock Interaction in Sub-Saharan Africa. Washington, D.C. The World Bank.

Mohammed, I., Bello, H.M. 1994. Transhumance: a rangeland resource management strategy in an endangered ecosystem, in Range development in the endangered Sudan savanna in Sokoto State. Unpublished report, Giessen, 73-80.

Mortimore. M., Adams, W.M. 1998. Farming intensification and its implications for pastoralism in northern Nigeria. In: Hoffmann. I. (ed.). Prospects of pastoralism in West Africa. Giessener Beiträge zur Entwicklungsforschung. Reihe I. Vol. 25. 262-273.

Mortimore. M., Adams, W.M. 1999. Working the Sahel: Environment and society in Nigeria. Routledge, London, UK.

Mortimore. M., Essiet. E.U., Patrick. S. 1990. The nature. rate and effective limits of intensification in the small holder farming system of the Kano Close-Settled Zone. Federal Agricultural Coordinating Unit. Ibadan.

Oltersdorf, U., Weingaertner, L. 1996. Handbuch der Welternaehrung - Die zwei Gesichter der globalen Nahrungssituation Verlag J.H.W. Dietz Nachfolger GmbH, Bonn - Hrsg. Deutsche Welthungerhilfe.

Powell, J.M. and E. Taylor-Powell 1984. Cropping by Fulani agropastoralists in central Nigeria, *ILCA Bull.* 19: 21-27.

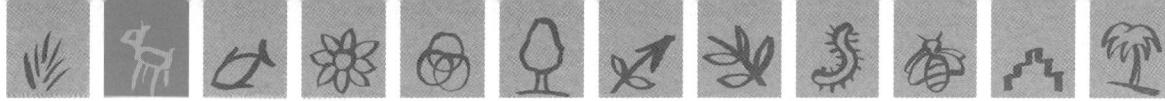

Powell, J.M. 1986. Manure for cropping: A case study from central Nigeria. *Expl. Agric.* 22, 15-24.

RIM (1991). Woody vegetation cover and wood volume assessment in northern Nigeria. Report for the Federal Government of Nigeria. St. Helier. UK.

Sandford, S.G. 1989. Crop Residue/Livestock Relationships. In: Soil, Crop, and Water Management in the Sudano-Sahelian Zone. ICRISAT Proceedings of International Workshop, 11-16 Jan. 1987 in Niamey/Niger, pp. 169-182.

Schaefer. C. (1998). Pastorale Wiederkäuerhaltung in der Sudansavanne: Eine Untersuchung im Zamfara Forstschutzgebiet im Nordwesten Nigerias. Cuveillir. Göttingen. 217pp.

Scoones, I. (Ed.), Living with uncertainty. New directions in pastoral development in Africa. London, pp. 1-36.

Sutter, J.W. 1987. Cattle and inequality: herd size differences and pastoral production among the Fulani of northeastern Senegal, *Africa* 57 (2): 196-217.

The Forestry Ordinance (Chapter 75) Sokoto Native Authority Forest reserve No. 9 (Zamfara Forests) (Amendment) Order 1957, in *Northern Regional Public Note* 38 of 1957.

Vabi, M.B. 1993. Fulani settlement and modes of adjustment in the Northwest Province of Cameroon, ODI *Pastoral Development Network Paper* 35d. 10 pp.

Van Raay, H.G.T. 1975. Rural planning in a savanna region. University Press, Rotterdam.

RESPONSIBLE
TECHNICAL DIVISION

Fishery Resources Division
Inland Water Resources and Aquaculture Service

Matthias Halwart

TRADITIONAL USE AND AVAILABILITY OF AQUATIC BIODIVERSITY IN RICE-BASED ECOSYSTEMS

TRADITIONAL USE AND AVAILABILITY
OF AQUATIC BIODIVERSITY
IN RICE-BASED ECOSYSTEMS

I. Kampong Thom Province, Kingdom of Cambodia*

AUTHORS

Peter Balzer
German Development Service
Phnom Penh, Cambodia
p.balzer@mobitel.com.kh

Tonette Balzer and Sibura Pon
(Consultants)

EDITORS

Matthias Halwart
FAO Inland Water Resources
and Aquaculture Service
matthias.halwart@fao.org

Devin Bartley
FAO Inland Water Resources
and Aquaculture Service
devin.bartley@fao.org

GUEST EDITOR

H. Guttman
Mekong River Commission, Cambodia
guttman@mrcmekong.org

ACKNOWLEDGEMENTS

The support of the following people in
the identification and verification of
species is gratefully acknowledged:
Walter Rainboth, University of
Wisconsin-Oshkosh, USA (fishes),
Bryan Stuart, Field Museum, Division
of Amphibians & Reptiles, Chicago,
USA (reptiles, amphibians),
Ruth O'Connor, FAO-Participatory
Natural Resource Management in Tonle
Sap Region, Cambodia (insects), and
Robert Cowie, University of Hawaii,
USA (molluscs). Thanks also to
Patrick Evans of the FAO-Participatory
Natural Resource Management in Tonle
Sap Region Project in Siam Reap and
Colin Poole of the Wildlife
Conservation Society in Phnom Penh
who facilitated contacts.
Ms. Georgia De Clancey Eva provided
assistance in documentation and
dissemination of study results.

CONTENTS

* Supported by the FAO/Netherlands Partnership Programme

INTRODUCTION

Food production in Kampong Thom Province is dominated by rice production and inland fisheries, both of which are heavily influenced by the Great Lake, also known as the Tonle Sap. Both systems contain a wealth of biological diversity that contributes to ecosystem processes, food security and cultural heritage.

The great lake ecosystem

The Tonle Sap, also known as the Great Lake in central Cambodia, is the heart of Cambodia's freshwater fisheries[1]. It is the largest freshwater lake in Southeast Asia and one of the richest inland fishing grounds in the world, with an estimated annual catch of 250 000 tonnes (Van Zalinge, personal communication). Its hydrological characteristics are mainly influenced by the Mekong River which, via the Tonle Sap River, fills the lake during the rainy season (May–October) and drains it for the rest of the year. The Tonle Sap basin contributes only 6.4 percent of the annual flood waters[2] (MRC, 1998). In the dry season, the lake covers an area of 2 500–3 000 km^2 with an average depth of between one and a half and two meters. During the rainy season the area increases to 9 000–14 000 km^2 and the depth reaches 9–11 metres.

A belt of freshwater mangroves known as the "flooded forest" surrounds the lake. This gradually changes to bushes and finally grassland with increasing distance from the lake. The floodplains are surrounded by low hills, which are naturally covered with evergreen or deciduous dry Dipterocarp forest. The figures below show transects from the lake to the surrounding hills at the end of the wet and the dry season respectively.

The lake's flooded forest and the surrounding floodplain are of great importance for Cambodia's freshwater fisheries. At the beginning of the flooding, many fish species leave the lake and the larger ponds for the now flooded forest to spawn. The young fish then move out into the floodplains to feed. The inflow of Mekong River floodwater brings with it large amounts of fry and fingerlings which also find shelter and food in the flooded forest and surrounding flood plains. At the end of the flooding, many fish follow the receding waters back to the lake and through the Tonle Sap River to the Mekong River.

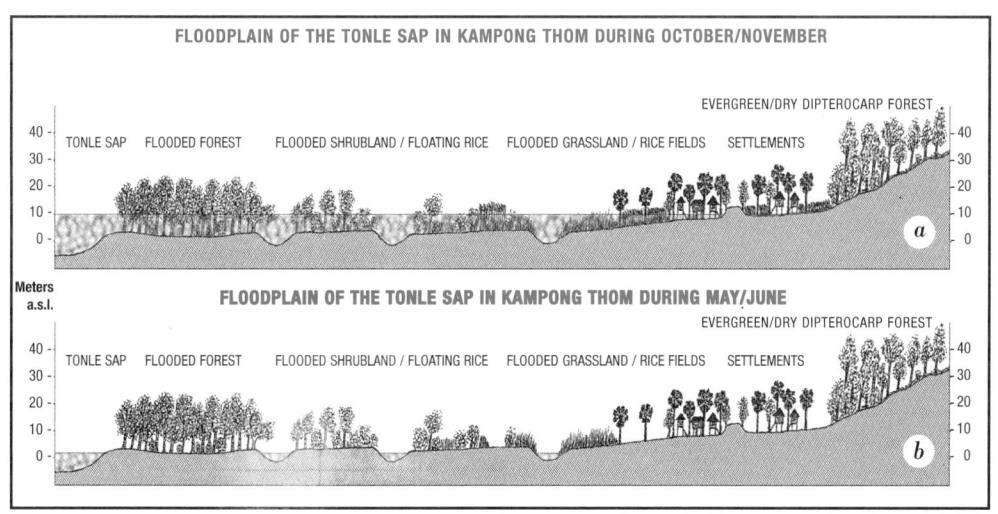

FIGURE 1

Floodplain of the Tonle Sap in Kampong Thom during (a) October/November and (b) May/June

---

1 The coastal fisheries are relatively minor in Cambodia, in contrast to Viet Nam, Thailand, Philippines and Indonesia.

2 It should be noted though that Tonle Sap River, which drains Tonle Sap Lake in the dry season, contributes 16 percent of Mekong River's flow in the dry season.

Ricefield fisheries in Cambodia

Traditionally the people living around the lake in areas subject to deep flooding have planted rice varieties that could cope with the high water levels by elongating their stems up to five meters with a maximum growth of ten centimetres per day. This floating rice is broadcast at the start of the rainy season on previously ploughed fields in areas naturally covered with bushes or grass. At greater distance from the lake where the flooding is not as deep, normal wet rice varieties are transplanted into the fields once the flood has reached them.

In some areas rice is planted in the receding water as the floodwaters recede. The water present in the soil has to be enough for the development of what is known as "receding" (recession) rice. With the introduction of improved rice varieties maturing quickly many areas formerly planted with deep water rice are now growing a recession rice crop instead. Controlled irrigation, which in Cambodia was present during the Angkorian period of the 13th Century, is nowadays the exception.

Rice is the staple of rural Cambodia and Ovesen *et al.* (1996) identified the deep cultural significance of eating (white) rice, going as far as being one of the things that separates humans from wild animals[3]. Fish and fishing are also central to rural households. In Cambodia there is a saying, "where there is water there is fish" *(mien tuk mien trey)* and considering that over a third of Cambodia is a seasonally inundated flood plain the significance is obvious. The importance of ricefield fisheries[4] is, however, generally underestimated.

Ricefield fisheries in the context of fishery activities in Cambodia

Rice field fisheries is only one chapter in the story of fishery in Cambodia. The Department of Fisheries (DoF) has traditionally concentrated on the management of the licensed "fishing lots" in the Great Lake and the Mekong and Bassac Rivers. This focus is mainly due to the revenues generated (> US$ 2 000 000 in annual tax revenue) through the auctioning of the lot licences to professional fishers and business ventures for commercial exploitation. This has a long history as the fishing lot system goes back to the early 1900s, when the French colonial government established the lot system[5].

These fishing lots occupy the most productive fishing grounds in Cambodia. In a recent move the Prime Minister has ordered the size of the fishing lots to be reduced and the parts removed opened to the general public for fishing. This was carried out by the Department of Fisheries of the Ministry of Agriculture, Fishery and Forestry, and the size of the fishing lots has countrywide been reduced to about 50 percent of their previous area.

A look onto the map reveals that the most productive areas in terms of fish catch –the rivers and streams that drain the floodplains during the onset of the dry season and channel the fish into the Tonle Sap and the Mekong - still remain within the fishing lot area.

In those cases where a more important fishery resource has been released to the public conflicts arose immediately about the control of these areas. Powerful persons within the areas tried to take control of these places, often supported by the local authorities in a move to supplement their meager salaries.

---

3 Or perhaps a distinction between civilised people and savages.
4 Defined as fishing in and around the rice fields.
5 Specifically the 1987 Fiat Fisheries Law formalised fisheries practices, such as the lot system, established in 1908.

Productivity of the fishing grounds

Statistics on the productivity of the lots for the entire country reveal that they are very productive, but precise estimates are not available. In recent years the estimates are improving and the official production from the commercial operations was reported to be 442 000 tonnes in 2 000, up from 75 000–85 000 in previous years. For unlicensed fisheries, such as ricefield fisheries, no such annual statistics are collected. Estimates of ricefield catch (from surveys) differ widely depending on the area and the year of the study. Regional figures vary from a low of 25 kg in Northeast Thailand (Spiller, 1985) through 125 kg in South-eastern Cambodia (Gregory *et al.*, 1996) up to 150 kg per hectare and year in Malaysia (Ali, 1990). Ahmed *et al.* (1998) used figures ranging from 25-61 kg per hectare and year to estimate the annual production of Cambodia's ricefield fisheries. Multiplied with the 1.8 million hectares of Cambodian rice fields they reach an annual production of 45 000 to 110 000 tons, amounting to 15-25 percent of Cambodia's total annual fish catch. This extrapolation, however, is a rather unreliable method of assessment, since the heterogeneity in agro-ecological conditions and differences in the micro-environments of rice fields result in a variable productivity (Little *et al.*, 1996). Generally speaking, there is a tendency to underestimate the importance of ricefield fisheries since they tend to yield only small amounts of fish at a time, but on a regular basis and for many people involved.

Also the diversity of organisms found in the ricefield ecosystem varies according to the place and the year of the study. Heckman (1979) found 19 aquatic animal species being used by farmers in Thailand, whereas Shams and Hong (1998) found 35 species in a study conducted in Kampong Thom Province, and Gum (1997) found 39 species during a survey in Battambang Province[6].

THE STUDY

The study was conducted in co-operation with the Provincial Department of Agriculture, Fishery and Forestry, the counterpart for the Natural Resources Management Component of the Rural Development Program (RDP-NRM) in Kampong Thom and Kampot, where one of the researchers works as an advisor. Most of the villages included are also target villages for RDP-NRM.

Scope of the study

This study was conducted from the middle of September to the middle of December 2001 in Kampong Thom Province, Cambodia[7]. Fish and other organisms were collected from the ricefield ecosystems at eight different locations in three districts of Kampong Thom Province: Stoung, Stueng Saen and Santuk. Two of the places were located closer to the Great Lake with direct access to the flood plain, the other ones at various distances from it along highway No. 6. The maps in the annex show the places where the samples have been taken.

In addition to the information on the availability of the organisms, the tool used for collecting and the ultimate use of each species, was collected along with people's preferences among the different species. Any additional information found has been added under the heading of traditional knowledge and observations.

Limitations of the study

The abundance of fish and fishers is extremely seasonal in the area. The study began in the middle of September and thus

---

6 For comparison, not restricted to ricefield ecosystems Rainboth (1996) reports 500 species of fish for Cambodia also stating that the real number of species is certainly higher.

7 It should be noted that in 2000 Cambodia was struck by the severest floods in over 30 years, leaving large areas inundated for months. The floods in 2001, albeit less severe, were still substantial, consequently one would expect a lot of aquatic organisms in 2001.

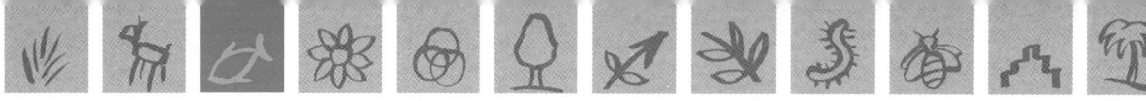

after the beginning of the fishing season. No attempt has been made to weigh the organisms caught by the fisher-folks involved in the study, since the day-to-day variability of the catches was too great and the scope of the study limited to three months and eight locations. This presents a challenge for further studies: to develop a method to reliably estimate the amount of fish caught within rice field ecosystems.

In order to obtain reliable data, the researcher himself will have to weigh the daily catches since the farmers/fishermen are often unwilling to do this by themselves, for a variety of reasons. Since a researcher can only do this with a limited number of farmers/fishermen, the data can only be valid for a small area and the year the study was conducted. As mentioned above, extrapolation of data generated on a small-scale are likely to give a wrong picture of the overall situation.

In terms of overall abundance and availability, certain groups of animals are under-represented in this study since, according to the local people, their "season" is either before the start or after the end of the study. Important groups include water insects, crabs, some snails, and frogs, which are mainly caught in the dry season and early wet season. Initially the study focused on aquatic animals collected from the ricefields to which aquatic plants were added later.

Methods used

To collect information from the local people, several different methods were used sequentially. The study was initiated by conducting Participatory Rural Appraisals (PRAs) in three villages. The second step was collection of information on the organisms caught by the local people. At the end of the study, single and group interviews were used to verify the information previously collected.

Participatory Rural Appraisal

As the first step, PRAs were conducted in the villages of Doun L'a in Stoung District, Panha Chi in Stueng Saen District and Tboung Krapeu in Santuk District. People were asked, during a village meeting, to enumerate the aquatic animals they collect from their ricefields, their uses, etc. At the same time the PRA served as an introduction to the people to ensure that they understood the purpose of the subsequent regular visits in their village.

Species collection

From the end of September 2001 to the beginning of December 2001[8], the researchers went to the field almost every day. The maps in the annex show the villages where the collections were made. The collection points were the sites where people went to fish in or near the ricefield ecosystems. The drawing below shows a typical situation in Kampong Thom: the road is built on a dam. Soil for it has been excavated on both sides, forming canals left and right of the road. During the rainy season these canals are filled with water and directly connected to the surrounding rice fields. People gather to catch fish near bridges and culverts, which are like a bottleneck for water and fish.

At points like this, as well as within the ricefields, specimens were collected and pictures taken of the various species caught. Samples of every organism smaller than 15 cm were collected and preserved for later identification.

---

8 It should be noted here that the period is covers only the mid of the wet season to the early dry season, but in 2001 the floods were extensive and thus there was much water even in the early dry season.

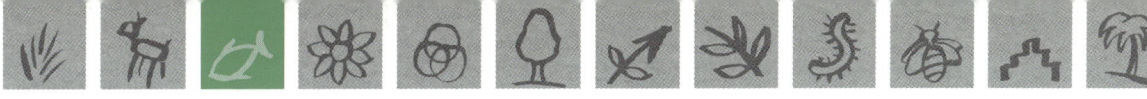

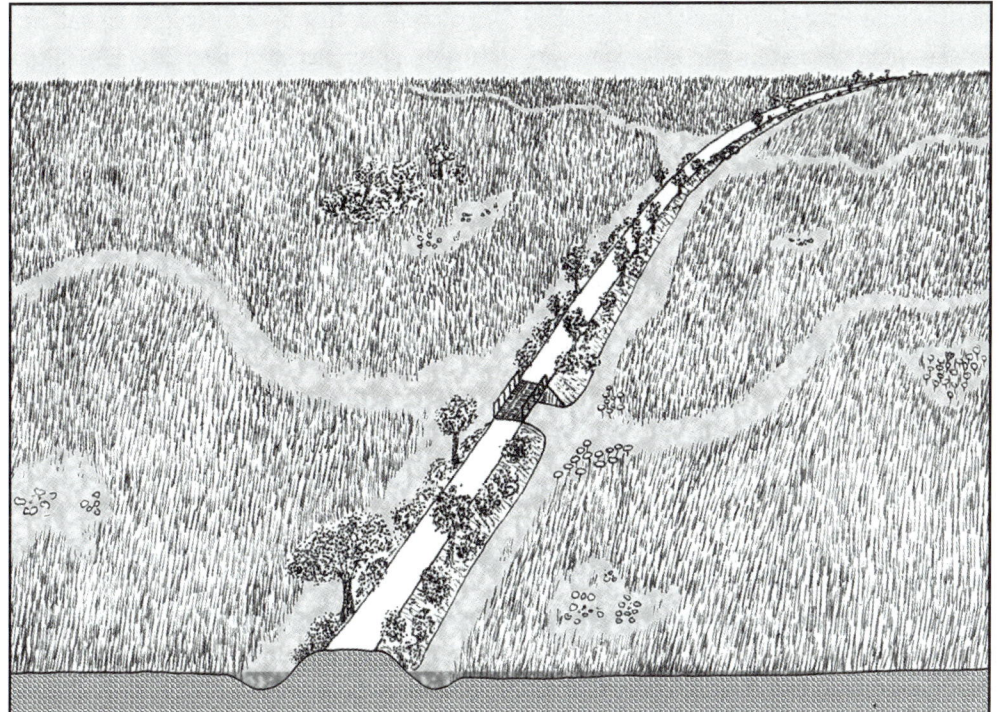

FIGURE 2

*ricefields and canals
all become inter-
connected in the
rainy season in
Kampong Thom
Province.*

N O . 3

The pictures were developed locally, then scanned and computer processed. Fish were identified as far as possible, using the field guide by Rainboth (1996).

While collecting the specimens, the fisher-folks were asked to give information on:

- the availability of the species;
- their uses in the rural community;
- the preferences of the people for them; and
- the various fishing tools used to harvest them.

Interviews

At the end of the fishing season, the information collected previously was consolidated and verified in single and group interviews conducted in the villages where the collections were made. Information on preferences was obtained during samplings, PRAs and group interviews and ranked on a scale from 1 = not liked, 2 = liked, to 3 = highly esteemed. Availability was ranked on a scale from 0 = absent, 1 = rare, 2 = little, 3 = medium, to 4 = abundant, all information obtained in group interviews.

Since the people were by then already familiar with the researchers, no initial shyness had to be overcome. People were talking freely about the aquatic animals they used to collect and also about the difficulties and problems encountered.

RESULTS

Species found

The species found have been categorized into seven groups: fishes, reptiles, crustaceans, amphibians, molluscs, insects and plants. They are discussed separately below.

Fishes are by far the most important group, both in species diversity and importance for the local population. For

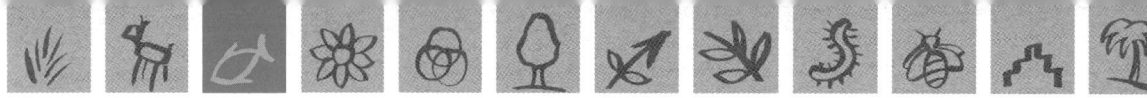

most Khmer people fish is the primary source of animal protein and is part of every meal in either fresh or processed form. Catching and collecting animals in and around the ricefields goes on all year round, however most fishing efforts in Kampong Thom start in August and end in December when the water recedes and leaves the fields dry[9].

In the course of this study 70 different species were found in the ricefield ecosystem. Of the 70 species, 25 were considered abundant and another 12 species were still commonly seen in the catches. Twenty-four species were rated as highly favoured. However, only four of them were abundant as well: the chevron snakehead *(Channa striata)*, broad-headed catfish *(Clarias macrocephalus)*, walking catfish *(C. batrachus)* and the swamp eel *(Monopterus albus)*. Eleven of the most favoured species are considered rare. The availability of the fishes changed during the course of the fishing season; some were more common at its beginning, others were found later. Some fishes came 'in waves', abundant one day and rare the next. The local people claim that they have seen the eggs of four species within the ricefields, three of them among the most favoured fish: the two catfishes mentioned above and the chevron snakehead, the latter being an opportunistic species that can breed all year round. The fourth species is the climbing perch *(Anabas testudineus)*, found abundantly but not as well liked as the other three (usually due to its small size, when caught large it is a high value species).

Most fish is eaten fresh, but there are a number of ways to preserve fish. A typical Khmer fish preserve is *Prahoc*, fermented fish paste. It is often made from the least favoured fish or from fish left over that cannot be sold fresh any longer. Other ways to preserve fish are to place it in salt, to dry it or to smoke it over wood fire. The latter method results in a highly priced product which is seen less and less in the villages and markets due to the lack of firewood needed for smoking. Two other types of processed fish command high prices in the markets: *mam* and *trey ngiat*. Both are made from the filet of large fishes like the snakeheads, marble fishes, catfishes and other favoured species. *Mam* is made by fermenting the filet while trey *ngiat* is sun-dried.

Reptiles: Seven snakes and one species of turtle were found during this study. No proper identification could be made of reptiles since no literature was locally available. Snakes are well liked for food and in some areas have been seriously reduced through over-collection, often for sale to urban markets. Of the snakes found only one was found in abundance, two were seen commonly, especially in the deep water ricefields of Roluos. Snakes are usually eaten fresh, only one snake has been reportedly used as a traditional medicine: preserved in alcohol it is said to enhance the appetite. Snakes can be found in the ricefields throughout the year.

The turtle found is considered uncommon, it was seen only twice during the course of the study and only in Roluos. It is considered a delicacy and is also reportedly used as a traditional medicine. A very destructive way of hunting turtles has been described for the dry season: the flooded forest – then dry – is set to fire and the turtles would come out of their hiding places to escape. In the ricefields, turtles can only be found in the months from August to December, during the rest of the year they are found in the flooded forest.

---

9 During the dry season toads, crabs and snails are dug out from the dry rice fields and in the early wet season insects and small frogs add to the catch.

Crustaceans: Five species of crabs and one shrimp were collected during the study. In his study on ricefield crabs Van Amerongen (1999) suggests that all ricefield crabs found belong to the same genus *Somanniathelphusa* only differing in colour and size with no morphological differences. In this study at least some morphological differences can be seen suggesting that more than one species were found. All but one of the different "types" of crabs were abundant, however, they are not well liked. Commonly used as bait or as feeds for pigs, people would eat them in time of scarcity. In other areas they are often an important source of food at the end of the dry season when they are dug out of the dry ricefields. Crabs are generally considered a pest in the ricefields where they feed on the rice plants and can do considerable damage to a newly planted field. They are chiefly collected from June to December.

Also the shrimp[10], *Macrobrachium lanchesteri*, is found in abundance. It is used for food either fresh or dried or processed into shrimp sauce. Shrimps can be found in the ricefields from September to December.

Amphibians: Two amphibians were found in the course of the study, one toad and one frog. Since literature on amphibians is not available locally, no identification could be made. Both species are reportedly abundant particularly early in the rainy season (June to September). The frog is very much liked for food and during the season commonly seen in the markets. The toad is used as an anthelmintic for cattle but it is also eaten. It is sometimes exported to China.

Molluscs: One snail was found within the ricefield ecosystem, other snails and also shells were only collected from the rivers and the lakes in the study area. While snail is liked better then crab it cannot compete with fish and is often used as bait to catch fish or fed to pigs. No species identification could be made due to the lack in literature available locally. Snails are collected in the ricefields from June to December, but in other areas, much like crabs, are an important dry season food dug out of the ricefields.

Insects: Two water insects were found to be of importance for the local people: a giant water bug from the Belostomatid family and a water beetle probably from either the Dytiscid or the Hydrophilid family. Both of them are used as finger foods and both command good prices in the markets following the season for frogs (September to October). The giant water bug is also used as traditional medicine: mixed with alcohol it is given to women after birth. In other areas dragonfly nymphs and other larger water insects are important sources of food in the early wet season (June-August). The water beetle is abundantly available; the giant water bug is still considered common.

Plants: Apart from the rice itself there are a number of other plants found within the ricefield ecosystem, which are used by the people in the area. Thirteen species were recorded, six of which were marketed. The other seven species are chiefly used as feeds or consumed locally and have no or a very low market value. All plants recorded were found in abundance, some during the time of the flood, others more towards its end. Of particular importance is morning glory *(Ipomoea aquatica)*.

The plants could be identified using the Dictionary of Plants used in Cambodia (Dy Phon, 2000) and the handbooks of the PROSEA series (2002).

---

10 Toft Mogensen (2001) identified the shrimp caught in ricefields as *Macrobrachium lanchesteri*.

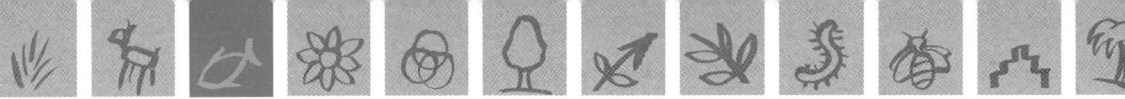

Fishing or harvest methods used

Farmers and fisher-folks of Kampong Thom Province use a wide variety of implements and techniques to collect fishes and other aquatic organisms from the ricefield ecosystem. In total 26 techniques to catch fish were recorded[11]. They can be sub-divided into four main categories:

Baited hooks are rather selective since some species are more attracted to a certain kind of bait then others (e.g. crab eggs are usually good for catching climbing perch). Four different types can be distinguished here, two of which are used actively, and the other two are attached to shrubs or sticks and checked at regular intervals. Baited hooks are also used to catch frogs.

Traps are usually less selective and apart from fish a variety of other aquatic animals like frogs, snakes, crabs and shrimps can be caught with them. Often made from woven bamboo the selectivity is defined by the distance between the bamboo strips. Traps for specific fish exist like the different eel traps. All together, six different types of traps can be distinguished.

Nets, like the traps, are not very selective and the main selectivity is determined by the size of the mesh. Six different types of net were observed. A distinction can be made between nets used actively, like a cast net, and nets placed and left, like gill nets. It was noted that gill nets used in Roluos had a larger mesh size probably since plenty of fish is still available in this area. In areas further away from the flooded plains, like Tboung Krapeu or Tuol Vihear, smaller mesh sizes were employed to catch all but the smallest fish. In these areas many juveniles were part of the catch, which can lead to over-use of the resource. Another particular problem is that old unusable gill nets are left at their position in or near the water becoming a

hazard to many water birds, which get caught and die in them.

Other: Here all techniques and implements that do not fit into the three categories above have been grouped together. They are usually characterized as active techniques such as digging, emptying depressions and catching fish with a spear or by hand.

An additional, illegal and very effective technique has become established during the last decade, i.e. electro-shock fishing. It is not a very selective method (though it tends to select larger organisms) stunning or killing all nearby animals. It is blamed, in addition to the destruction of the flooded forest, for the great reduction of the fish catches during the last years.

The implements used to catch aquatic animals can also be grouped according to their use in either shallow or deep water. They can thus also be assigned to periods of time when they are commonly used: a succession of different fishing tools could be observed along with the changing levels of the flood waters in the area.

Most of the implements used to catch fish are traditionally used either by men or women. The elder children, mostly boys, would already support their father in the use of typical male implements, smaller children and girls would accompany the mother in her collecting activities. While both men and women catch fish and crustaceans, plants, snails and insects are rather collected by women and children. Since reptiles and amphibians are typically caught with fish traps, men are the ones collecting them.

Some farmers have established ponds in or near their ricefields. These are not usually stocked actively with fish but the fish enter them from the surrounding ricefields when they are drying up[12]. Some farmers would also place small fish caught during the end

---

11 The Mekong River Commission's Fisheries Programme is in the process of compiling a complete catalogue of the different fishing tools used in Cambodia. A previous study by Gregory *et al.* (1996) recorded 23 techniques.

12 For further reading on this please refer to Gregory (1997) and Guttman (1998 and 1999).

of the flooding into their pond to allow them to grow bigger. However, most farmers do not have such a pond, because the soil in their area is not suitable or they cannot afford the time needed to dig such a pond deep enough to retain water throughout the year. They declared that if an organization would support them they would readily agree to have a pond[13].

Traditional knowledge and observations

Over time people have accumulated a profound knowledge about fishes and their behaviour. They have very detailed understanding about what kind of fish can be found where and when[14]. A common observation is that many fishes lay their eggs in the flooded forest or in the flooded shrubs surrounding their rice fields. The fingerlings then come to look for food in the ricefields and the flooded grasslands. Another observation is that once the trees and shrubs are gone in an area, the abundance of fishes is reduced.

People in the area have observed that over the last two decades fish catches have been greatly reduced, and some fish species have disappeared altogether. This is blamed partly on the increasing use of illegal fishing tools like electro-shock but also on the destruction of the flooded forests surrounding Tonle Sap Lake[15]. The increased number of people living in the area, and thus fishing in the area, is also given as a reason for the diminishing catches. In fact it is likely that the total catch in the area has actually increased as a result of increased fishing pressure[16] (i.e. a greater number of

people fishing), but the individual catch as well as the average size of individual fish have dropped.

On average, a family of five persons would consume about one kilogram of fresh fish every day during the fishing season, and the same family would need about 20 kg of *Prahoc* to eat during the dry season, amounting to a total of some 200 kg of aquatic animal products annually for the family. Other studies (e.g. Ahmed *et al.*, 1998) indicate that fish consumption around the Great Lake is around 60-70 kg per person per year (fresh weight equivalent) totaling an annual consumption of about 350 kg for a family. In other areas of Cambodia the consumption is estimated at around 40 kg per person per year (Gregory *et al.*, 1996, APHEDA, 1997, Shams and Hong, 1998, Gregory and Guttman, 2002a).

A more market-oriented approach drives people to catch more fish for sale in the markets. Depending on the fishing tool employed, a farmer-fisherman can catch 15-20 kg of fish on a good day. The average fish catch during the fishing season is less than 10 kg per day. However, this is still very small-scale in the Cambodia context where a fishing lot operator can catch about the same amount of fish in less than ten minutes by blocking the migration route of fish in a medium sized river. Recent changes in the Cambodian fisheries legislation are dismantling a number of the fishing lots to be converted to co-management operations by surrounding communities.

The effects of the disease epizootic ulcerative syndrome (EUS), locally known as *dambao*[17], which spread throughout

---

13 It should be noted that the Family Food Production (FFP) programme dug some 20 000 ponds in Cambodia during the early 1990's.

14 The MRC Fisheries Programme has compiled a study on local knowledge of fisheries in the Mekong River basin (Bao *et al.*, 2001).

15 It should be noted that in the beginning of the 1980's the area of flooded forest cover was greater than in the late 1960's, due to the reduction in rice cultivating areas around the lake during the 1970s

16 As a comparison the population in Cambodia has grown from an estimated 8.5 million in the early 1980's to just over 11 million in 2000, an increase of almost 30 percent.

17 This is what local people call it; the word actually means that the fish has skin ulcers and wounds.

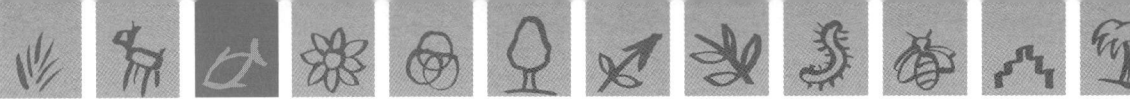

Southeast Asia in the mid 1980s are still noticeable and the disease is reported to occur every year with varying strength. Most noticeable in the period January to March, fishes (most susceptible are snakeheads and catfishes) in ponds and lakes start dying of big ulcers, but by the onset of the rains the disease disappears. Local people say that these fish cannot even be used to make *Prahoc*. The worst year in memory was 1995 when most lakes were transformed into stinking pits[18]. People claim that the increasing use of fertilizer and pesticides in the receding rice cultivation is to blame for this phenomenon. Epidemiological studies show that the disease spread to Cambodia in 1984 and has been endemic since then (Lilley *et al.*, 1992).

Local people believe that on Buddhist prayer days many fish can be caught. Since Buddhist prayer days coincide with the phases of the moon, this is supported by another observation: fish like the moonlight, they are playful in moonlight and are easily caught with gill nets at full moon. When rain is coming up, however, no fish can be caught. Only when the rain starts falling the fish would come out of their hiding places. During certain times of the day, very little fish is caught. Asked for the reason, a fisherman told that the fishes are now in the ricefields looking for food. They would come out later to play in the canal where they can be caught with the cast net.

CONCLUSIONS AND DISCUSSION

The ricefield ecosystem is of major importance to the local population for the supply not only of rice but also of animal protein and vegetables. Development that only focuses on increasing yields of rice through intensification and the use of chemical fertilizers and pesticides may possibly give the people more rice to eat, but it may, as pointed out by Guttman (1999) and Gregory and Guttman (2002a), take away much of the aquatic animals and vegetables also harvested from and around the ricefields. This is an important point since the current agricultural policy is to further develop rice production in Cambodia[19]. Without a sound understanding of the other components of the ecosystem and careful preparation of suitable extension there is a great risk that the aquatic animal production is severely affected. Importantly, it will be the poorer segments of rural society who will suffer most from the negative impacts of such development.

With the increased fishing pressure there is a growing perception that what is needed is the introduction of pond culture stocked with (often exotic) fishes.

However, caution is advised here (Halwart *et al.*, 2002). Small-scale pond culture is unlikely to produce large amounts of fish to fill the increasing (or developing) gap between supply of naturally available fish and demand of the growing population. In the study areas it is at best difficult to maintain a fishpond during the flooding period when most of the area surrounding the Great Lake becomes flooded. The amount of fish produced in small-scale low input ponds is correspondingly low, averaging a few kilograms to a few hundred kg per family. As an illustrative example, the development of

---

18 Also in other areas of Cambodia the years 1995 and 1996 were seen as especially bad with respect to EUS incidence.

19 A very understandable goal since on average the yields are poor, averaging around 1.4 tonnes per hectare per year (Javier, 1997), well below the world average of 3.8 tonnes per hectare (IRRI, 2002).

small-scale aquaculture took off dramatically in the southern province of Svay Rieng in the mid-1990s with the support of intensive aquaculture extension. After five years the number of households farming fish had risen from virtually none to a total of 1 300 households producing some 50 tonnes of fish (Gregory and Guttman, 2002b). This is less than one percent of the estimated provincial production from ricefields, swamps and rivers, estimated at 5 000-10 000 tonnes (Guttman, 1999). Small-scale aquaculture has a great potential in providing fish in areas where fish is scarce[20], and can successfully address needs of poorer sections of rural population, but it cannot provide large amounts of fish for a growing population (although commercial intensive aquaculture is capable of this).

A more promising approach seems to be a participatory development approach that addresses all the needs of the local people through locally developed natural resources management plans and a more holistic view of a system that has catered the needs of the people for many generations[21].

The recent development of legislation in the forestry and fishery sectors points into that direction. A new fishery law is under development, as is a sub-decree on community fishery which allows for the creation of community fishery associations and the management of fishery resources by these associations.

However, until now a rather centralistic approach has prevailed and the political will to hand over full responsibility for the management of an area to the communities is not reflected in either the draft of the new law nor in the sub-decree. The communities are rather restricted to a "watchdog" function, acting as an extended arm of the authority which grants the newly elected commune councils the right to develop rules and regulations in accordance with existing laws and to enforce these rules within the area of the commune. Starting on the village level the people receive assistance to develop a set of Village Regulations including a list of fines specifying the amount of fine to be paid to the village NRM Committee for the violation of the Regulations. Central to this approach is that the villagers by themselves control the observance of the regulation and have the authority to collect fines in case of a violation.

A problem until now is the development of management plans for the fishery resource. While the programme can assist the villagers in producing maps of the resources and evaluate the present availability of fish and other aquatic resources through studies such as the one presented here, true management measures in this open system remain difficult. The difficulty is perhaps to know where to start. A few pointers have emerged over the years. Firstly, it is important to make sure that enough adult fish survive the dry season. In a monsoonal flood pulse environment like the Great Lake (and much of the Mekong River Basin) this means that enough dry season refuges are available, either in the lake or in swamps and ditches around the lake. This is sometimes contrary to current practice since the best places to fish are in the dry season refuges as they tend to have a lot of fish and little water. In the target villages of the NRM Component it was experienced that to protect these ponds is extremely difficult. The declaration of a protected pond will keep the majority of the villagers away from it, leaving the field clear for illegal fishing, often with electro-shock equipment since this appears to be the most effective way of collecting a lot of fish in a short time. Guarding these ponds is difficult for the villagers, since they do not receive pay for their activity and have to spend their time fishing in areas that are not protected.

---

20 Demaine and Halwart (2001) discuss various forms of interventions in an overview of rice-based small-scale aquaculture.

21 An overview on thoughts and lessons collected by DFID and FAO in 2000 with regard to participatory approaches for aquatic resources management and development is given in Halwart and Haylor (2001).

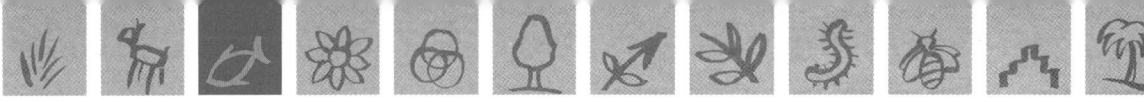

A second point of importance is the protection of the flooded forests around the lake which serve as spawning ground for many fish species. A lot of this forest is being destroyed for the production of charcoal or cleared for the establishment of mung bean or deep water ricefields.

Thirdly there is a need to make sure that the migration routes out of the refuge are unhindered, so the fish can reach the spawning grounds or fry can reach the nursing grounds and the adult fish can reach their dry season refuges[22]. Also this proves a difficult task. At the onset of the flooding, catching fingerlings is a very profitable albeit illegal business. In the target area of RDP NRM the people involved in this activity are equipped with radios and in some cases protected by local authorities, backed by district officials.

Streams are blocked completely by lot operators as well as by the local people at the time the flood water recede, giving the fish no chance to reach the Tonle Sap or the Mekong, their major dry season refuge. Also this practice is illegal but law enforcement remains difficult.

Finally there is a need to determine the aim of the fishing effort, equity or efficiency. If efficiency is sought (i.e. as much fish for as little effort as possible) it is often better to have a smaller group of specialized fishermen undertaking the fishery. If, however, equity is desired then small 'inefficient' gears are to be favoured making it possible for a greater number of people to be involved in the fishing.

More practically resources must be managed with a more holistic view. Hoggarth *et al.* (1999) provides general advice on how to manage flood plain river fisheries, and identifies several levels of management units (national, catchment, intermediate and village level). The same paper also identifies the different management categories; environment, who has access, amount and type of fishing and fish stock enhancements. One successful strategy has been to encourage villagers to protect their dry season refuges to ensure that there are sufficient stocks for the following year. This is an intuitive and practicable solution that has been successfully introduced in some areas. Other solutions are to agree on certain fish catch methods during certain periods (this is practiced traditionally in some areas), to reforest/improve spawning and nursing habitats (possible if the areas are also under some environmental protection) as well as making sure that infrastructure development (such as roads) are not blocking important wet season migration routes for fish.

Finally, the ricefield ecosystem is a traditional modified ecosystem, its diversity and productivity is high making it a suitable system for low input farming. Intensification and specialization of the system will be associated (in most cases) with losses in some of the other products. It is therefore especially important to assess what those changes will be, who will benefit and who will loose, and to try to find ways to minimize the losses and maximize the gains.

---

22 Guttman (1998) provided a schematic illustration of the migrations of fish in a lowland flood pulse environment.

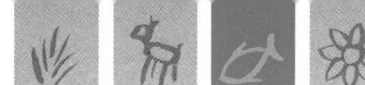

LITERATURE

Ahmed, M., Navy H., Vuthy, L. & Tinogco M. 1998. *Socioeconomic assessment of freshwater capture fisheries in Cambodia: Report on a household survey.* Mekong River Commission, Phnom Penh, Cambodia. 186 pp.

Ali, A.B., 1990. Some ecological aspects of fish populations in tropical ricefields. *Hydrobiologia* 190: 215-222.

APHEDA, 1997. Baseline Survey Report (AngkorChey, BantemayMeas, Chhouk and KompongTrach District), Report prepared by N.C. Paul, Domestic Fish Farming Program, Australian People For Health Education and Development Abroad (APHEDA)-Department of Agriculture, Forestry and Fishery Kampot Province, Cambodia, 29 pp.

Bao, T.Q., Boukhamvongsa, K., Chan, S., Chhoun, K.C., Phommavong, T., Poulsen, A.F., Rukawoma, P., Suntornratana, U., Tien, D.V., Tuan, T.T., Tung, N.T., Valbo-Jorgensen, J., Viravong, S. & Yoroong Y., 2001. *Local Knowledge in the Study of River Fish Biology: Experiences from the Mekong.* Mekong Development Series No. 1, Mekong River Commission, Phnom Penh. 22 pp.

Demaine, H. & Halwart, M., 2001. An overview of rice-based small-scale aquaculture. In IIRR, IDRC, FAO, NACA, and ICLARM. Utilizing different aquatic resources for livelihoods in Asia: a resource book, p. 189- 197. IIRR, Silang, Cavite, Philippines. 416 pp.

Dy Phon, 2000. *Dictionary of plants utilized in Cambodia.* Imprimerie Olympic, Phnom Penh.

Gregory, R., 1997. *Ricefield Fisheries Handbook.* Cambodia-IRRI-Australia Project, Phnom Penh, 38 pp.

Gregory, R., Guttman, H. & Kekputhearith, T., 1996. Poor in all but fish: *A study of the collection of ricefield foods from three villages in Svay Theap District, Svay Rieng.* Working Paper C-5, AIT Aqua Outreach (Cambodia), Asian Institute of Technology, Bangkok, 29 pp.

Gregory, R., & Guttman, H., 2002a. The ricefield catch and rural food security. *In* P. Edwards, D.C. Little & H. Demaine, eds. *Rural Aquaculture,* p. 1-13. CABI Publishing, Wallingford, 358 pp.

Gregory, R. & Guttman, H., 2002b. Developing appropriate interventions for rice-fish cultures. *In* P. Edwards, D.C. Little & H. Demaine, eds. *Rural Aquaculture,* p. 15-27. CABI Publishing, Wallingford, 358 pp.

Gum, W., 1997. *Consultancy report on fisheries development in Northwest Cambodia.* Cambodia Rehabilitation and Regeneration, CARERE UNDP/RGC, Phnom Penh.

Guttman, H., 1998. Rice and Fish. AARM *Newsletter* 3(3): 6-7.

Guttman, H., 1999. Rice Field Fisheries: A Resource for Cambodia. *NAGA the ICLARM Quarterly* 22(2): 11-15.

Halwart, M. & Haylor, G., 2001. Participatory approaches for aquatic resources management and development. In IIRR, IDRC, FAO, NACA, and ICLARM. *Utilizing different aquatic resources for livelihoods in Asia: a resource book,* p. 87-94. IIRR, Silang, Cavite, Philippines. 416 pp.

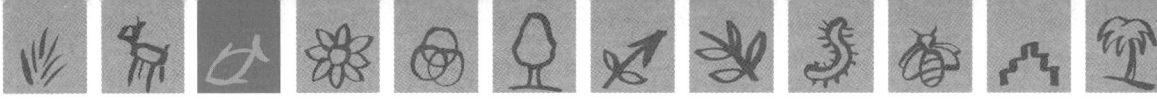

Halwart, M., Funge-Smith, S. & Moehl, J., 2002. The role of aquaculture in rural development. *In* Review of the State of World Aquaculture. FAO *Fisheries Circular No. 886* (Rev. 2). FAO, Rome.

Heckman, C.W., 1979. *Rice field ecology in Northeastern Thailand: The effect of wet and dry seasons on a cultivated aquatic ecosystem.* Junk Publishers, The Hague, The Netherlands.

Hoggarth, D.D., Cowan, V.J. & Halls, A.S., 1999. Management guidelines for Asian floodplain river fisheries. *FAO Fisheries Technical Paper Nos 348/1, 348/2.* FAO, Rome.

IRRI, 2002. *Rice Production: Europe, Australia, USA, World.* (available at IRRI webpage URL http://www.riceweb.org/riceprodleurope.htm, visited 23/06/2002).

Javier, E.L., 1997. Rice ecosystems and varieties. *In* Nesbitt, H.J., ed. *Rice production in Cambodia,* p. 39-81. International Rice Research Institute, Manila.

Lilley, J.H., Phillips, M.J. & Tonguthai K., 1992. *A review of Epizootic Ulcerative Syndrome (EUS) in Asia.* Aquatic Animal Health Research Institute and Network of Aquaculture Centres in Asia-Pacific, Bangkok. 73 pp.

Little, D.C., Surintaraseree, P. & Innes-Taylor N., 1996. Fish culture in rainfed ricefields in northeast Thailand. *Aquaculture* 140: 295-321.

MRC, 1998. *Lower Mekong Hydrological Yearbook 1998.* Mekong River Commission, Bangkok. 499 pp.

Ovesen J, Trankell I-B & Ojendal J., 1996. *When every household is an island. Social Organisation and Power Structures in Rural Cambodia.* Uppsala Research Reports in Cultural Anthropology, No 15.

PROSEA, 2002. *Plant Resources of South East Asia.* PROSEA Foundation Bogor, Indonesia and Pudoc, Wageningen, Netherlands, 1989-2002 (publication ongoing).

Rainboth, W.J., 1996. *Fishes of the Cambodian Mekong.* FAO species identification field guide for fishery purposes. FAO, Rome. 265 pp.

Shams, N. & Hong, T., 1998. *Cambodia's rice field ecosystem biodiversity – resources and benefits.* Deutsche Gesellschaft für Technische Zusammenarbeit (GTZ), Kampong Thom Provincial Development Programme, Phnom Penh. 60 pp.

Spiller, G., 1985. *Rice cum fish culture: Environmental aspects of rice and fish production in Asia.* Report FAP/WP-15. FAO Office for Asia and the Pacific, Bangkok. 48 pp.

Svedrup-Jensen, S., 2002. *Fisheries in the Lower Mekong Basin: Status and Perspectives.* MRC Technical Paper No. 6. Mekong River Commission, Phnom Penh. 103 pp.

Toft Mogensen, M., 2001. *The importance of fish and other aquatic animals for food and nutrition security in the Lower Mekong Basin.* The Royal Veterinary and Agricultural University, Copenhagen. 140 pp. (M.Sc. Thesis)

Van Amerongen, S.R., 1999. *The rice field crab* (Somanniathelphusa sp.) *– a short study on its role as a pest to rice culture.* FAO-Participatory Natural Resource Management in the Tonle Sap Region Project GCP/CMB/002/BEL, Siam Reap, Cambodia. 21 pp.

ANNEXES

ANNEX 1 DISTRICTS OF KAMPONG THOM PROVINCE, KINGDOM OF CAMBODIA

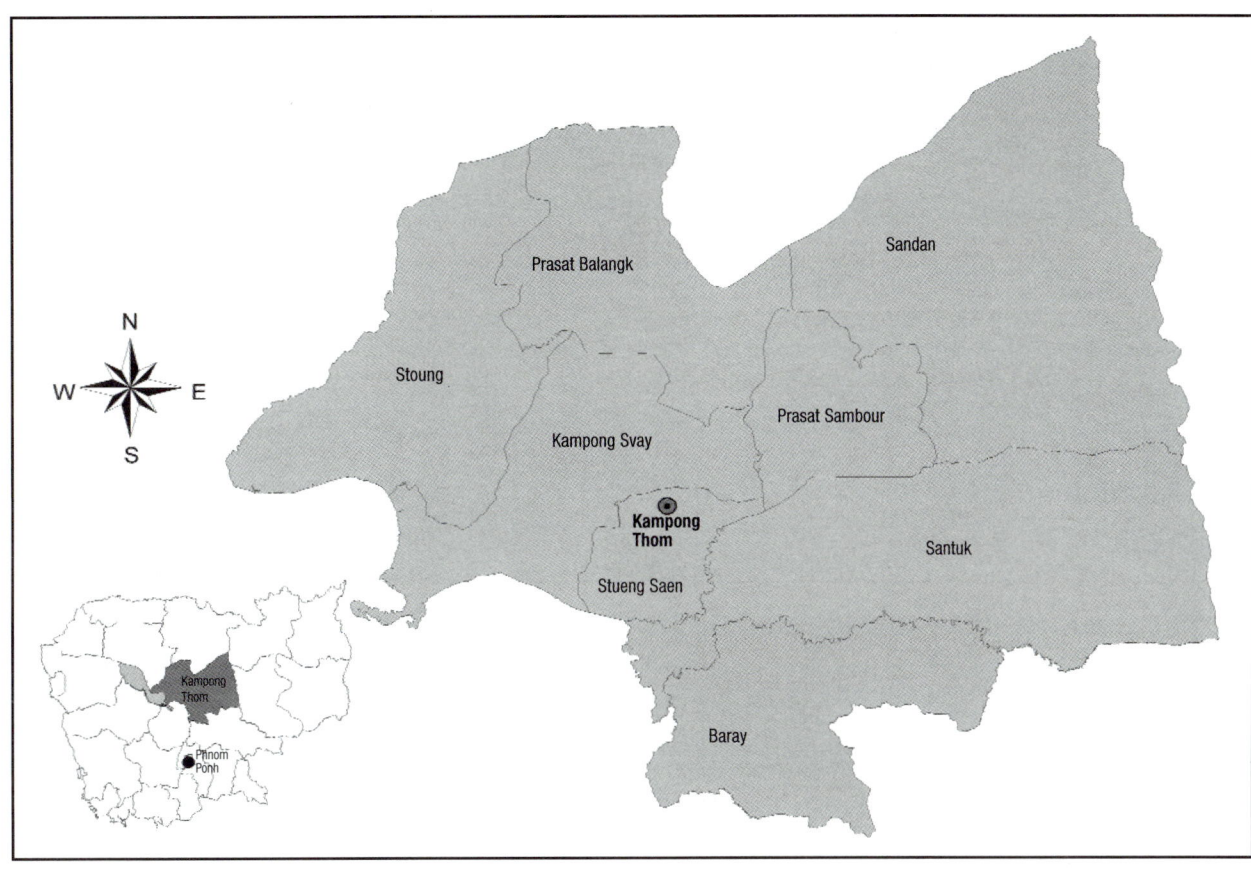

Landuse

ricefields

Water features

large river or stream

lake or pond

temporal lake or pond

Water courses

—— river or stream

-------- intermittent river or stream

—— small canal

══ large canal

ⅿⅿⅿⅿ levee

Settlements

built-up area

village (high density)

village (medium density)

village (low density)

floating village (high density)

floating village (medium or low density)

Settlements

══ all weather, hard surface, two lanes

══ all weather, hard surface, one lane

══ all weather, loose surface, one lane

—— dry weather, loose surface

------- cart track

············ footpath

—— urban street

◉ Provincial Capital

● District Capital

■ Target Villages

For full colour image please access Ecoport at
http://www.ecoport.org/EP.exe$PictShow?ID=26731&Subj=E113

ANNEX 3 MAP OF STOUNG DISTRICT (FOR LEGEND PLEASE REFER TO ANNEX 2).

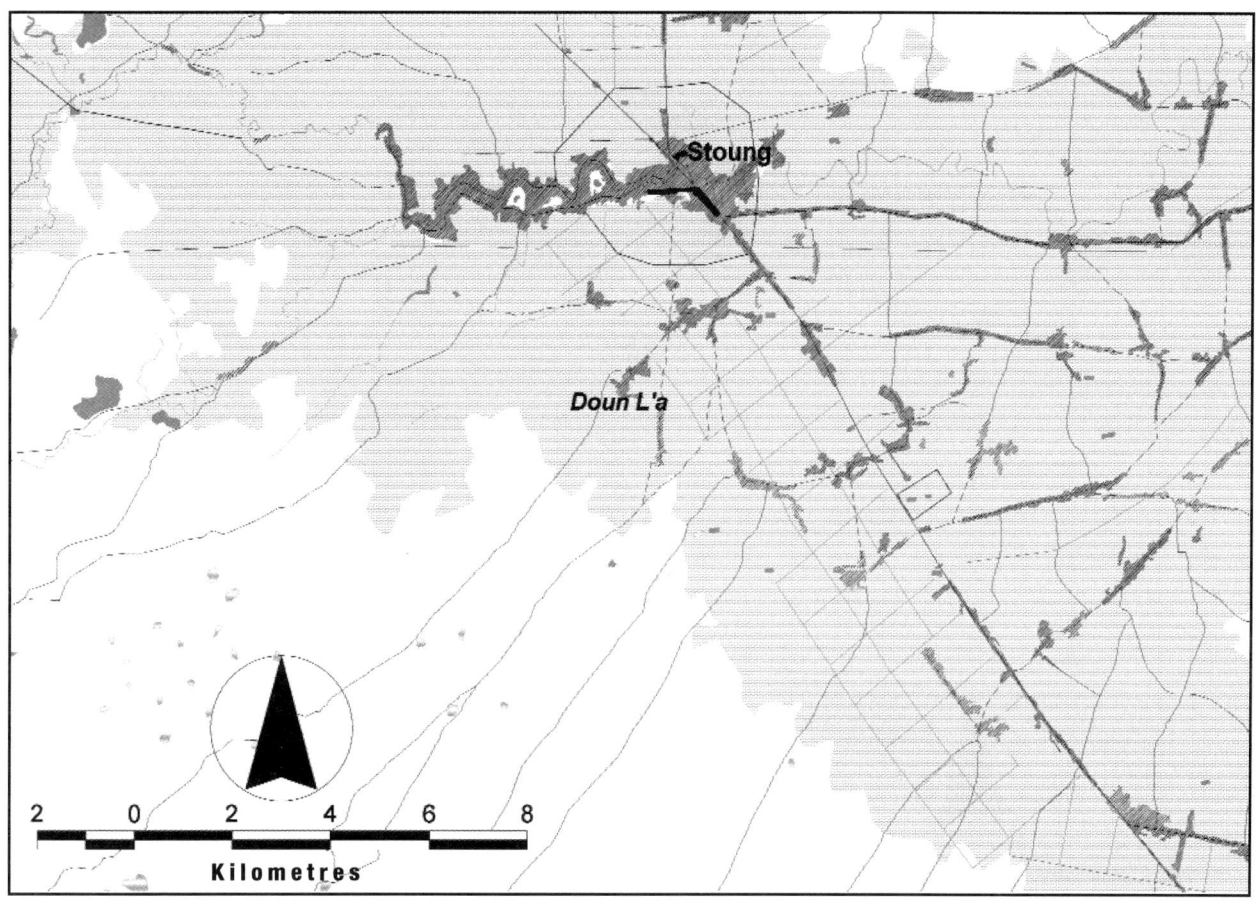

For full colour image please access Ecoport at
http://www.ecoport.org/EP.exe$PictShow?ID=26732&Subj=E113

ANNEX 4 MAP OF STUENG SAEN DISTRICT (FOR LEGEND PLEASE REFER TO ANNEX 2).

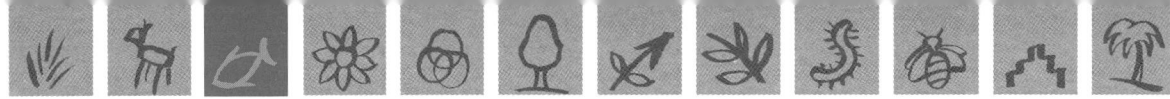

Kampong Thom
Stueng Saen
Roluos
Tboung Krapeu
Kiri Voan
Puk Yuk
2 0 2 4 6 8
Kilometres

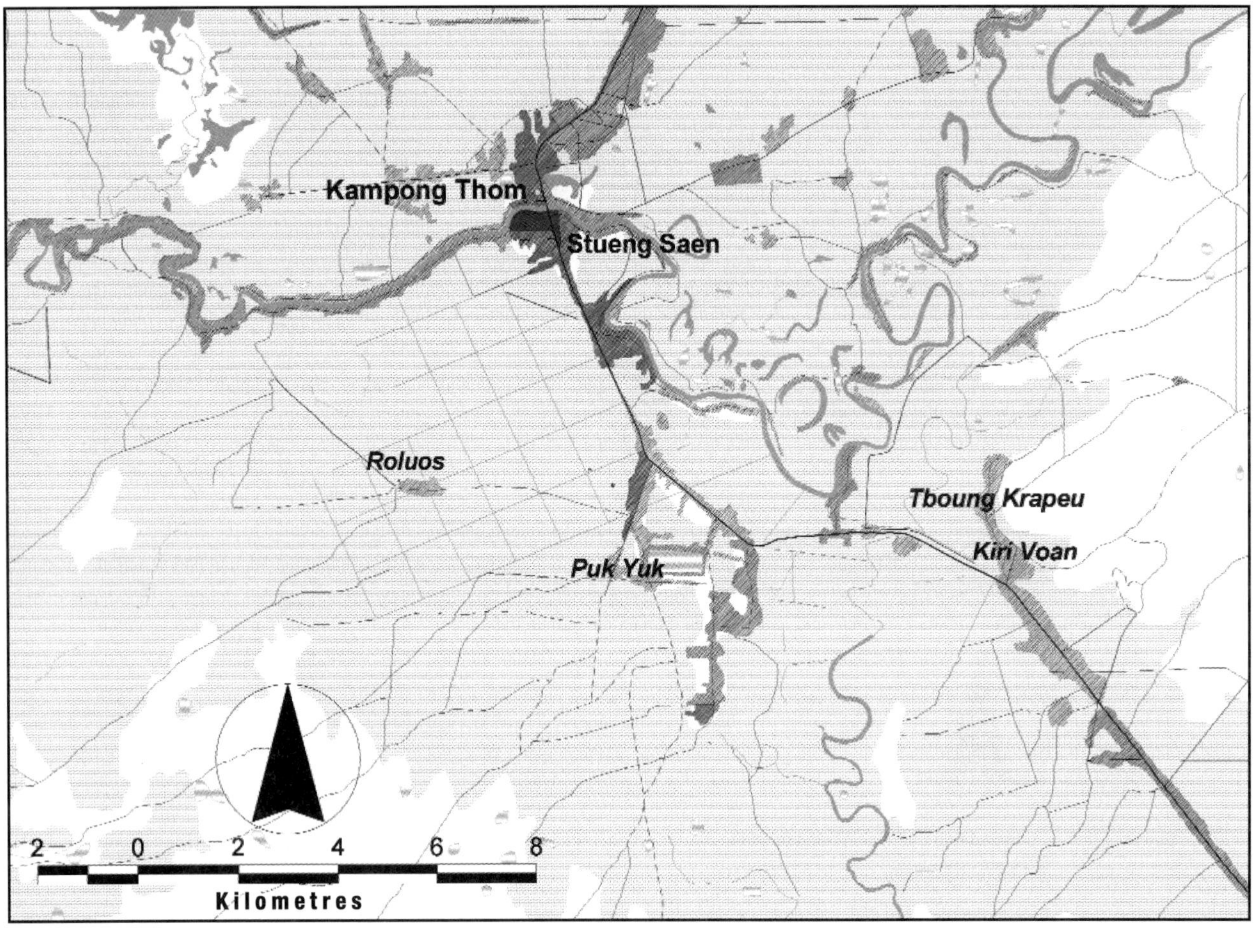

For full colour image please access Ecoport at
http://www.ecoport.org/EP.exe$PictShow?ID=26733&Subj=E113

ANNEX 5 MAP OF SANTUK DISTRICT (FOR LEGEND PLEASE REFER TO ANNEX 2).

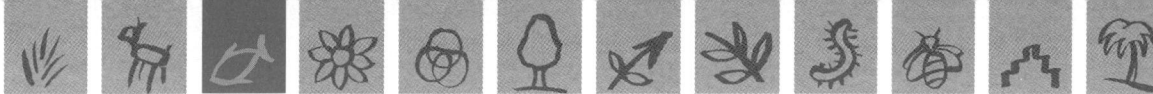

Kiri Voan
Tuol Vihear
Santuk
Ta Nhaok
2 0 2 4 6 8
Kilometres

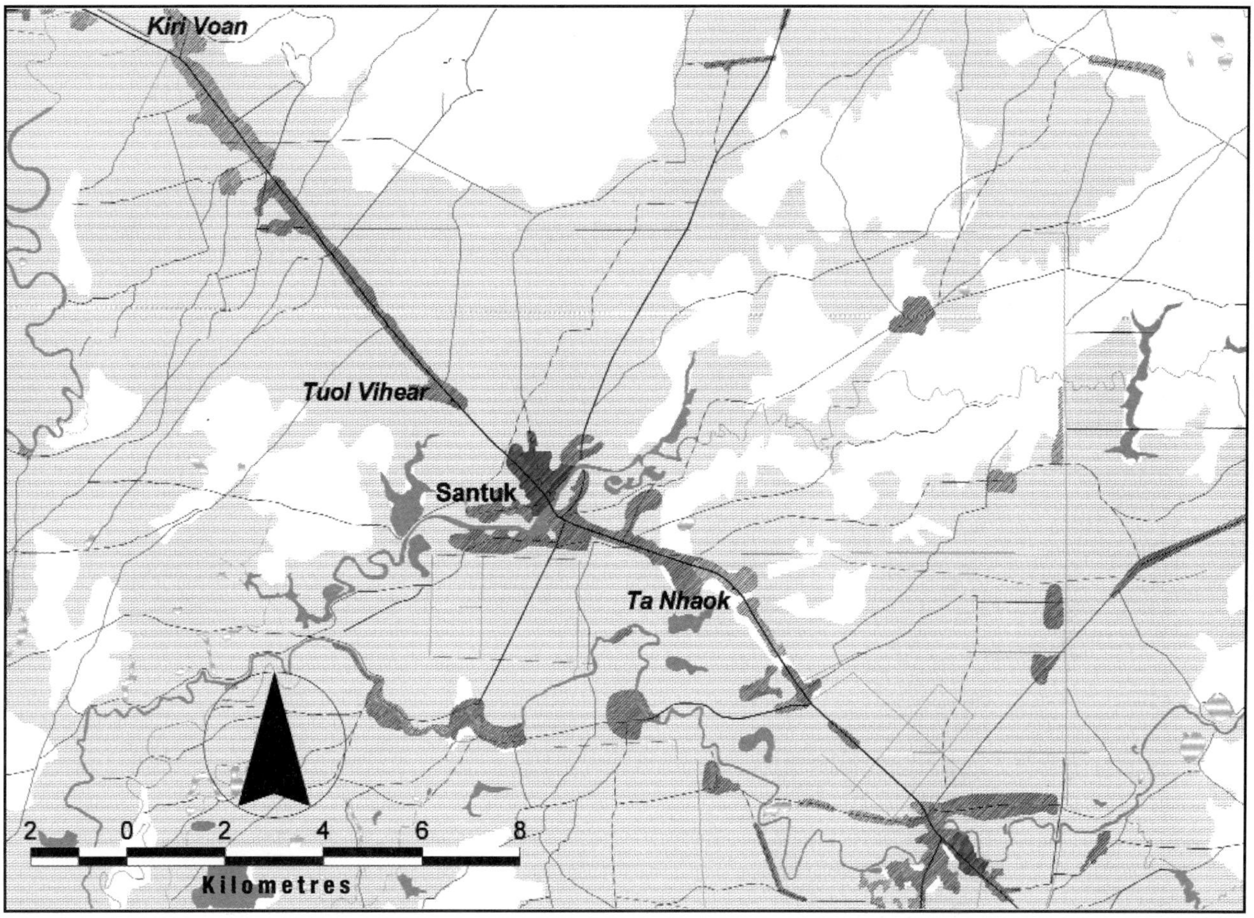

For full colour image please access Ecoport at
http://www.ecoport.org/EP.exe$PictShow?ID=26734&Subj=E113

BIODIVERSITY AND THE ECOSYSTEM APPROACH IN AGRICULTURE, FORESTRY AND FISHERIES   FAO INTER-DEPARTMENTAL WORKING GROUP ON BIOLOGICAL DIVERSITY FOR FOOD AND AGRICULTURE

72

RESPONSIBLE
TECHNICAL DIVISION

Research, Extension and Training Division
Environment and Natural Resources Service

Nadia El-Hage Scialabba

ORGANIC AGRICULTURE

AND GENETIC RESOURCES FOR FOOD AND AGRICULTURE

4

ORGANIC AGRICULTURE
AND GENETIC RESOURCES FOR FOOD AND AGRICULTURE

AUTHORS

Nadia El-Hage Scialabba

FAO Environment and Natural
Resources Service
nadia.scialabba@fao.org

Cristina Grandi

Associazione Italiana per
l'Agricoltura Biologica (AIAB),
Rome, Italy
c.grandi@ifoam.org

Christina Henatsch

Association for Biodynamic
Vegetable Plant Breeding
(Kultursaat), Germany
christina.henatsch@gmx.de

ACKNOWLEDGMENTS

The authors are very grateful to
Caroline Hattam for translation
of the Spanish text.

CONTENTS

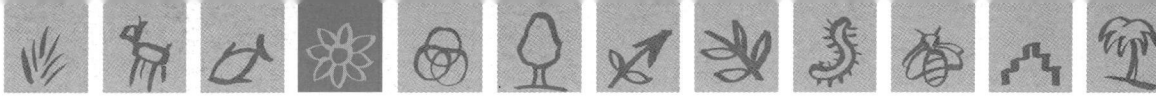

INTRODUCTION

Scope of this paper

Farm specialization and the general abandonment of mixed farming is a significant factor in the decline of biodiversity, including genetic resources for food and agriculture and wildlife, and of the disintegration of traditional and community-based management. The adoption of high-yielding, uniform cultivars has led to a considerable reduction in the number of genetically viable species used in agriculture. Many food crop varieties have virtually disappeared from their centres of diversity.

There is now an increasing body of evidence that organic agriculture supports a much higher level of biodiversity than conventional farming systems, including species that have significantly declined. This paper addresses the contribution of organic agriculture to agricultural biodiversity, including genetic resources for food and agriculture. The positive impact of organic agriculture on wildlife (e.g. soil organisms, arable flora, predatory invertebrates, pollinators, birds) and in creating and connecting habitats that enhance nature conservation is outside the scope of this paper. The adverse impact of genetically modified organisms on organic agriculture is also outside the scope of this paper.

The cases presented in this document illustrate how organic agriculture, to be viable, is reversing the decline in species diversity as well as abundance of each species. If biodiversity is to be maintained, it should be an integral part of a healthy landscape where not only diversity but also abundance is a fundamental factor.

What is organic agriculture?

The close relationship between organic agriculture and agricultural biodiversity is expressed at the philosophical and theoretical levels in the basic principles, standards and regulations that govern organic agriculture and by the practical experiences of organic farmers around the world.

According to the FAO/WHO Codex Alimentarius Guidelines for Organic Food "Organic agriculture is a holistic production management system which promotes and enhances ecosystem health, including biological cycles and soil biological activity ... The primary goal of organic agriculture is to optimise the health and productivity of inter-dependent communities of soil life, plants, animals and people".

The adoption of organic agriculture methods requires farmers to follow a series of agronomic practices (e.g. essentially rotations and associations of a large number of plants and animals) that make organically managed systems biologically much more complex than conventionally managed systems. The organic agriculture system relies on the creation and maintenance of conditions that positively nurture the health of crops and livestock and on harnessing of natural processes (instead of using artificial inputs). Many involve the positive use of biodiversity, thus making the conservation of biodiversity an integral part of the farming activity.

Reasons for the adoption of organic agriculture

There are millions of small farmers in developing countries who do not have the economic means to buy high yielding seeds or the fertilizers and pesticides necessary for their cultivation. Many of them have opted for

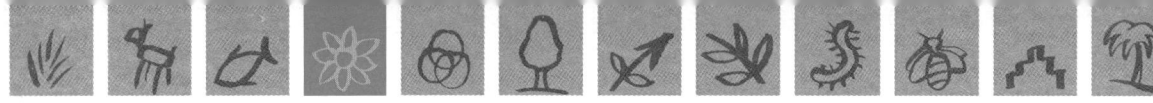

the maintenance or re-introduction of organic systems based on traditional forms of agriculture. These promote the use of varieties that are better adapted to local biotic and abiotic conditions (e.g. biological control of pests and diseases, climatic stress).

Convincing farmers to adopt organic agricultural techniques is not a difficult task. Many farmers, besides cultivating high yielding varieties that require large amounts of off-farm inputs, have continued to plant a part of their land with crops for self-consumption, using natural methods.

In developed countries, the role of organic agriculture in the restoration and maintenance of species, varieties and breeds risking extinction has also led many organizations or individuals to adopt this system to save genetic, species and ecosystem diversity. Conservation efforts, through organic management, are however even more important in centres of diversity. In these areas, cultivation of populations with high genetic variability is an indispensable inheritance for agriculture and as such, for food security for future generations.

The price premiums which organic products command (and in Northern countries, payments received for organic agriculture) give an economic value to biodiversity and to the efficient use of resources. There are farmers who farm organically because they are attracted by the strong demand for organic products. For these farmers, conversion to organic management is a way of adding value to production and obtaining better prices on the market. These farmers are required to implement a minimum level of biodiversity in order to be granted the organic label. In fact, crop rotation is the first step towards improving agricultural biodiversity. This is one of the methods required by organic certification bodies, as well as by financial support programmes. For

the organic system to be economically viable, market-driven farmers are led to use local species, varieties and breeds that are more resistant to disease and local environmental conditions in order to compensate for the restriction on synthetic input use.

By choice or by necessity, organic farmers do not make use of synthetic agricultural inputs. They rather rely on the "natural inputs" by enhancing biodiversity through *in situ* conservation and sustainable use of genetic resources. Independent of the motives that farmers have for the adoption of organic agriculture, in every case a marked increase in biodiversity is visible. Many of these empirical systems have given satisfactory results and have since become the focus of study in research centres.

Why is biodiversity important in organic systems?

When diversity is encouraged, locally adapted plant and animal breeds which are more appropriate to local ecosystems can be used. Most importantly, agricultural genetic diversity is a basic insurance against crop and livestock disease outbreaks.

Organic farmers breed varieties for quality, nutrition, resistance and yield, in reduced input growing conditions. Research has shown that these characteristics are more likely to be found in older native cultivars. In particular, open pollinated varieties offer diverse and regionally adapted characteristics that are better suited to organic agriculture. Open pollinated varieties, which represent an important gene pool for resource-poor farmers living in marginalized and stress-prone areas, are rapidly vanishing. They are replaced by very few hybrid varieties which require inputs not available to poor farmers and which entail dependence on large seed

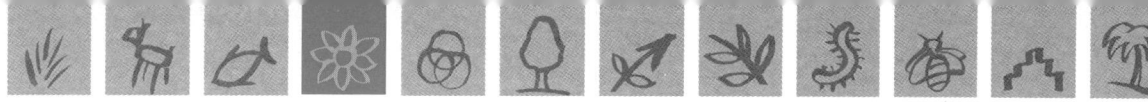

companies. A significant proportion of local breeds remains in the care of pastoral people and traditional livestock owners in developing countries (e.g. pigs in China, cows in India and poultry in Asia and Latin America). Local breeds are robust and suitable for free ranging; however, two local breeds are becoming extinct every week.

Organic systems encourage the preservation and expansion of older, locally bred and indigenous varieties and breeds. Farmers who save their own seeds can gradually increase crop resistance to pests and diseases by breeding for "horizontal resistance". Horizontal resistance is the ability of a crop to resist many or all strains of a particular pest (which differs from breeding for "vertical resistance" to have a gene to resist one specific strain of a disease). By exposing a population of plants to a certain disease or pest (or to several pests at one time), then selecting a group of the most resistant plants and inter-breeding them for several generations, a given population becomes more resistant than the original population. Horizontally resistant cultivars are well adapted to the environment in which they were bred, but may be less suitable for other growing conditions.

Many indigenous groups have an expert knowledge of biodiversity in their own areas. Traditionally, these groups have conserved, improved and shared genetic resources according to their food preferences and socio-economic and environmental conditions.

Case studies

Following are a series of 16 case studies, selected from published materials from diverse sources including inter-governmental, governmental and private organizations. In all of these, the close relationship between the introduction of the organic agricultural system and the maintenance of biodiversity is evident, as is the resulting improvement in the socio-economic conditions of the farmers.

The case studies are presented according to the main contribution they make to the different aspects of agricultural biodiversity, including:

- traditional and community-based management;
- *in situ* conservation and sustainable use of centres of diversity;
- rescue of under-utilized species and varieties for quality diets and culinary traditions;
- selection of biodiversity adapted to local ecological conditions and resistant to disease;
- alternative breeding criteria and participatory research.

TRADITIONAL AND COMMUNITY-BASED MANAGEMENT

In some communities, the adoption of conventional agriculture has substituted traditional cultivation systems with high biodiversity for monocrops of genetically similar individuals. In a relatively short period of time, such systems have led to environmental degradation, social disintegration and misery within the communities. However, many traditional agricultural systems that have been the basis of food security and community cultures have since been saved through organic agriculture approaches.

The introduction of organic management, based on traditional experiences and new knowledge of natural processes, has allowed

the maintenance of the agro-ecological systems and has improved socio-economic conditions of rural communities, especially in environmentally vulnerable areas. These agricultural systems are also based on strong farmer participation in the decision-making process, exchange of information and distribution of benefits.

The examples below illustrate the community-based rehabilitation of abandoned and degraded agro-ecosystems, through organic agriculture, in the flood plains of Bangladesh and mountainous areas of Indonesia and Mexico. The polycultures systems established by these communities (like many others around the world) are characterized by highly diversified ecosystems and an improved agricultural biodiversity. The good markets associated with organic products have not only provided food but have also generated further community services.

Nayakrishi Andolon: a community-based system of organic farming, Bangladesh[1]

In the flood plains of Bangladesh, community-based organic agriculture resulted from an increasing awareness of the harmful effects of the Green Revolution. The latter was showing a tremendous decline in crop yields despite an enormous increase in the need to apply fertilizers and pesticides. Groundwater was less available, livestock and fish populations were diminishing, the health situation was worsening (including gastric, skin and respiratory diseases) and exogenous varieties were gradually replacing traditional varieties. This forced many poor farmers to sell their land and other productive assets, shifting from farming to non-farming occupations.

Following particularly terrible floods in 1988, some farmers, together with UBINIG (Policy Research for Development Alternatives), a non-governmental organization, gathered together to seek an alternative – not just an alternative method of farming, but also community-based work, which is organic in nature. They named it Nayakrishi Andolon (New Agriculture Movement). The rationale for such a name was to indicate that this method is not "old" in a backward sense; but is a newer method, incorporating traditional knowledge and wisdom and appropriating newer ideas and "scientific" innovations that are suitable for farmers and the environment.

Initially, the peasant women took the lead in stopping the use of pesticides, mainly for health reasons. Then, a group of farmers organized themselves to experiment with green manure and compost. Compost made of water hyacinth, available in abundance, became quite popular and soon Nayakrishi Andolon spread from village to village. As experience and confidence grew, the farmers developed a set of ten simple principles for Nayakrishi farming, all focusing on the use of locally available resources to enhance the efficiency of land, water, biodiversity and energy as well as the control over seeds within the farming community.

In addition to chemical-free agricultural practices, the production of biodiversity is built-in within the Nayakrishi method of food production. As a fundamental principle of agricultural practice, Nayakrishi farmers reject monoculture and base their practices on mixed cropping and crop rotation. This has an immediate effect in overcoming the present narrow genetic base, but is also a highly effective method for pest management and the nutritional health of the soil.

In Nayakrishi villages, farmers derive more varieties of fish, together with a wide range of uncultivated crops, which either

---

1 Sources: UNDP, 2000. The Nayakrishi Andolon: A Community-Based System of Organic Farming – Innovative Experiences; Farhad Mazhar et al., 2001. Nayakrishi Andolon: Recreating Community Based Organic Farming. Low External Input Sustainable Agriculture; Mark Lynas, 2001. A Message from Bangladesh: Recipes Against Hunger. Greenpeace International.

come as accompanying crops due to multiple cropping in the fields, or grow on the common land as no more herbicides are used. Livestock and poultry also develop more rapidly, thereby enriching the food security of the people. Similarly, the planting of local-variety trees is an integral part of the practice in Nayakrishi villages, which, in turn, attracts birds, butterflies and other pollinators and predators.

The local species, varieties and breeds are always preferred to those that are introduced. The strategy of Nayakrishi Andolon for the maintenance and regeneration of biodiversity and genetic resources is based on some simple rules and obligations between members. The strategic importance is in the conservation and regeneration of species and the genetic variability of the cultivated crops and homestead forestry. However, there are a large number of species and varieties that are not cultivated. The conservation and regeneration of biodiversity for these species and varieties are mainly maintained by the overall organization of Nayakrishi Andolon.

Every village where Nayakrishi is actively adopted has its own *gram karmi* (extension workers). Apart from networking and campaigning for Nayakrishi, *gram karmi* maintains audits of the natural resources of the village. This information is pooled collectively and is a vital practice in maintaining and managing the local biodiversity. The Nayakrishi farmers can easily be put on alert if it appears that any "land race" or "wild" species or variety is disappearing or being lost.

Around 65 000 families from all over Bangladesh now follow Nayakrishi principles and the movement is spreading fast. Most important is the general confidence among farmers that Nayakrishi is "economically viable", but the ecological situation is also improving, the land is regaining fertility and biodiversity is being strengthened.

Ladang cultivation of organic spices in Sumatra, Indonesia[2]

On the island of Sumatra, Indonesia, ruthless exploitation by forestry, fishing and extractive industries in the last decades has decreased the rich biodiversity of the region. Despite this, some areas still survive in a state close to that of pre-European colonization, mainly because of their mountainous, remote locations. Some of these areas form part of national parks that still support the rare Sumatra tiger, rhinoceros and elephant, as well as some native people following traditional lifestyles in the forest.

Many poor farmers live around these areas and use slash-and-burn techniques for the production of crops for self-sufficiency and for the market. These practices threaten the remaining forest areas and even the national parks on whose boundaries they encroach.

Thomas Fricke, a former advisor to the United Nations, World Wildlife Fund and Indonesian national parks on sustainable agriculture projects, aimed to find a solution to that problem. In 1995, he and his wife Sylvia Blanchet founded ForesTrade, an international company dedicated to preserving biodiversity through responsible trade.

In 1996 ForesTrade, in collaboration with local NGOs and the National Parks of Indonesia began the Indonesian Cassia Cinnamon Project, encouraging local farmers to stop clear-cutting the rainforest. The project focused on land bordering a national forest park, providing a buffer zone for the protection of biodiversity in the rapidly disappearing forested areas.

Some of the local people joined forces with ForesTrade, creating grower groups for the production of organic spices for the European and the United States markets. Good prices, generally a little better than could be expected

---

2 Sources: Tim Marshall, 2000. The Spice of Organic. The Organic Gardener; Tim Marshall, 2000. Responsible Approach to Spice Production. Acres Australia, The National Newspaper of Sustainable Agriculture.

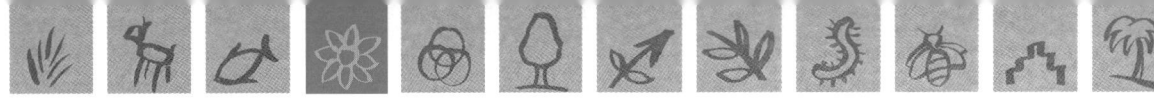

from the conventional market, are paid to the farmers, together with a bonus to the local community. The community bonus is used to run training centres, nurseries and other community services.

These farmers agree to follow organic agriculture practices, avoiding the use of chemical pesticides and fertilizers. They rely instead on composting, crop rotation and biological pest and disease control. Organic growers are not permitted to use fire for clearing their plots. All slashing and weed control is done by hand, using simple tools such as axes, hatchets, machetes or knives. Slashed matter is then reduced to mulch. Farmers also agree not to poach rainforest resources, where some farmers previously clear-cut slopes to plant crops, spoiling the environment and causing widespread erosion.

Crops are produced in a modified, traditional "home garden" or "shifting cultivation" situation. Each grower operates one or more traditional gardens (or *Ladang*) in which a variety of annual plants (such as potato, aubergine and onion), short-lived plants (such as cassava, banana and yam) and longer-lived plants (such as cloves and cinnamon) are produced. As the *Ladang* matures, the longer-lived trees dominate shading under-storey crops. These trees can be either selectively felled (e.g. cinnamon) or left to produce during their mature phase (e.g. cloves and coconuts) before the cycle is started again.

Certified-organic growers in Sumatra produce a variety of spices and essential oil crops, such as chilli, turmeric, ginger, vanilla, cloves, allspice, cardamom, nutmeg (and mace), black and white pepper, patchouli and cassia (cinnamon). They are inspected and certified by the National Association for Sustainable Agriculture (Australia), by the Dutch certification body SKAL or by Oregon Tilth.

In a short time approximately 3 000 Sumatran farmers have begun producing organic spices for the world markets. This has led to improved socio-economic conditions for the communities while at the same time preserving biodiversity both in the national parks and in the local agro-forestry systems (garden/forest plots).

Organic coffee production and biodiversity management, Chiapas, Mexico[3]

At the end of the 1980s, small coffee producers (most of them Tzotzil and Tzeltal indigenous people) in San Cristóbal de Las Casas, the highlands of Chiapas, southern Mexico, faced a deep economical crisis due to the drop in coffee prices on the world market. Together with the disappearance of direct support for coffee growers in terms of technical assistance, marketing and financial support, this resulted in the abandonment of practices for maintaining the crop and processing the beans, leading to lower yields and product quality.

Many of these farmers organized themselves in 1983 into the Beneficio "Majomut" Coffee Growers' Union of Ejidos and Communities, a grassroots social organization with 1 450 members in 25 communities. The organization was created as a means to bring together farmers in the processing and direct marketing of their coffee. Members work an average of two hectares and cultivate corn, beans and coffee. As coffee is sold, it forms the main source of family income. Gradually, work has expanded to include the entire productive process and it has become a means for organizing, managing and carrying out integral development projects for the communities.

---

3   Source: Walter Anzueto Anzueto and Alberto Ortiz Gutiérrez, 2001. Organic coffee production and its contribution to natural resource management and conservation. Growing Diversity.

To fight the declining price crisis, farmers were compelled to find alternatives to conventional coffee production and marketing, so they decided on organic coffee production. The conversion to organic agriculture began in 1992, and by 1995 the first organic certificate was granted. The introduction of organic techniques has been carried out through the training of community promoters who create experimental organic lots in each community as a base for the learning process and research.

Activities are based on the exchange of experiences through a farmer-to-farmer approach including: development and evaluation of agro-ecological practices, participatory research through farmers experimentation, and training of community promoters and community participation. The agents participating in the process include: communities associated with farmers' organizations, the network of promoters, the coordinating network of small coffee growers' organizations and international cooperation agencies. The Majomut Union is also promoting work with women to strengthen participation in the organization's internal democracy.

Farmers' extension covers the following themes: soil conservation; production of organic fertilizers; coffee pruning; management of the diversity of animal and plant species; natural control of pests and diseases; organic production of crops for self-sufficiency in maize, beans and other food species; and internal control for supervision of the organic certification of coffee and quality control of the product.

The management of biodiversity within the coffee production systems and other cultivations constitutes an example of a rich local germplasm and of knowledge applied in the design of the stratification of the vegetation. This is knowledge transmitted from generation to generation resulting from a continuous process of adaptation.

In 1995, a census was carried out on the species found in the organic coffee production systems of La Unión. The study demonstrated that in addition to coffee, there were more than 30 associated plant species of agronomic interest: fruit trees (loquat, mango, lime, guava, peach and orange), shade trees (eucalyptus, ash and pine), horticultural crops (tomato, chilli and beans), as well as medicinal and other plants used for the prevention of erosion. The benefits generated by the higher organic coffee prices were therefore accompanied by an improvement in the biodiversity of the coffee production system.

IN SITU CONSERVATION AND SUSTAINABLE USE OF CENTRES OF DIVERSITY

The continued cultivation of crop species within their centres of diversity plays a fundamental role in the maintenance of genetic diversity. As such, centres of diversity represent a fundamental resource not only for farmers of the present but also of the future. It is this genetic variability that allows populations to adapt to changing environmental conditions.

The introduction of organic agriculture in the two case studies below has increased the economic value of cocoa in Mexico and cotton in Peru and as such, has provided livelihoods to peasants and indigenous communities. This is a necessary basis for the maintenance of agricultural production in the centres of diversity. The creation of a market outlet for indigenous products represents, together with sustainable use, a viable way of maintaining, *in situ,* a diverse genetic heritage.

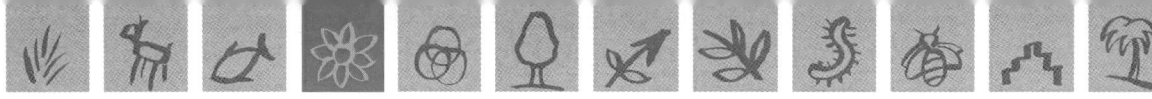

Organic farming for the Mayas' chocolate, Tabasco, Mexico[4]

Tabasco is a very fertile, mainly flat region in the southeast of Mexico, bordering the Gulf of Mexico. It was there that the Spanish conquistador Hernán Cortez landed and cocoa was discovered together with the new continent. Tabasco is in fact the first place in the world where cocoa was cultivated. It was the Mayas who originally cultivated cocoa, a species that has its biological origins in Amazonia. Tabasco now produces 80 percent of Mexican cocoa. The remaining 20 percent come from nearby Chiapas.

Cocoa has a deep bond with Mexico and its culture. Today, however, its cultivation is suffering a period of crisis. The brokers of multinational companies, who have the monopoly on purchases, establish low prices from which small farmers do not earn enough to live and maintain their families. In desperation, farmers have been cutting down the tallest trees in order to earn minimal incomes from the wood. Together with this, emigration and abandonment of the countryside are undermining the socio-economic and environmental situation in the traditional cocoa-growing areas of both Tabasco and Chiapas.

In 1984 some farmers began producing organic cocoa but they faced marketing problems. In 1993 a group of women headed by Doña Sebastiana (Slow Food Award) decided to process the organic cocoa grown by their husbands in the traditional manner of the women of Tabasco and the Mayan women before them. These women began making chocolate to sell to tourists in hotels and at Villahermosa airport. The idea was a winner.

In 1997 two biologists decided to form a non-profit association, the Asesoria Técnica en Cultivos Orgánicos, to promote a complex cocoa-farming project, involving both men and women. The men would be responsible for the land, producing organic cocoa and developing a system of reforestation, recreating ideal conditions for shade production of the crop. In fact, men's cooperatives provide a very important ecological function; in addition to farming organically, they also implement a programme of reforestation. As cocoa requires shade, there must be taller trees in the plantation. This system assures the conservation of the Tabascan ecosystem. Today, the cooperatives are carrying out a "progressive level" reforestation plan. Timber is obviously important but the trees are also valued for their tropical flowers and fruits, which the women preserve and sell with the chocolate. Women's main role is to produce chocolate in the traditional manner.

In this project, the added value to the organic crop and processed chocolate boost family earnings and make it possible to purchase drying equipment. The result is that today there are seven cooperatives, four men's and three women's, directly involving 200–300 people and indirectly involving thousands of farmers and processors.

Organic and naturally coloured cotton, Peru[5]

With a unique agricultural history, the cotton plant was domesticated independently in four different geographical regions, giving rise to four distinct botanical species: *Gossypium arboreum* (northern Africa), *G. herbaceum* (India), *G. hirsutum* (Central America) and *G. barbadense* (Peru). Organically cultivated, naturally pigmented cotton is one of the oldest industrial crops of humankind and still survives as a backyard cultivar among many peasant and indigenous peoples of the tropics. Natural cotton colours recovered from

---

4   Source: Carlo Bogliotti, 2002. Doña Sebastiana Juárez Broca - Slow Arch. Magazine of the International Slow Food Movement.

5   Source: Vreeland, J. M., 2000. Systems and Genetic Diversity in the High Jungle of Peru. In: The Relationship between Nature Conservation, Biodiversity and Organic Agriculture. IFOAM/IUCN/AIAB.

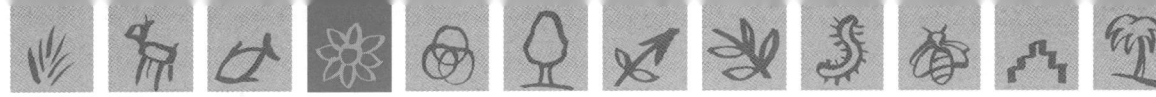

Indian communities include colours ranging from white, beige, brown, chocolate, green to purple.

Since 1982, the Native Cotton Project of Peru has aimed to identify, recover, multiply and redistribute seeds from indigenous landraces of coloured cotton to local farmers and artisans. With the support of the American Indian Institute, Aid to Artisans, the Peruvian Science and Technology Council and other organizations, the project has facilitated the reintroduction and commercial revival of naturally coloured cotton as an organic cash crop providing much needed income to thousands or rural farmers and weavers.

The project concept holds an efficient, economical and lasting strategy for the maintenance of genetic diversity, which sustains indigenous agricultural systems. It promotes local consumption and good linkages with external markets, placing added value on organic textile fibres.

As a result of the agro-ecological farming systems of the jungle Indians and the conservation of ecotypes adapted to moist tropical conditions, native farmers constitute the most qualified and legitimate guardians of this unique cotton germplasm. With a high degree of natural resistance to insects, disease and drought stress, native cotton can also provide valuable genetic materials for improving other cultivars and commercial cotton varieties.

RESCUE OF TRADITIONAL AND UNDER-UTILIZED SPECIES AND VARIETIES FOR QUALITY DIETS AND CULINARY TRADITIONS

In the past, agriculture played an important role in the maintenance of genetic diversity. The substitution of a large quantity of species for only a few and the adoption of high yielding and uniform varieties from a genetic point of view, has caused a significant reduction in the genetic inheritance of cultivated species. Many agricultural species, varieties and breeds which have played an important role in the human diet and traditional cultures have practically disappeared over the last century.

In the last decade, the adoption of organic agriculture has indirectly established a rescue process of species, varieties and breeds threatened by under-use or extinction. Stronger collaboration has been evident among movements aiming to defend biodiversity (such as the Slow Food Movement) and the organic agriculture movement. This is especially the case now that there is interest in traditional, speciality and organic products. For the rescue of varieties threatened by extinction, the development of a market is fundamental and it is here that organic agriculture plays an important role as the price premium gives an additional value to the product.

As illustrated by the case studies below, the restoration and enhancement of under-utilized species and varieties has been motivated by a food demand concerned with health and culinary traditions. The first two cases illustrate the discovery of the nutritional value of the gluten-free quinoa in Peru and Saraceno grain in Italy. The case of rice in Indonesia shows the role of local varieties in traditional diets and cultures. Consumer

No 4

demand for speciality products, such as the Garfagnana spelt in Italy, has re-established this product's economic viability. In all these cases, organic agriculture has allowed the maintenance and improvement of species and varieties that otherwise would suffer strong genetic erosion or extinction. Although not illustrated under this heading, similar examples exist for animal breeds and races.

Organic quinoa from the Cotahuasi river basin, La Unión, Peru[6]

At least since 3000 BC, if not longer, the seed of the plant *Chenopodium quinua* has been a vital part of the Andean diet, used as a grain in baking, as well as served in numerous dishes prepared by Aymara, Quechua and other indigenous peoples throughout the Andean region. Yet, in spite of its nutritious value and hearty growth, the arrival of the Spaniards led to change. Farmers were sent into the mines of Peru and Bolivia and non-native crops were introduced for consumption by the Spaniards. During the colonial period, quinoa use was associated strictly with native populations, leading to an undesirable perception of the seed as belonging to the lower class. In the last ten years, however, there has been an increasing interest in quinoa. The absence of gluten makes it ideal for sufferers of *Celiacs* disease.

In 1994, following the new interest in quinoa production, a group of farmers from La Unión, the northern part of the department of Arequipa, Peru, decided on a scheme for the promotion of farming within the province, adjusting their activities to the requirements of the international market. One of the initial strengths was considered to be that of agricultural production without the use of synthetic

organic chemicals, following their rich cultural tradition.

The Association of Organic Crop Growers (APCO) was then formed in 1996. Its social objective was to promote and develop organic agricultural production among the farmers of La Unión and encourage the conservation of biodiversity, research, commercialization and other activities that would lead to an improvement in the productivity and quality of the produce. By the end of 2001, there were 238 farmers belonging to the association and 350 enrolments were being processed for inclusion in the 2001-2002 agricultural campaign.

The territory coincides approximately with the Cotahuasi River Basin. With valleys between 2 400 and 3 400 metres above sea level, organic farmers produce quinoa associated with maize in a wide rotation that includes alfalfa, potatoes, peas, wheat and other crops, depending on the altitude of the farm. Land is fertilized directly through bovines that are allowed to graze on the crop residues or pastures. Control of pests and diseases is carried out through modifications in the sowing season and through other agro-ecological practices. An Internal Control System Committee ensures, with the participation of the farmers, that the norms of organic production are followed. APCO organizes the certification of its produce with the certification agency, Biolatina.

With the help of the Specialised Sustainable Development Association (AEDES), a development NGO that advises the local government and the population of La Unión Province, awareness has been raised amongst the APCO members of the role of organic crop production as a method of conserving and protecting the environment and biodiversity of the Cotahuasi Basin. This has included the creation of a micro-regional rural development

---

6   Source: Jordan Erdos, 2002. Quinoa, Mother Grain of the Incas. In: Sacred Food of the Incas from the World's Deepest Valley. APCO/AEDES web site: www.aedes.com.pe

plan integrating within the management of the Cotahuasi river valley, the local knowledge of the Andean cultures, in order to achieve sustainable management of biodiversity. This has been carried out according to the guidelines of the local Agenda 21 and has improved the socio-economic conditions of the farmers of the valleys.

Slow Food and the Ark of Taste: the case of Saraceno grain and Zolfino bean, Italy[7]

In 1996, the International Slow Food Movement (an association of some 80 000 gourmets worldwide, with objectives ranging from gastronomic education of consumers to the promotion of biodiversity for food and agriculture) launched the idea of an Ark of Taste. The objective of this project is to document, catalogue and safeguard small and quality agricultural diversity threatened, or potentially threatened, by extinction. The products chosen to be safeguarded include plant species, varieties and ecotypes, as well as autochthonous or well-adapted animal populations in a specific territory.

In order to promote these products, guaranteeing them an economic and commercial future, preserving degraded territory and creating new employment opportunities, an organizational instrument called "Presidia" was created. These are formed by local producers and are backed by local public organizations. "Presidia" elaborates production regulations for each of the products that it intends to save. These take into consideration not only the cultural-historical aspects and biodiversity, but also environmental problems and small-scale economies, proposing agronomic and livestock practices that are not aggressive to the natural environment. In some (but not all)

cases the production regulations are explicitly of organic nature and prohibit the use of synthetic fertilizers and pesticides, as in the case for Saraceno grain from Valtellina and the Zolfino bean from Pratomagno, Italy.

Saraceno grain (*Fagopyrum sculentum*) was one of the fundamental foods in the diet of the poor farmers from Valtellina (Piemonte Region, Italy) and of the whole Alpine region until the last century. It was used to make black polenta and *"pizzoccheri"*, a type of home-made pasta made with flour from the Saraceno grain and wheat. However, its labour intensive and costly cultivation is a factor that has led to its decline. The Saraceno grain is now only cultivated on the slopes of the high valleys due to its rusticity and resistance to cold climates, and where it is lightly attacked by parasites. However, as the Saraceno grain does not contain gluten, the last few years have seen an increased interest in its cultivation as it is an ideal food for people affected by *Celiacs* disease.

The Slow Food Presidia, supported by the Municipality of Teglio (Piemonte) and by the Mountain Community of Valtellina de Tirano, proposed the reintroduction of the cultivation of Saraceno grain using organic production techniques and has since elaborated production regulations that only allow for organic fertilization and prohibit pesticides. The land used for Saraceno grain cultivation is put through a rotation of two or three years.

Another Presidia that has adopted organic regulations is that of the Zolfino bean, a dwarf variety *(Phaseolus vulgaris L.)* cultivated in an area of Pratomagno, Tuscany, Italy. It is characterized by certain peculiar organoleptic qualities that make it especially suitable for some gastronomic preparations typical of the Tuscan cuisine.

This variety is particularly adapted to low-potential lands in the hills and mountain

7   Sources: L'Arca, 2001. Il grano Saraceno della Valtellina. Quaderni dei Presidi Slow Food 2001; Di Napoli Raffaela and Davide Marino, 2001. Biodiversità e sviluppo rurale.

areas and is frequently cultivated on terraces in association with olives. The drastic reduction in the farming population that occurred in the region after the Second World War has caused the progressive abandonment in the cultivation of this bean[8]. It is only the patient work of a few farmers over the last few years that has allowed the Zolfino bean to avoid extinction, stimulating its revival and development.

Recovery of local varieties of rice through organic methods, Indonesia[9]

Rice is the staple food of not only 95 percent of the Indonesian population, but of Asia as a whole. In 1967, paddy rice varieties in Indonesia were diverse, including over 7 000 varieties. However, in 1965, the planting of local rice varieties was prohibited, resulting in their near extinction.

The changes in the Indonesian agricultural system over the past 25 years have had many deleterious effects. Among them are soil infertility, the appearance of new pests, the economic dislocation of many poor and food insecurity. After some years of adoption of modern farming techniques, the productivity of rice dropped, even though the use of fertilizers increased.

Due to the near extinction of many local rice varieties, in 1997, "Pusspaindo" (a private organization that focuses on biodiversity) launched a project for the recovery of local rice varieties. The main objective was to promote farmers independence through the use of local varieties, conbined with local wisdom and traditional production systems.

In cooperation with poor farmers and farmhands, one kilogram of a local rice variety was obtained. This was planted, multiplied and distributed amongst the farmers. In 1997, several local rice varieties (e.g. Siyem Putih, Rajalele, Nongko Bosok), which are top "yielders" and resistant to pests, have been found in East Java and are being planted in East and Central Java.

Aside from the revival of local rice varieties, Pusspaindo is also promoting the production of rice using organic methods of pest control. Pusspaindo has carried out experiments proving that local rice varieties can achieve higher yields than the newly introduced varieties. Yields of 10–14 tonnes per hectare have been achieved.

Indigenous agricultural knowledge, from land preparation to planting, harvest, processing, rituals and prayers and recipes for medicine, has also been documented. Local rituals have been staged to make farmers realize the importance of local rice varieties.

Other activities carried out in relation to local rice varieties and the indigenous knowledge that accompanies them include: organization of farmers into groups; establishment of demonstration plots; production and dissemination of educational materials; organization of seminars and training workshops; opening of a consultation service; development of a marketing system; and establishment of a network and advocacy campaign.

According to Pusspaindo, local rice is superior compared to modern rice. It has a better flavour, higher production, is more nutritious, can be grown continuously throughout the year, is easier to plant and more economical, especially if grown organically. Some local varieties also have medicinal properties for common diseases such as stomachache, cough, metabolic acceleration, and others.

---

8   In the Regional Register (Tuscany) of Autochthonous Genetic Resources LR 50/97 it is shown at risk of genetic erosion.

9   Source: Sismanto Joseph, 2002. Management of Diversity of Local Rice and Organic Agriculture for Strengthening Indonesia's Food Security. Centre of Study and Development of Indonesian Rice (Pusspaindo). Paper presented to the GRAIN International Workshop on the Local Management of Agricultural Biodiversity, Rio Branco, Brazil, 9-19 May 2002.

Most importantly, local rice is part of Indonesian culture. For thousands of years, Indonesians have been growing local varieties of rice, developing their own technique for the production of high yields, providing for their own needs, as well as a surplus. This they did for a thousand years without damaging the soil, at the same time developing and conserving their varieties. In fact, Java was previously known as *Jawa Dwipa* (*Jawa* meaning island and *Dwipa* meaning rice) and used to be famous for its tradition as an exporter of delicious rice.

In this context, many communities who care about the future of Indonesia, especially in terms of food security, have been bound together in the study of the traditional knowledge surrounding rice production and natural agricultural ecosystems, understanding that their futures could be dependent upon them.

Protected Geographical Indication and organic production norms for the Garfagnana spelt, Italy[10]

Spelt wheat (diverse species from the genus *Triticum* and *Spelta*) is the oldest known cereal. It was first cultivated by the Babylonians and later by the Egyptians. For centuries it was the staple food of Asian and Mediterranean populations. According to some studies, its centre of origin is in Palestine, from where it was spread by nomads. In Italy it has been cultivated since the Bronze Age and later spelt became one of the principle foods of the Romans. Its decline began in the Middle Ages when other cereal crops of greater yield and easier working started to be cultivated.

For these reasons, spelt wheat remained in limited marginal areas of altitude between 500 and 1 500 metres above sea level. This includes the Garfagnana (an area of high hills and mountains in the Lucca Province, Tuscany) where the difficult geo-pedological and climatic conditions allow the plant, thanks to its rusticity, to vegetate and grow.

Spelt persistence in the Garfagnana area depends above all on its links with local traditions. It is the fundamental ingredient in some traditional dishes such as soups and savoury cakes. Since the early 80s, spelt wheat has seen a return in various regions in central Italy, as the healthy properties of this cereal attract consumers. Spelt wheat contains high levels of fibre and is cultivated traditionally, without the use of synthetic pesticides or fertilizers.

Consumers' interest has determined an increase in price of the cereal and consequently, the diffusion of its production to the plains. Here yields are higher, but cultivation practices do not always follow traditional methods, threatening the production on the hills that has been maintained for centuries.

In order to overcome this situation and give value to local production, the Mountain Community of Garfagnana applied for, and obtained, European recognition for Protected Geographical Indication (PGI) in 1996. The regulations drawn up for Garfagnana spelt common variety *(Triticum dicoccum)* and the description of the genotype that through the years has adapted to the local climate and terrain, prescribe agronomic practices for its production as "organic". These include rotations with meadows, the prohibition of the use of chemical pesticides, herbicides and fertilizers and the mandatory use of seeds coming from local populations. Compliance with these regulations is guaranteed by the activities carried out by the Italian Association for Organic

---

10 Sources: Rossi Asanella and Massimo Rovai, 1999. La valorizzazione dei prodotti tipici: un analisi secondo l'approccio di network. Rivista di Economia Agraria N° 3/1999, INEA; Di Napoli Raffaela, Davide Marino and Paolo Foglia, 2001. Biodiversità e sviluppo rurale. Quaderno informativo INEA.

Agriculture (AIAB), under the authorization of the Ministry of Agriculture.

These measures have stimulated an increase in the production of Garfagnana spelt and in the value of the production. The area under cultivation has practically doubled in the last three years, reaching about 200 ha, the producers obtaining a price of between 25 and 30 percent higher that spelt without a geographical indication label.

The adoption of the PGI for Garfagnana spelt is important for the preservation of local varieties that have been selected throughout centuries. The "organic" regulations (including the prohibition of chemical pesticides and fertilizers) afford greater value to production and provide a guarantee to the consumer with an interest in food with decreased pesticide residues. Production also favours the local economy of Garafagnana which, like all mountain areas of Italy, has suffered a decrease in population over the last decades and a progressive socio-economic marginalization of its inhabitants.

SELECTION OF BIODIVERSITY ADAPTED TO LOCAL ECOLOGICAL CONDITIONS AND RESISTANT TO DISEASE

The necessity for organic farmers to find methods for obtaining quality products with good yields and limited production costs is greater than for other farmers. Besides the fact that organic farmers cannot apply synthetic inputs, their use of organic fertilizers, natural pesticides and other permitted substances is uneconomical in the long-term. External inputs as such are relied upon mainly during the conversion period to organic agriculture or under exceptional circumstances. The comparative advantage of certain breeds to withstand local natural stress, especially in marginal areas, leads organic farmers to adopt biodiversity management as their main productive strategy.

This has stimulated organic farmers to use species, varieties and breeds better adapted to local climatic conditions and more resistant to pests and diseases. This has led to the restoration of many traditional varieties and breeds that are better adapted to local environments and has indirectly generated a significant contribution to agricultural biodiversity. A few centres of public and private research are now collaborating with farmers in the search for such species, varieties and breeds.

The first case illustrates the rescue, in Germany, of an old variety of wheat with a vegetative cycle that allows the absorbency of nitrogen available in the soil, improving the quality of the grain. The second example describes the rescue of the Maremmana cattle, on the verge of extinction, due to its suitability to grow in marshy environments of Italy. The third case describes the re-

establishment of native chicken in South Africa, due to their resistance to disease. The fourth case describes the techniques adopted and the complexity of the permaculture system that provided a solution to desertification in Brazil.

Selection of quality wheat varieties for organic agriculture on sandy soils, Germany[11]

Sandy soils are quite common in Lower Saxonia, Mecklenburg-Vorpommern and Brandenburg (around Darzau), Germany. Combined with spring rains, soil nitrogen is easily washed away. As this cannot be compensated for in organic systems by the use of chemical fertilizers, organic farmers find it difficult to produce the quality standards for wheat (especially gluten content) demanded by mills and bakers for the production of organic foods.

In order to find a solution to the problem and improve the quality of organic wheat produced in the area, the Cereal Breeding Research Station Darzau is carrying out a project entitled Quality Wheat Project for Organic Agriculture on Sandy Soils. Under the umbrella of the Association for Goetheanistic Research, the criteria for cereal breeding at the research station have been developed following biodynamic farming techniques. Specific attention is paid to soil fertility, natural manures, weed competitiveness and seed-corn diseases, together with nutritional quality. Once important old or new characteristics of cereals have been identified, these varieties or species are entered into breeding programmes, aiming to enhance production under modern farming practices.

The research station is undertaking field experiments involving many varieties, several selected from the gene bank at Braunschweig-Voelkenrode. Field trials have shown that the loss of nitrogen during the winter months can be minimized by early sowing in autumn. For this purpose it is necessary to have wheat varieties that do not develop too quickly before winter (as this would lead to losses during winter), but would on the other hand, develop sufficiently to suppress competing weeds (such as *Apera spica-venti L.*). Fast development in spring, as observed with the old Austrian variety "Staatzer", resulted in successful suppression of competing weeds as wheat shadowed the soil already early in the year. Field experiments showed that with this variety, even under poor conditions, it was possible to obtain the qualities that the bakers and millers request.

Breeding activities also aim to find a more favourable relationship between gluten content, yield and sedimentation value (the sedimentation value quantifies the swelling ability and the ability of the proteins to aggregate). Selection of varieties with high sedimentation values could improve the gluten contents of wheat by approximately 22 percent.

The station's researchers have also networked with other breeders in order to develop, exchange and test locally adapted wheat varieties for organic conditions in different locations of Central Europe. For five years, five initiatives tested old and new varieties and developed new breeding lines, through an interchange at each location. These activities led to the development of new criteria for wheat varieties suitable for organic agriculture and to an exchange of information on breeding problems. It became soon apparent that it was impossible to obtain a single variety that was suitable for the diversity of organic growing conditions in the entire region. For example, the availability of manure depends on the number of cows per

---

11 Source: Cereal Breeding Research Darzau, 2000. The Quality Winter Wheat Project for Organic Farming on Sandy Soils; Mueller Karl-Josef, 2002. Developing Criteria for Breeding and Breeding Itself of Cereals for Organic Farming in Northern Germany. Cereal Breeding Research Darzau (www.darzau.de/en/projects/quality_wheat.htm).

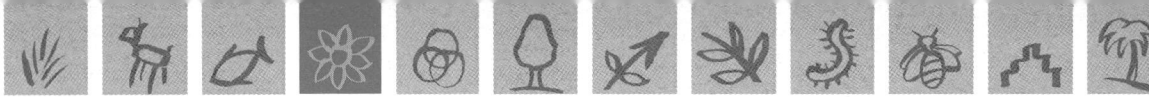

hectare, which in turn is dependent upon the fodder harvest that varies according to a specific location's rainfall, temperature and soils conditions. All theses factors affect wheat quality, directly or directly. The results of this collaborative effort suggest that more biodiversity is required for organic agriculture and that organic agriculture requires more varieties of locally adapted wheat.

The research station also aims to work on resistance or tolerance in relation to seed-corn fungal stinking smut or common bunt disease *Tilletia caries*. Other research in this matter suggests that conventional breeding programmes often pay little attention to the characteristics which can be of special importance to organic farmers. For example in the conventional "value for cultivation and use trials", breeders often choose wheat varieties that yield a few kilograms more, and not varieties which are less susceptible to, for instance, *Septoria,* that can be controlled by fungicides.

Contrary to the conventional approach, a healthy crop is of more importance when breeding for organic agriculture. Other characteristics of interest to organic crops and ignored by the conventional breeding programmes include cereal varieties with a reasonable length of straw; this is because soil born fungi (such as *Fusarium* and *Septoria*) are less likely to harm crop varieties with longer stems.

A further question that needs consideration refers to the protein quality of wheat for human nutrition. Which relationships between the different protein components should be aimed for? How can the quality demands of millers and bakers be combined with high value for human nutrition? These questions are currently under investigation.

Maremmana cattle with the organic rearing approach in marshy areas, Italy[12]

The Maremmana race is a robust breed of cattle, directly descended from *Bos taurus macreceros*, originating from the Asiatic steppes. Through the course of the centuries the Maremmana cattle have adapted to the marshy and malarial environment of Maremma (part of Tuscany and Lazio, central Italy). Thanks to its extraordinarily robust characteristics it has been used above all for working in the fields and for the transport of marble from the mines of the area.

This animal has the capacity to feed and develop in areas where other races have difficulty in surviving. It is few-troubled by diseases, is long lived and is characterized by cows with good maternal instincts. It is not uncommon to find cows over 20 years old that have produced between 13 and 15 calves during their productive life. The animal lives outdoors all year round and is sufficiently nourished by grazing herbaceous plants and bushes.

With additional payments available for areas subject to flooding, the mechanization of agriculture and the appearance of races more specialized in meat production, the Maremmana cattle began to face a crisis and was on the verge of extinction.

Thanks to the suitability of the Maremmana race to marshy environments, the situation is now changing. Following the entry of agri-environmental measures in 1992 (EU Regulation 2078/92), incentives are provided to protect animals threatened with extinction and to ecological agriculture. Livestock owners and regional governments have therefore began to revive interest in the development of Maremmana cattle. The race has now been included in the rural development plans of the regions

---

12 Sources: Slow Food, 2001. La vacca Maremmana. *In:* L'Arca, Quaderni dei Presidi; Giannone Mario, 2002. La ferrea dieta della Maremmana. In: A-Z Bio.

of Tuscany and Lazio. Livestock keepers raising this cattle now receive financial compensation the reduction in meat yield of this animal, which is bonier compared to other types of cattle.

Farmers can also benefit from the European Union Community Regulation 1804/1999 that regulates the production of organic livestock. The raising of Maremmana cattle easily falls within this regulation. The exponential demand for organic meat after the "mad cow" crisis has also further stimulated livestock owners to raise cattle in an organic manner.

One of these organic farms is "Alberese Natura", in the Uccellina Park, extending over 3 000 ha (of which 1 000 ha of forest) and raising in complete freedom 500 Maremmana cows. This farm offers an ideal performance test centre for the production of Maremmana cattle. Together with cattle, the farm produces organically: forage barley, maize and beans for feeding the cattle, durum whead for the production of pasta, sunflowers, flax, millet and horticultural crops.

Following the adoption of the EU Regulation for organic livestock (1999), the number of heads of Maremmana cattle has increased: in 2000, 5 840 cattle heads were registered in the "genealogical book" of Lazio, Marche and Tuscany. Despite the fact that the yield of meat of this animal in the abattoir is less than for other races, rearing of Maremmana cattle is highly profitable. It achieves higher market prices than conventional breeds as its meat is sold as organic meat while at the same time the cost of production is much lower.

Rearing native chickens through organic agriculture, South Africa[13]

In South Africa, an estimated 70 percent of the rural population are classified as poor and many are locked into poverty and subsistence farming. Poultry production is a very important source of animal protein in subsistence agriculture.

In the past few years, programmes aimed at increasing animal protein production have proposed intensive poultry breeding as the solution. Besides causing a reduction in native and locally adapted breeds, they have generally failed at household level due to high losses (often reaching 80 percent) of chicks before they reach maturity. Newcastle disease is prevalent in the area and cyclic outbreaks have had devastating effects. Although vaccines are available, no organized or systematic vaccination campaign has been undertaken. Poor nutrition, lack of protection and predation are also contributing factors to high losses. In South Africa, Newcastle disease, which caused serious problems in 1994, regressed in 1995 and again in 1996. However, it has still not been eradicated and isolated outbreaks have been observed in chicken and ostrich production units in several regions of the country[14].

Although many native birds grow more slowly, they are good layers, have genes adapted to survival in extreme conditions, are less choosy about what they eat and are more resistant to disease; these characteristics make them more suitable for poor farmers. Their meat has a good flavour and texture. Following the devastation caused by Newcastle disease, these indigenous birds were used to re-establish the poultry population.

---

13 Sources: FAO, 1999. Programme Framework for the Republic of South Africa for Food Security. Special Programme for Food Security; Frei Berthould Annette, 2002. Noel Honeyborne. *In:* Slow Ark, Magazine of the International Slow Food Movement.

14 United States Animal Health Association, 1997. Animal Disease Status Worldwide Reports. Encyclopaedia of Sustainability, 2001. Polyculture in the Brazilian Drylands.

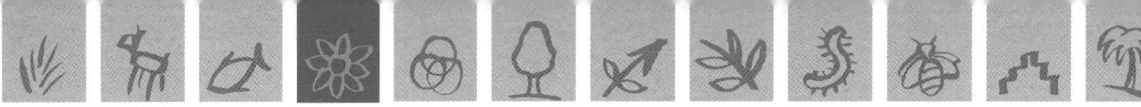

These native properties, of the birds acquired over hundreds of years, are important for future breeding and should be conserved. In 1994, the Animal Improvement Institute of the Agriculture Research Council founded the Fowls for Africa project, which neatly links the task of saving old breeds with the fight against hunger in southern Africa. The project is based on the idea of "conservation through utilization", combining the production of poultry birds with know-how and research. With the help of regional Poultry Supply Centres the Fowls for Africa project focused on supporting farmers. The primary aim was to produce protein at low cost through the utilization of suitable breeds and appropriate technology.

The breeding programme follows organic approaches and birds are free-range. The system also includes the use of movable chicken coops, allowing them to be moved frequently. Where the chickens leave their manure, the soil is well fertilized and vegetables are planted. Four poultry breeds *(i.e. Potchefstroom Koeloek, Ovambo, Venda and Naked Neck)* and two other poultry breeds (i.e. *Black Australorp* and *New Hampshire*) were identified as suitable for small farmers. Theses breeds are in fact well adapted to survival under harsh, low-input conditions with only basic requirements of shelter, feed, water and hygiene.

Anyone interested in taking part in the programme can attend a training course and obtain credit from a recognized financial institution. The Fowls for Africa network supplies the software, the necessary background information, training and veterinary care, and also the hardware (i.e. the poultry and additional materials such as chicken coops).

Initially, the project had the objective of saving the biodiversity of autochthonous races of chicken and combating hunger in the rural population by giving support to development projects. However, it also proved equally useful for the provision of breeding animals for organic farms, for eco-tourism and in educational and research institutions.

Restoring drylands with permaculture, Bahia, Brazil[15]

In the North East of Brazil there is a large semi-arid region of 900 000 square kilometres. Much of this region is severely degraded due to large-scale deforestation, ploughing and goat herding beyond the carrying capacity of the land.

Rain in this region is erratic, often coming in downpours followed by long dry periods even in the rainy season. In spite of this, farmers plant corn and beans which are dependent on rain, practically programming themselves for agricultural failure. Agriculture in this region is collapsing and many men and women have migrated to the slums of Sao Paulo in search of employment. This physical and economic collapse has been accompanied by a general depression and disbelief in the potential of the region; unless one has sophisticated irrigation equipment, however, this is also unsustainable due to high costs and dwindling water reserves.

Marsha Hanzi is a professional permaculture teacher and consultant, with long experience in agroforestry. When travelling out to the drylands of Bahia, she pointed out that the problem of crop failure was not lack of rain but rather lack of strategies to maintain the water on the land. Castor (*Ricinus communis*), a major crop of the area, was planted with very large spacing. The constant dry wind blew unimpeded down these corridors. At the same time, she noticed that other fields, where pigeon peas (*Cajanus cajan*), cassava, elephant grass (for fodder) and

---

15 Encyclopaedia of Sustainability, 2001. Polyculture in the Brazilian Drylands.

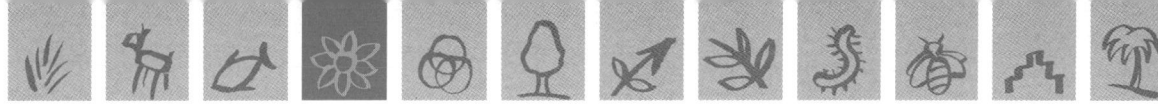

local fruit trees had been planted, survived under extreme conditions.

The idea emerged to plant all of these crops together in one field, maximizing the use of space. This improved the performance of the castor crop thanks to greater soil moisture conservation and fertility, due to inter-cropping with leguminous plants. The soil was covered at all times, combining high and low growing plants, short-and long-season crops as well as trees to protect the land during the long dry season.

From these ideas, the Polyculture Project was born in 1999, run by the Bahian Permaculture Institute and VITA, an agency involved in sustainable agricultural development. In 2001 the project boasted 47 demonstration fields on farmers' plots and worked directly with 1 000 farmers.

The polyculture model adopted is based on castor bean and sisal *(Agave sisalana)* associated with maize, beans, sesame, sunflower and pigeon peas. To these are added short, medium and tall legumes for nitrogen fixation and production of organic mass, such as jackbeans *(Canavalia sp.)*, *leucena* and *gliricidia* trees. Finally, native and adapted fruits and lumber trees are planted together with native legumes and fodder plants, guaranteeing that even in the worst years the farmer will harvest something from the fields.

The results were far beyond expectations, doubling castor production in the first year, as compared to neighbouring fields, and offering a number of food products for the farmers from space that otherwise would have been empty. Farmers are now organizing the sale of surplus food products. These have the advantage of being organic and as such, are more appreciated by the consumer in the markets of the cities of Santana and Salvador.

ALTERNATIVE BREEDING CRITERIA AND PARTICIPATORY RESEARCH

The objectives of selection for the development of varieties for organic agriculture differ from those for conventional agriculture. It is of crucial importance to utilize the genotype potential for an improved adaptation of varieties to the low-input conditions prevailing in organic agriculture.

Breeding for high performance and selecting for early maturity have led to increased susceptibility to infectious diseases, joint inflammation and mastitis as well as circulatory, metabolic and fertility problems of livestock. Loss of breeds is exacerbated by the narrowing genetic base of modern breeds and hybrid lines. The trend towards inbreeding increases the degree of genetic uniformity in the animals and hence, susceptibility to infection, parasites or epidemics.

The majority of crop varieties available on the commercial market are not suitable for organic cultivation methods as they have been selected for production dependent on large quantities of synthetic fertilizers and pesticides. Many of these are also hybrids and are not open pollinated. In the last few years, the problem has worsened with the arrival on the market of genetically modified varieties.

Limitations and threats associated with crops has stimulated many organic farmers, especially in the horticulture sector, to produce their own seeds. In order to do this, they have often had to rescue local varieties and develop their own systems of selection and distribution. These empirical systems have been based on the selection of individuals better adapted to the local environment and more resistant to pests and diseases. In many cases, the systems include

the exchange of seeds between farmers as a fundamental instrument (e.g. organic seed fairs). In other cases, the tastes of the consumers have also been taken into consideration, as is the case of La Verde Cooperative, Andalusia, Spain.

Many of these systems have demonstrated interesting results in the selection of varieties suitable for low external input situations. The selection systems developed by organic farmers have restored and improved local varieties. These varieties often present a high degree of genetic variability and as such, the systems have played an important role in the *in situ* conservation of agricultural biodiversity through cultivation and production. This has since been studied by research centres specialized in the selection of seeds and plants. For example, in Cuba, such studies are being used as the basis for the refinement of methodologies for the selection of varieties for low input situations.

Historically, farmers have managed many varieties and breeds according to agronomic and culinary properties. Considering the need for a wide gene pool to improve and multiply genetic resources for food and agriculture, seeds and breeds from the formal and informal sectors should be included harmoniously in local and national breeding programmes. Benefits derived from new varieties bred by farmers require a legal system of common ownership that allows equitable access and benefits sharing. The biodynamic network of farmers and breeders in Germany provides an example of how such a system could be organized.

In situ restoration of local varieties through organic agriculture, Andalucia, Spain[16]

The cooperative "La Verde" in Villamartin, Cadiz, Spain, was founded in 1987 when a group of day labourers linked to the farm workers syndicate decided to organize themselves to overcome the precarious labour situation in which they found themselves.

From the outset they followed methods of production that respected the natural environment and which equally integrated women in the cooperative. The main concern was the restoration of traditional cultivation techniques and the transmission of the knowledge generated. The decision to cultivate following methods of organic agriculture was due, on one hard, to the negative effects of conventional agriculture on health and the natural environment and on the other, to the few market opportunities that conventional agriculture offered.

In 2000, the cooperative cultivated 13 hectares of horticultural crops and employed seven people. The cooperative sells the majority of it products directly on the local market, complementing this with sales through consumers' associations.

One of the problems that the cooperative has faced since the beginning is the lack of commercial varieties adapted for organic agriculture. For this reason, one of the practices adopted was the restoration of local varieties of horticultural crops, better adapted to the system of organic agriculture. This has involved the exchange of seeds with other farmers and the selection of seeds on the basis of a number of criteria, principally resistance to distinct adversities and the taste of local consumers.

Other organic cooperative activities highlighted the necessity for a long-term project for the restoration of traditional

---

16 Sources: Soriano Niebla Juan José, 2000. *In situ* on Farm Conservation of Vegetables Landraces; Figueroa Zapata M., *et al.*, 2000. Recuperación de variedades locales para la agricultura ecológica; Soriano Niebla Juan José, 2001. Manejo agroecológico de recursos genéticos.

varieties. With the help of the Institute of Sociology and Farming Studies at the University of Cordoba, the Council of the Assembly of Andalusia and the Syndicate of Farm Workers of Andalusia, a research project was initiated in 1998. The project, entitled "Study of the potential use of local varieties of horticultural crops for organic agriculture", concentrated on the process of agro-ecological use of biodiversity by local farmers in two zones of Andalusia: Sierra de Grazalema (Cadiz) and Antequera.

The project was carried out in a participatory fashion and following objectives linked to the management of plant genetic resources in line with the necessities of the cooperative and the rest of the organic agricultural sector. The project included 63 of the most common local varieties of horticultural crops in the area (e.g. tomatoes, aubergine, bell peppers, water melon, melon, lettuce, courgette, squash and carrot) and involved:

- strengthening farmers' network for the exchange and conservation of local varieties of seeds;
- diversifying the financial resource of the cooperatives and creating alternative markets for organic seeds;
- taking advantage of knowledge and genetic material restored during the years that the cooperative has been functioning, strengthening their use,
- elaboration of an inventory and characterization of varieties following series of botanical and agronomic descriptions, with a view to restore the knowledge of use linked to them;
- increasing on-farm biodiversity, including the use of wild plants from the area as forage and green manures;
- offering to the organic sector the possibility to use seeds of traditional varieties aiming to guarantee their long-term survival in the area and allowing the restoration of organoleptic and cultural characteristics associated with these varieties.
- introduction of varieties obtained from the area or restored through germplasm banks;
- agronomic valuation of the varieties with the help of agronomic experts of the area;
- participatory evaluation of the quality of the varieties, following criteria required by farmers and consumers.

The strategy followed takes into consideration not only the characteristics of the varieties but also the values of the consumers and the contribution of farmers. Members of the cooperative are responsible for deciding what varieties should be used directly for production and above all, what varieties and characteristics they will continue working with in their selection.

The activities of La Verde cooperative in the restoration and use of suitable local varieties of horticultural crops for organic agriculture has formed the basis of a project for the *in situ* restoration of the genetic biodiversity of distinct varieties of horticultural crops. The empirical selection criteria adopted by the organic farmers has allowed the maintenance in cultivation of local varieties, but has also permitted researchers to elaborate a valid system of *in situ* conservation that includes an active role for farmers and consumers.

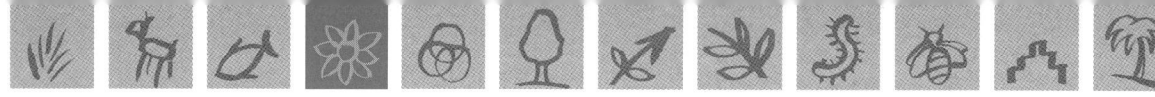

Changes in plant breeding of pumpkins as response to socio-economic limitations, Cuba[17]

Since the beginning of the economic crisis in 1989, the Cuban Government has attempted to reduce the negative impact of the lack of inputs for agriculture. National strategies have been implemented to accelerate research and its application in areas, including biological control, crop rotations and polycultures. This has resulted in major changes in some of the goals of Cuban plant breeding and a search for more appropriate methods of participatory plant breeding.

In Cuba pumpkin *(Cucurbita moschata)* is very popular thanks to its culinary and medicinal properties, taste, beta-carotene content and use in some religious ceremonies. With the drastic reduction of chemical inputs and artificial irrigation, the abrupt decrease in productivity resulted in pumpkins disappearing from the market.

Research into the pumpkin provides a clear example of how plant breeding systems have changed in response to these constraints. Initially the use of conventional seeds was maintained, but substantial yield reductions necessitated a change in response. After exploring new seed varieties from international seed companies to little avail, landraces from different sources in Cuba were investigated. Cuba then began a system of participatory plant breeding whereby trials were conducted by farmers themselves on their own land in collaboration with researchers.

This change in approach stimulated discussions on the efficiency, advantages and weakness of chemical as compared to organic inputs and on their application in approaches to plant breeding within the country. In terms of energy consumption, inputs used on farms and farmers' participation, the collaborative effort towards crop improvement under low input conditions was much more efficient in terms of energy use. Notably, the yield obtained under the low input system was comparable to yields under the conventional, high input technology package (i.e. 6-8 tonnes per hectare).

Pumpkin varieties were maintained and their seed multiplied through cross-pollination (rather than isolation) and honey bees were frequently used. Farmers' participation allowed on-farm selection of half sib families (rather than contracted seed production), screening germplasm, facilitating availability of new germplasm and evaluation of varieties with farmers.

Working on the farm with farmers provided two important insights. First, wide genotype variability of useful traits exists and has been documented among pumpkin landraces grown under low input conditions. Second, it is possible to increase production by selecting directly for fruit yield under low input conditions. Under this experience, plant breeders offered a bridge between the plant genetic resources conserved in gene banks and the farmers, and the opportunity to screen those resources. Clearly, farmers' agricultural knowledge and skills were an inspiration to develop a new, collaborative approach towards a more efficient use of inputs such as energy, more profitable crop production and maintenance of greater genetic diversity *in situ*.

A Participatory Plant Breeding for Strengthening Agrobiodiversity is now on-going to investigate how such alternative practices can rebuild, improve and distribute biodiversity in Cuba. Interesting results of farmers' experimentation are already apparent in research for maize resistant to

---

17 Rios Labrada H., D. Soleri and D.A. Cleveland, 2002. Conceptual Changes in Cuban Plant Breeding in Response to a National Socio-Economic Crisis: the Example of Pumpkins. *In:* Farmers, Scientists and Plant Breeding, CAB International 2002. pp 213-237.

fallarmy worm *(Spodoptera frugiperda)* and beans with good productivity under low input conditions.

Network of biodynamic seed production and plant breeding, Germany[18]

The disappearance of open-pollinated varieties and, more recently, the development of gene technology and its multinational structure are the main drives for establishing breeding methods and cultivars suited to organic agriculture and organic markets.

For over 15 years now, the Association for Biodynamic Vegetable Plant Breeding (Kultursaat) in Germany has been working on biodynamic plant breeding and seed production. This is done through a network of farmers, breeders, a seed company and the Kultursaat Association (see graph attached). The objectives of the Network are the following:

- good plant development and root growth (capable of relating to beneficial soil organisms);
- growth through organic fertilizers (i.e. energy and nutrient efficiency);
- ability to interact with the environment;
- tolerance and resistance to adversities (disease resistant or tolerant);
- development of species-typical growth patterns and maturation processes;
- good, species-typical taste and nutritional quality.

The activity of plant breeding is returned to farmers themselves. Practical care of plants, their propagation, testing of new varieties and maintenance are best applied by farmers. Farmers build up on their experience and knowledge of crops

and cultivation methods. Breeders and farmers practice, at the farm level, breeding methods that both achieve quality and respond to the specific needs of organic agriculture. Since maintenance and breeding are integrated in the vegetable production cycle and the most applied breeding technique is positive mass selection, the additional work to the farmer is relatively small.

Breeding and selection skills are provided by the Kultursaat Association through regular meetings and, if necessary, through individual training by experienced farmers working on the same crops. Meetings are held once a year at the regional level and three times a year at national level. During these meetings, farmers share knowledge on botanical and breeding issues, develop new ideas and exchange experiences and breeding lines. International exchange is starting within Europe and there are plans to further extend it.

Since the association receives financial support from donations, some of the farmers are paid for full-time breeding activities. This allows establishing broader breeding programmes and research. The new varieties generated from these programmes can be registered directly or can be given to other breeders/farmers in different areas for adaptation to different growing conditions.

With regard to conservation of biodiversity, it is not sufficient to preserve varieties in seed banks, or just on-farm. A constant and proper selection effort is necessary to maintain varieties value as well as adaptation to specific growing conditions. The maintenance of one variety in different places will create new varieties within a few generations. Therefore, more diversity is created, according to different biophysical conditions.

Research experience has so far demonstrated that most varieties (e.g. cereals) do best under the conditions under

---

18 Henatsch C., 2002. Organic Farming Needs Organic Plant Breeding: a Network for Independent Seed Production and Plant Breeding. *In:* Cultivating Communities. Proceedings of the 14th IFOAM Organic World Congress, Victoria, Canada, August 2002 (p. 300).

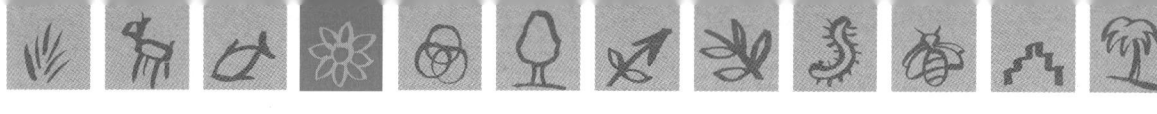

which they were bred. This is especially true in terms of resilience to adversities which are bound to the place of origin. Cultivation under different natural conditions takes three to five generations to regain resilience. The general condition of the variety, however, is not to be discounted when evaluating its likely adaptability.

Breeding methods are based on diversification and regional adaptation, including:

- consistent and rigorous selection from a large stock base;
- single plant selection;
- cross-fertilization;
- creating varieties and developing special characteristics through: geology, geography and mineral provision; effects of planetary influences and the biodynamic preparation; and influence of human and cultural conditions.

The organic seed company was founded and is partly owned by the farmers/breeders. The seed company looks after cleaning, quality testing and distribution. The Kultursaat association coordinates plant breeding and provides financial support, payment of registration and testing fees. To date, Kultursaat has bred more than 20 new (registered) varieties (e.g. carrots, cabbage, spinach). Most importantly, the association is the owner of new varieties and ensures common ownership and benefit sharing among all participants.

Recently, the breeding and selection network is expanding to other European countries (i.e. Austria, Italy, Switzerland and the United Kingdom): ideas, experiences, varieties and breeding lines are shared to provide a widely available open-pollinated assortment of vegetable seeds of high quality.

CONCLUSIONS

As demonstrated by the case studies, organic agriculture, almost without the help of governmental institutions, is providing an important contribution to the *in situ* conservation, restoration and maintenance of agricultural biodiversity. The spontaneous establishment of participatory systems of research and development is shaping a simple and practical system of equitable sharing of benefits derived from genetic resources for food and agriculture.

The growth pattern shown by the conversion to organic agriculture throughout the world suggests that this contribution is likely to increase even further. Considering the role that organic agriculture plays in the maintenance of agricultural biodiversity, public institutions, especially research centres and universities, are recommended to carry out actions that could include:

- documentation of existing cases and the systematic diffusion of data concerning agricultural biodiversity on organic farms;
- development of participatory approaches to the evaluation, selection and multiplication of varieties and breeds that better adapt to situations of low external input and low potential areas, directly on organic farms;
- establishment of proper systems for registering and determining access to genetic resources developed for organic agriculture;
- establish proper systems for compensating the maintenance activities and breeding efforts of organic farmers.

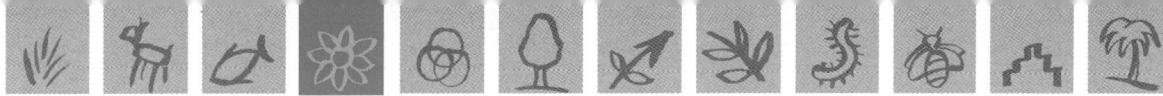

INITIATIVE FOR BIODYNAMIC VEGETABLE SEEDS NETWORK
ORGANIC PROPAGATION AND BREEDING

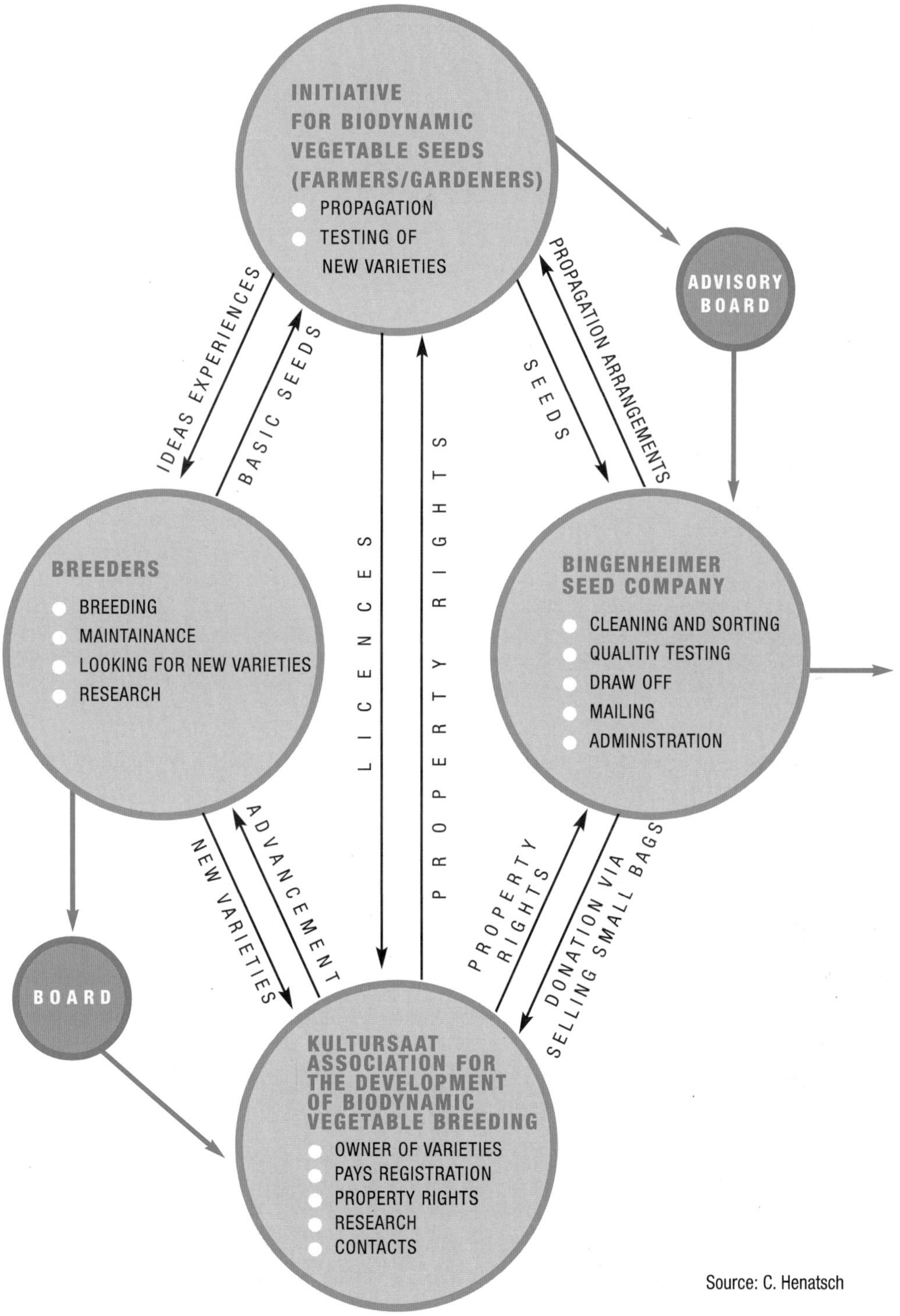

INITIATIVE FOR BIODYNAMIC VEGETABLE SEEDS (FARMERS/GARDENERS)
- PROPAGATION
- TESTING OF NEW VARIETIES

ADVISORY BOARD

IDEAS EXPERIENCES

BASIC SEEDS

PROPAGATION ARRANGEMENTS

SEEDS

BREEDERS
- BREEDING
- MAINTAINANCE
- LOOKING FOR NEW VARIETIES
- RESEARCH

LICENCES

PROPERTY RIGHTS

BINGENHEIMER SEED COMPANY
- CLEANING AND SORTING
- QUALITIY TESTING
- DRAW OFF
- MAILING
- ADMINISTRATION

BOARD

ADVANCEMENT

NEW VARIETIES

PROPERTY RIGHTS

DONATION VIA SELLING SMALL BAGS

KULTURSAAT ASSOCIATION FOR THE DEVELOPMENT OF BIODYNAMIC VEGETABLE BREEDING
- OWNER OF VARIETIES
- PAYS REGISTRATION
- PROPERTY RIGHTS
- RESEARCH
- CONTACTS

Source: C. Henatsch

No. 4

BIODIVERSITY AND THE ECOSYSTEM APPROACH IN AGRICULTURE, FORESTRY AND FISHERIES

FAO INTER-DEPARTMENTAL WORKING GROUP ON BIOLOGICAL DIVERSITY FOR FOOD AND AGRICULTURE

RESPONSIBLE
TECHNICAL DIVISION

Forest Resources Division
Forest Conservation, Research and Education Service

Douglas Williamson

EFFECTIVENESS

OF BIODIVERSITY
CONSERVATION

5

EFFECTIVENESS
OF BIODIVERSITY
CONSERVATION

AUTHORS

Martin Jenkins
UNEP-WCMC Cambridge, UK
martin.jenkins@unep-wcmc.org

Douglas Williamson
FAO Conservation, Research and
Education Service
douglas.williamson@fao.org

CONTENTS

INTRODUCTION

There is an important difference between the way people relate to agriculture and the way they relate to the conservation of the natural world [hereafter referred to simply as 'conservation']. The need for food is an empirical fact, so nobody questions the need to produce it. But there is no general consensus about the natural world and attitudes towards it are by no means universally benign, as the following quotations illustrate:

Not everybody cares about the fate of wild animals or the state of the natural environment. I met a lady who said she wouldn't worry if all the wild animals in the world disappeared overnight. She was a city person, she said.
Melvin Bolton [1997]

From a past replete with legends of man-bears, Montgomery brings us to the grim present: the shuddering horror of bears with paws cut off for soup – one at a time to keep the rest fresh. The animals' vocal cords are then cut so that as they walk on the bloodied stumps their cries don't disturb the tourists.
Adrian Barnett [2002]

The ways in which humans as a species relate to the natural world lie along an axis ranging from unthinking cruelty and destructiveness to complacent indifference to informed dependence to committed engagement, and the way people are distributed along this axis obviously has profound implications for its conservation. In particular, it means that the level of official financial support for conservation is much lower and less secure than funding for agriculture, including the conservation of agricultural biodiversity.

Attitudes to conservation by those not involved in it are an external consideration, which often takes the form of a constraint. This paper mainly concerns conceptual and practical issues within the domain of conservation which influence its effectiveness. It is based on a recent analysis of these issues by Jenkins (2002) .

Being "effective" is taken to mean "producing a desired effect". In the context of biodiversity conservation it is often difficult to reach a consensus either on what desired effects or ends should be or how they should be produced. Jenkins [op. cit] provides some insight into the difficulty of reaching consensus with the following argument:

There may be broad agreement that the ultimate goal of conservation is to maintain as much as possible of the world's existing biodiversity

BUT
- There is no consensus on what attempting to achieve this goal actually means.

BECAUSE
- Conservation always involves costs at least in the short-term and at least for some people.
- There are limited resources to offset these costs.
- Agreement on using scarce resouces is difficult.

BECAUSE
- No universal value system exists which could lead to complete agreement.
- Even if there is agreement in principle, scientific uncertainty makes it difficult to reach a consensus on what should be done in practice.

Another factor which probably contributes to the difficulty of reaching

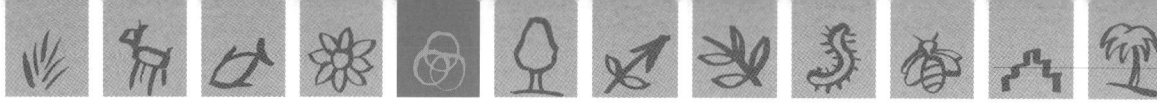

consensus is the sheer complexity and immensity of the natural system we are trying to understand and manage. It is inevitable that most of us only see and work with a tiny part of it, and that the overall enterprise of conservation is therefore highly fragmented.

A convenient way of exploring the issue of consensus about conservation is to address three fundamental questions:

> **Why** should we conserve biodiversity?
> **What**, exactly, should we be conserving?
> **How** should we conserve biodiversity?

Why should we conserve biodiversity?

This question has been at the root of conflicts over environmental and conservation issues since the earliest stirrings of the modern conservation movement. In the United States of the late nineteenth and early twentieth century, for example, differences in values were personified by John Muir, leader of the movement to preserve wilderness and elected as first president of the Sierra Club in May 1892, and Gifford Pinchot, described as America's first professional forester. Muir's relationship with the natural world was essentially a spiritual one. He claimed that the "demands and the discontents of modern American civilization... were so great and the rewards were so fraudulent that wilderness preserves were a spiritual and psychological necessity" (Turner, 1997, page 312). Pinchot on the other hand was a utilitarian. In his view: "The object of our forest policy is not to preserve forests because they are beautiful or wild or the habitat of wild animals; it is to ensure a steady supply of timber for human prosperity. Every other consideration is secondary. ...no

lands will be permanent reserves which can serve the people better in any other way." (Turner, 1997, page 323)

In the arguments that raged during the first decade of the twentieth century over plans to build a dam to provide water for San Francisco that would flood the Hetch Hetchy valley in California's Yosemite National Park, the two men were predictably on opposing sides, Muir vehemently against, Pinchot resolutely for.

During the latter part of the twentieth century environmental economists attempted to capture the spectrum of values attached to biodiversity with categories such as: direct use, indirect use, option, existence and bequest. In these terms, the most marked difference is between those who allocate a high existence (or existence and bequest) value to the components of biological diversity in principle and those who do not. Preservationists are those who believe that biological diversity has intrinsic value and should be conserved for its own sake to the maximum extent possible, regardless of whether any given component can be shown to produce tangible economic benefits. They are effectively giving priority to existence and bequest values. Utilitarians, like Gifford Pinchot, attach low existence value to the individual components of biodiversity and hold that it is only justifiable to expend serious effort in maintaining those that can be shown to produce tangible benefits for humans, or conversely, that conservation actions are only justified if they do not entail any appreciable costs.

Despite the efforts of environmental economists to articulate the different values that may motivate conservation, both academic opionion and current events suggest that the conflict of values over why biodiversity should or should not be conserved has not abated.

In terms of academic opinion, Norton (2000) expresses the view that:

"Recent international discussions of biodiversity policy have established two points:

- there is growing international commitment to sustain and protect biodiversity; and
- there is little agreement regarding why this should be done. Thus, while a significant international consensus regarding policy has apparently emerged, this consensus is not grounded in a consensually accepted value theory to explain why biodiversity protection, however strongly supported, should be a top priority of environmental policy."

In terms of current events, there is the ongoing argument in the United States over the Arctic National Wildlife Refuge, environmentalists arguing that the refuge should remain inviolate, other constituencies arguing that the national interest requires that it should be opened to drilling for oil prospecting and development. The current controversy is strongly reminiscent of the controversy over the Hetch Hetchey valley almost a century ago.

Is there a way of transcending or at least ameliorating the apparently intractable difficulty of reconciling opposing positions on divisive environmental issues? Two ideas come to mind.

Firstly, it is obvious that in the modern world with its growing population, rising material aspirations and increasing appetite for energy and other critical resources, almost any substantial tract of land may attract the interest of a diversity of constituencies, which could include environmentalists, developers, international donors, national government departments and officials, civil society, local residents and communities, and so on. It is surely necessary for all these constituencies to recognize the legitimate interests of other stakeholders and to make what concessions they can without compromising their own needs, rights and principles.

Secondly and more tentatively, there seems to be some scope in such situations of conflict for deploying the expertise in decision analysis that has been developed over more than half a century, mainly in relation to business management but also in relation to thinking about values (e.g. Keeney 1994). Finding new ways of understanding and articulating old problems and exploring alternative solutions in ways that are rigorous and systematic but also imaginative, could reveal ways of reconciling seemingly conflicting interests.

What, exactly, should we be conserving?

The Convention on Biological Diversity (CBD) has been ratified by more than 180 states and can thus be taken to represent, at least nominally, the views of the vast majority of the world's governments. The CBD, along with others (e.g. Mangel *et al.*, 1996), in effect advocates conservation of genes, species and ecosytems, but it is rarely, if ever, a practical possibility to operate at all three levels, so decisions still have to be made about which aspects of biological diversity deserve priority. There is little consensus on this, and the absence of authoritative and usable guidance on what exactly we should be conserving has led to a diversity of approaches to the question of how biodiversity can best be conserved (e.g. IUCN, 1994, Noss, 1996 at page 574, Janzen, 1998, Angemeier 2000, Parks Canada, 2000, Myers *et al.*, 2000).

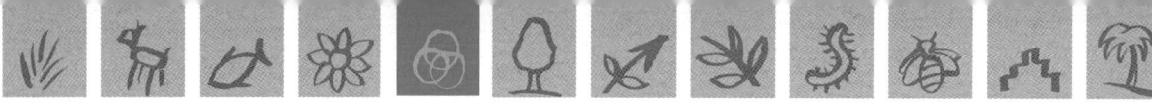

How should we conserve biodiversity?

If one examines existing practice in terms of the genes, species and ecosystems approach it becomes evident that the influence of the CBD is as yet rather limited.

Conservation of genetic resources, in the sense of intra-specific genetic diversity, is seldom a feasible activity in field-based work, but it is a central concern in captive breeding projects and is also a consideration in reintroduction projects.

Since it is clearly impossible to conserve all species, decisions have to be made about what species to prioritize. There are two different approaches: one that concentrates on identifying individual species of importance, and one that identifies important areas where it is hoped that actions will benefit a significant number of species.

SPECIES-BASED CONSERVATION

There is no global consensus as to what constitutes an important species, but species may be singled out for conservation action if they fall into one or more of the following categories:

- Threatened species
- Ecologically important species
- Species useful to humans
- Species with non-use value.

Threatened species

Threatened species are those that are believed to be in danger of extinction. Threatened species listing systems, such as the US Endangered Species Act and the IUCN system, give clear guidance as to which species are believed in most urgent need of conservation action. But the guidance is incomplete because only a small proportion of the world's species has yet been assessed in terms of extinction risk. Information is most complete for birds and mammals and is very incomplete in most invertebrate and plant groups.

It is widely agreed that it is unrealistic at present to expect concerted conservation efforts to be undertaken for each of these threatened species individually, but there is not much agreement on which threatened species are most deserving of attention. For example, it is often assumed that the most threatened species are those that should be accorded highest priority and should therefore be the principal focus of action. But it can also be argued that some of these species are lost causes and that resources are better spent elsewhere.

Ecologically important species

It can be inferred from basic ecological understanding that keystone species which play a crucial role in ecosystems should be considered to be a high priority for conservation, and the thinking of ecologists is being vindicated by the results of so-called 'small world' analyses.

"The true keysones in an ecological community are the most highly connected species, the hubs of the network. The keystones are the ecological control centres, so to speak, and clearly the most important targets for preservation." (Buchanan, 2002, page 154). But there are no rules for determining which are likely to be the keystone species, so identifying them can be difficult and demanding.

Species useful to humans

These include wild species that are harvested for food, medicines, clothing, building materials or other purposes, wild relatives of domesticated species or wild species (chiefly bacteria) with biochemical attributes that can potentially be harnessed industrially. In addition, some species are subject to non-consumptive use that can be expressed in economic terms. These are chiefly species that play an important role in tourism.

Species with non-use values

A number of species have an importance that cannot easily be quantified in economic terms. That is, a significant number of stakeholders ascribe a non-trivial existence or bequest value to them. Globally, the most important of these are the so-called "charismatic" species, particularly the charismatic megafauna, including large carnivores and birds of prey, cetaceans, sea-turtles, elephants, rhinoceroses and the great apes, but also some groups of smaller species such as other primates, parrots, large butterflies, and even some plants. Species may also be important for religious, spiritual or scientific reasons.

And of course, a species may be important for more than one reason. For example, the Malagasy Indri (*Indri indri)* is regarded as a sacred species by local people in much of its range, is decidedly charismatic, with great popular appeal, and is of considerable scientific interest to primatologists (Harcourt, 1990).

With so many reasons for attaching importance to species, it is not surprising that there is little agreement on which species merit special conservation action.

Local priorities may be different from national ones, which may in turn be different from global ones. For example, a species that is considered of high priority by some biologists because it is both taxonomically distinct and threatened, may be of little interest or concern to local people within its range. Conversely, a population of a widespread, non-threatened species regarded by conservation biologists as of low priority internationally may be of considerable local importance.

AREA-BASED CONSERVATION

Area-based approaches are widely advocated for planning in species' conservation. They are based on the observation that some parts of the world have far more species than others. Areas with large numbers of species, especially endemics, are often referred to as "hotspots" (Myers *et al.,* 2000). It is argued that by concentrating conservation efforts in these areas, a disproportionate impact can be had on the maintenance of global biological diversity. This approach can theoretically be applied at any geographical scale. It is widely accepted that such area-based approaches are the only realistic hope of maintaining a significant proportion of the world's biological diversity, but there are both practical and theoretical difficulties in identifying the most important areas.

On problem is that information on global species' distribution is very incomplete and heavily biased towards large, conspicuous forms, so identification of important areas has to be made on the basis of partial knowledge and is usually

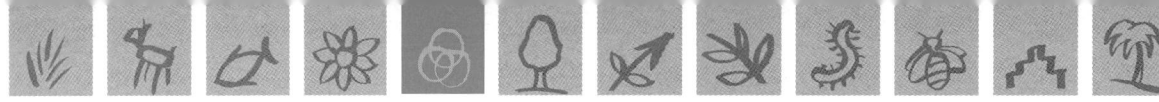

based on an assumption that areas important for well-known species are also important for others, that is that measures of diversity in different groups of organisms are highly correlated, but this may not necessarily be the case.

A further difficulty is that species' diversity may be important either for its richness or for its endemicity. There is notway of judging the relative importance of an area with high species' richness but low endemicity against an area with lower species' richness but higher endemicity.

Much also depends on the scale at which any assessment is made: a square metre of European chalk grassland will contain many more plant species than a square metre of tropical moist forest, while for an area of one square kilometre the reverse will be the case.

Concentration of conservation efforts on global hotspots of species' richness and endemism, assuming that these can be reliably identified, has a number of implications. Most importantly, it implicitly ignores the large part of the world that is not within a hotspot, and the high proportion of species not present in such areas. It therefore embodies what is ultimately a narrow conception of global conservation priorities.

The ecoregion approach avoids the problem associated with hotspot methods, of recognizing only a limited set of areas as of conservation importance. It combines analysis of biogeography, based on the distributions of species and species' groups, particularly narrow endemics, with an assessment of the dominant natural ecosystem or ecosystems in a particular area to divide the world, or part of the world, into a series of ecoregions. The identified conservation goal is then to maintain representative samples of natural areas in each of the identified ecoregions, or in those identified as of high priority because of their uniqueness or the urgency or scale of the threats they face.

But the ecoregion approach is based on the assumption that the world can usefully be divided into discrete, identifiable regions of this kind, and that these regions are suitable units for conservation planning. Both these assumptions are to some degree problematic. In the first instance, natural habitats and ecosystems seem generally to form part of a highly variable continuum, unless they are separated by very definite physical barriers, rather than form discrete entities. Dividing this continuum into units is to some extent therefore an artificial exercise. In addition, biogeographic patterns in different groups of organisms are not necessarily highly correlated with each other. Thus, a biogeographic analysis using vascular plant families will yield a different set of patterns from one using vertebrate families. Similarly, ecoregional classifications based on terrestrial ecosystems and biota have limited relevance, at least in continental regions, to freshwater systems where biogeography is very largely determined by drainage patterns.

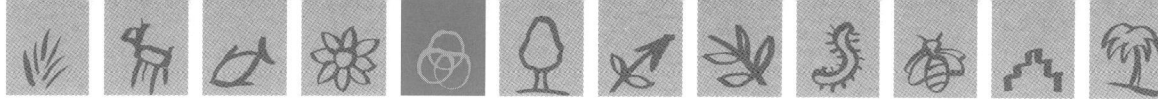

ECOSYSTEM APPROACHES

The Parties to the Convention on Biological Diversity have decided that the ecosystem approach should be the primary focus for actions undertaken to meet the objectives of the convention. They have subsequently devoted some effort to deciding what this actually means in practice. There appear to be two separate, though linked, concepts involved here. The first is the maintenance of particular ecosystems of importance. Implicit in this is the assumption that such ecosystems can be considered to be spatially distinct entities, so that this involves conservation of particular more or less well-defined areas. The second is the maintenance of ecosystem processes.

The CBD itself recognizes the following ecosystems and habitats as of importance (Annex I of the Convention). Those containing high diversity, large numbers of endemic or threatened species, or wilderness; required by migratory species; of social, economic, cultural or scientific importance; or, which are representative, unique or associated with key evolutionary or other biological processes. It further defines an ecosystem as "a dynamic complex of plant, animal and micro-organism communities and their non-living environment interacting as a functional unit".

Several of the categories of importance that are defined implicitly or explicitly with reference to species are effectively reformulations of the area-based approaches to species' conservation outlined above.

An alternative approach to conservation emphasizes natural processes rather than the particular entities (populations of various species) that mediate these processes. At the most fundamental level, these processes are energy fixation (almost entirely through photosynthesis), the cycling of that energy and of a range of organic and inorganic chemicals. They also include regulation of climate and water movement on land, soil formation and retention and, in the sea, formation of reefs and other zoogenic structures. These processes take place across all spatial and temporal scales.

Maintenance of these processes is seen as important for three main reasons. The first is to allow ecosystems to continue providing goods and services to humans; the second is to maintain or restore naturalness; the third is as a means of allowing populations of species to be maintained, or rather to maintain themselves. This last case can be seen as essentially a methodology for organism or species-based conservation, discussed above.

Major constraints on the effectiveness of ongoing efforts to conserve existing biological diversity

Given the uncertainties that arise from the lack of consensus about optimal approaches to conservation, it is hardly surprising that manifold difficulties are experienced in attempting to carry out effective conservation in practice. There are many reasons for this, ranging from the global to the local, and from the general to the highly specific. These are illustrated in Table 1.

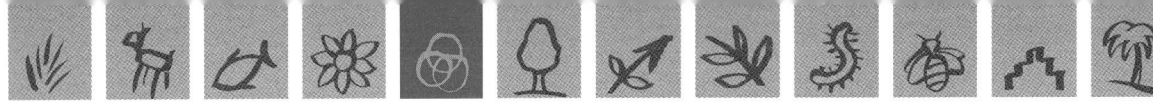

CONTEXT	CONSTRAINING MECHANISM
Global	Pressure from population growth and increased resource consumption, production of waste, breaking down of biogeographical barriers
Global	Lack of political will to prioritize or mainstream biodiversity conservation and to invest in its implementation
National	Lack of enabling policy, legal and institutional environment
Practical implementation	Failure to articulate clear goals, and failure to identify specific problems that need solving in order for goals to be met
Practical implementation	Inherent difficulty of changing behaviour that is having a negative impact on biodiversity
Practical implementation	Difficulty of maintaining conservation values while promoting development
Practical implementation	Simplistic assumptions about effect of poverty alleviation on attitudes to biodiversity conservation
Practical implementation	Simplistic assumptions about nature and functioning of communities
Multiple-use protected areas	Need to reconcile conflicting needs of users while also meeting conservation goals
Technical problems	Scientific uncertainty or inadequate capacity creates uncertainty about outcomes of actions
Alien species	Lack of capacity to control, selection of inappropriate control methods
Strict protected areas	Conflicting management goals, disagreement over permissible level of intervention
Highly threatened species	Uncertainty and conflict about optimal approach to conservation
Complexity of problems	Scarcity of skills in leadership, conflict management and applying a multi-disciplinary approach
Funding regime, project cyles	Conservation has no end-point, funding and project cycles are short-term – difficult to reconcile these time scales

Increasing the effectiveness of conservation

Because of the complexity and unpredictability of the world, there is no single, definitive approach to effective conservation.

Those implementing conservation activities have to deal with the complexities of the physical environment, of ecosystems, of populations of non-human species and of human society, and with the interactions between all these levels. Any given conservation scenario will almost certainly present a unique combination of these. Thus, even if it is possible to obtain a persuasive account of why certain approaches have succeeded or failed, it may be difficult to work out which of these may be widely applicable and which may be products of a particular and unique set of circumstances.

But of course as experience accumulates generalities do emerge and it would be extremely wasteful to ignore all previous activities. Success lies in learning from other experiences without being prescriptive and avoiding a "one size fits all" mentality.

A general issue that requires more attention than it has yet received is the opportunity that exists to conserve biodiversity in production landscapes of all kinds. Given the inexorable expansion and intensification of human activities, this is a crucial issue.

Those who fund and manage conservation can contribute to improved practice on the ground by working to create an supportive environment for conservation. Those who implement conservation on the ground are best placed to improve its practice.

CREATING A SUPPORTIVE ENVIRONMENT

Engaging with the biodiversity-related conventions

The processes of the biodiversity-related conventions involve the vast majority of the world's governments and have generated wide-ranging programmes of work that directly or indirectly involve conservation. Constructive engagement with these processes on the part of non-governmental and government organizations that are concerned with conservation is a way of working to make national and international government and policy arenas more amenable to and supportive of conservation efforts.

Achieving internal coherence

Organizations involved in conservation should articulate strategic goals and visions so that those working for them, particularly those on the ground, know what they are supposed to be aiming at. This would provide a vital wider context for local efforts, as well as help bring a sense of cohesion to what are often a highly dispersed group of people. It involves devoting considerable energy and resources to communications, which should be two-way, so that the practical experience of those working locally informs higher-level policy and advocacy work and vice-versa.

Moving away from the centrality of projects

It has become clear that if conservation is to be successful it has to be a sustained and continuing process, like providing health care, for example. This means modifying the time-scale over which interventions take place, accepting the possibility of long-term support, for example through trust funds and other means, and eschewing expectations of rapid results, both in terms of changes in human behaviour and in impacts on biodiversity.

It does not mean that good money should be thrown after bad, that once a commitment has been made, investment should continue regardless of outcomes. There should always be a preparedness to withdraw from a particular area or activity as a last resort. Indeed, any programme working in an area as complex as conservation should expect a certain percentage of failures. But this percentage can be kept to a minimum if great care is taken in deciding what to invest in and where to invest.

No less important than a decision to invest is a decision, after careful appraisal, not to invest in a particular area or activity - that is, a willingness to walk away if there is little realistic hope of any significant success.

A further requirement is that any long-term commitment should involve not just financial investment, but also the establishment of mechanisms to ensure that the activities undertaken are subject to continuing review and, if necessary, modification.

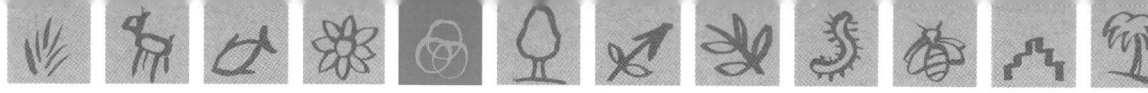

External communications

Obviously there are strong incentives for organizations and institutions involved in conservation to emphasize their successes in their external communications, but this can be counterproductive because as much can be learnt from where things go wrong as from successes. A more open and honest approach to the sharing of information and experience would enhance lesson-learning and help build capacity across the entire conservation community. It is, of course, unrealistic to expect conservation organizations and institutions to shift completely from what is still an essentially competitive approach to a wholly cooperative one. Such a shift would in any event almost certainly not be wholly beneficial to the practice of conservation, as it would be unlikely to encourage innovation. However, what is definitely needed is a shift in balance towards a more cooperative approach.

IMPROVING THE PRACTICE OF CONSERVATION ON THE GROUND

Clarify objectives

Just as any organization involved in conservation activities should articulate its visions and long-term goals, so field activities should have a clear objective or set of objectives that fit in with those visions and goals. These should be specified as precisely as possible.

A set of activities may, of course, have multiple conservation objectives. In this case the objectives should be set out and prioritized - it is very possible, or even likely, that the objectives may conflict with each other (e.g. restoration of natural ecosystem processes may be inimical to some threatened species' populations).

Interventions in any given area may equally have development objectives as well as conservation objectives. If this is the case it is vital that the relationship between the two sets of objectives is clarified and made explicit at the outset. Determining which takes precedence may have a profound impact on the nature of the activities undertaken.

Set targets

Once the conservation objectives of a set of interventions have been established, explicit targets for each of them should be set. These targets should relate to outcomes rather than activities. For example, the target should be that the population of a given threatened species has reached a certain threshold, or has increased at five percent per year for at least five years, not that hunting has been successfully controlled.

Targets should be realistic. They do not necessarily have to be quantitative - in many circumstances, particularly in tropical forest ecosystems where accurate monitoring of most components of biodiversity is extremely difficult, it may be counterproductive to set precise quantitative targets. Here, targets based on relative changes in abundance, say, or even qualitative measures based on expert assessment may be as valuable. It is equally important in these cases, however, that the targets, and the criteria used to judge whether they have been met, are made explicit.

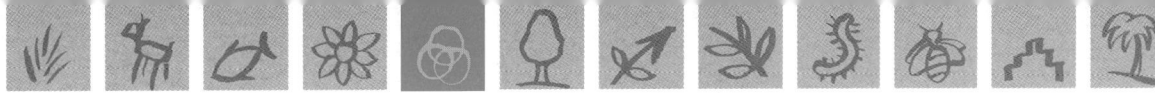

Identify impediments

Having determined objectives and set targets, the next step is to identify impediments to reaching those targets. These can be loosely divided into two categories: actual or potential threats to the components of biodiversity in question, and constraints on action.

Identifying actual or potential threats to particular components of biodiversity may be extremely complex. General categories of threat may be rather obvious, but determining the relative importance of different threats is often much more difficult. It is crucial that the assumptions on which threat assessment is based are carefully questioned. Without this a great deal of time, money and energy can be spent trying to address factors that are of little or even no immediate importance (e.g. brown tree snakes, cahows, logging in tropical forests).

There will always be limits to the kinds of actions that can be undertaken locally to mitigate threats. These limits may be imposed by the interests of different stakeholders, by technical constraints, limitations in available resources, or where the origin of the threat lies outside the geographical sphere of influence of those engaged in direct management. A clear, realistic understanding of these from an early stage in planning may prevent much wasted effort later on.

Develop an active management plan aimed at reaching desired targets

A plan of some kind is a fundamental requirement of virtually any conservation intervention. The plan itself needs to be unduly complex – indeed it should be as elaborate as

necessary and no more. At minimum it should set out what is intended to be done, how (and by whom) this is to be achieved and how the achievement is to be assessed.

Adaptive management

The original proponents of adaptive management (e.g. Holling, 1978) intended it to be management by active experimentation. This involves construction of a model of the system under consideration. Different management scenarios are applied to the system to generate a series of possible outcomes. The most favourable of these are then tested in real-life situations. Monitoring before and during the application of a particular management regime allows its impacts on the system to be assessed. A choice can then be made between different regimes. Implicit in this technique is the comparison between different approaches, with one (usually non- or minimal-intervention or business-as-usual) regarded as control. This entails operating different management regimes either in parallel, that is in two areas at the same time, or in series - that is one after the other in the same area.

For whatever reason, active adaptive management is evidently still an uncommon approach in management of biodiversity, even in North America, where the concept originated and where, at least potentially, the resources and legislative and institutional infrastructures are available to implement it.

A much more widely used approach is what may be termed passive adaptive management or "adaptable management" (Alexander, 2000). In terms of this, what is considered to be the best available management option is adopted. Its impact on the identified targets is subject to

continuous monitoring which is fed back so that modifications to the management regime can be made as appropriate. Implicit in this is a setting of thresholds in the status or trends in those aspects of biodiversity identified as priorities under the objectives. Reaching or crossing these (negative) thresholds will normally trigger modifications or at least detailed examination of the management regime.

Stakeholder involvement

Rarely will conservation entail a single individual acting entirely in isolation. In most cases more than one person, and often a large number of different interest groups, will be involved.

The perceptions of the different groups, the relationships between them and their respective responsibilities must be clearly understood if any other than the simplest of plans is to be successful. Clearly the better people understand what is expected of them, the greater the likelihood of success. As a general rule, this implies that the more participatory the approach, the better.

Thus, whenever management programmes are being planned, priority should be given at the start of the process to explicitly identifying different stakeholder groups, what their stake or interest in the process is, and how best they might be represented. Many different methods for ensuring participation have been developed. Most involve facilitated workshops or other kinds of meetings. Whatever system is adopted, it is important that at each step, clear individual responsibilities are allocated for carrying the process forward.

Establish monitoring protcols and regimes

Just as effective conservation can ultimately only be defined in terms of impacts on the actual components of biodiversity on which interventions are focused, so the effectiveness of management can only be assessed by monitoring those components of biodiversity. Establishment of protocols and regimes for this should be an integral part of all conservation interventions. Because of the complexity of most biodiversity issues, innovative approaches may have to be developed. These may include involvement of local people in surveys and monitoring (so-called "citizen science"), training of parataxonomists and use a range of remote-sensing techniques.

While the importance of monitoring cannot be over-emphasized, it is equally important that the protocols established should not be unrealistically complicated or expensive to implement. Long-term continuity is the most important aspect of any monitoring regime, and this should be the prime consideration in its initial design.

Reporting and documentation

For most people involved in conservation management, reporting is one of the most unrewarding and onerous tasks. Unfortunately it is also one of the most important tasks, particularly in complex situations, which most conservation activities are. The major reason for keeping records should be to improve conservation management in the area concerned. Conservation objectives are not, or should not be, time-bound, so that the need for management of some form should outlive the involvement of any given individuals. However, much successful conservation action

depends on the expertise of particular individuals. Ways need to be found of transmitting this expertise to succeeding generations of managers. Oral transmission and learning-by-doing are very often the most successful of these, but there are almost certain to be gaps and discontinuities at times when these forms of transmission are broken. Permanent records – written words, photographs, video or tape-recording – can play a vital role in filling this gap. This form of documentation does not necessarily have to be highly formalised.

Documentation in conservation management activities is also important in situations where different interest groups are involved. Sooner or later disputes are almost certain to arise over why a given set of actions has been carried out. The better the decision-making process has been documented, the more easily such disputes can be resolved. In this case, documentation should concentrate on what decisions have been made (i.e. actions to be undertaken and designated responsibilities) with brief justification. Concision and clarity are the two most important characteristics of this form of documentation.

As well as improving practice locally, reliable documentation serves many other purposes. In the first instance, it can be used to help disseminate lessons learned elsewhere, spreading knowledge and helping to overcome the barriers to learning from experience. Related to this, it should help in constituency-building and wider advocacy. Narratives of successful conservation action are some of the most powerful communication tools available (although the temptation, as noted elsewhere in this report, may be to present unduly positive versions of what has happened; this is likely to be counterproductive in the long-term).

CONCLUDING THOUGHTS

- Always consider the larger context.
- Be clear in your objectives.
- Look for charismatic leaders.
- Build alliances – be prepared for creative partnerships.
- Be patient – look to the long-term.
- Be realistic about costs and benefits and long-term sustainability of activities.
- Be prepared to walk away.
- Be as adaptable as possible, but not at the expense of your original objectives.
- Increase the effectiveness of conservation by improving the quality of thought.
- Thinking costs nothing.

No. 5

REFERENCES

Alexander, M., 2000. *The Guide to Management Planning in Nature Reserves and Protected Areas*. Conservation Management System Partnership. http://www.cmsp.co.uk/cmsmain/main_frame.htm

Angermeier, P.M., 2000. *The Natural Imperative for Biological Conservation*. J Conservation Biology 14(2), pp. 373 – 381.

Barnett, A., 2002. *Review of Search for the Golden Moon Bear: and adventure in Southeast Asia* – Sy Montgomery, Simon&Schuster. New Scientist 21 September 2002 page 53.

Bolton, Melvin ed., 1997. *Conservation and the Use of Wildlife Resourcs*. London: Chapman & Hall, (at page xiii).

Buchanan, M., 2002. *Small world: uncovering nature's hidden networks*. London: Weidenfeld & Nicolson.

Harcourt, C.S. 1990. *Lemurs of Madagascar and the Comoros – the IUCN Red Data Book*. IUCN, Gland, Switzerland and Cambridge, U.K.

Holling, C.S. ed., 1978. *Adaptive environmental assessment and management*. New York John Wiley and Sons, Inc.

IUCN, 1994. *Guidelines for Protected Area Management Categories*. IUCN: Gland, Switzerland.

Janzen, D., 1998. Gardenification of Wildland Nature and the Human Footprint. Science Vol. 279, pp. 1312 – 1313.

Jenkins, Martin, 2002. *Effective Conservation*. Working Paper WL/01, Forest Conservation, Research and Education Service, Forest Resources Division, Foresty Department, FAO, Rome.

Keeney, R.L., 1994. *Value-focused Thinking – A Path to Creative Decisionmaking*. Harvard University Press: Cambridge, Massachusets.

Mangel, M. *et al.*, 1996. *Principles for the Conservation of Wild Living Resources*. Ecological Applications 6(2), pp. 338 – 362.

Myers, N., Mittermeier, R.A., Mittermeier, C.G., da Fonseca, G.A.B. and Kent, J., 2000. *Biodiversity hotspots for conservation priorities*. Nature 403: 853-858.

Norton, B. G., 2000. *Biodiversity and environmental values:* in search of a *universal ethic*. Biodiversity and Conservation 9: 1029 – 1044.

Noss, R.F., 1999. *Conservation of Biodiversity at the landscape scale*. In Biodiversity in managed landscapes: Theory and practice. eds. R.C. Szaro and D.W. Johnson, pp. 574 – 589. Oxford University Press: New York.

Parks Canada, 2000. *Report of the Panel on the Ecological Integrity of Canada's National Parks*. Available at: http://parkscanada.pch.gc.ca/EI-IE/index_e.htm

Turner, F., 1997. *John Muir: From Scotland to the Sierra*. Edinburgh: (Canongate, First published in the USA in 1985 by Viking Penguin Inc.)

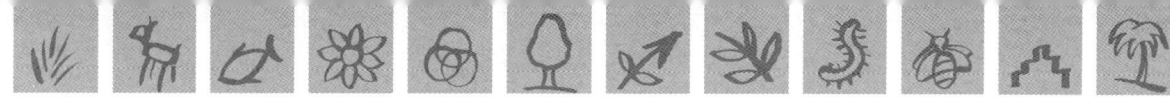

RESPONSIBLE
TECHNICAL DIVISION

Forest Resources Division
Forest Resources Development Service

Pierre Sigaud

CONSERVATION
AND USE
OF MAHOGANY IN FOREST
ECOSYSTEMS IN MEXICO

6

CONSERVATION AND USE
OF MAHOGANY IN FOREST
ECOSYSTEMS IN MEXICO

AUTHORS

Fernando Patiño Valera

Instituto Nacional de
Investigaciones Forestales y
Agrìcolas y Pecuarias

Roberto Centeno Erguera

Juana Marín Chávez

INIFAP, CIR Sureste, Mexico
cezohe@cirse.inifap.conacyt.mx

CONTENTS

INTRODUCTION

Mahogany (*Swietenia macrophylla* King) is one of the best-known and more frequently used tree species of forest stands in Latin America. Mahogany and other *Meliaceae* of *Cedrela genus* (*Cedrela odorata L.* and *Cedrela fissilis* Vell.) are both the backbone of the forestry industry in Latin America and the main sources of income for a large number of people in rural communities.

Due to its biological and commercial characteristics, mahogany has a large potential to become the basis for a sustainable use and management system of the tropical forest, applied in the framework of appropriate silvicultural practices. Nevertheless, former and current traditional use systems tend, in most of the cases, to harvest the best trees, regardless of forest regeneration and future growth.

In the last decades, the quality of *Swietenia* genetic characteristics has been severely affected and the number of individuals per hectare has decreased, due, among other reasons, to the selective exploitation of the best trees (dysgenic selection); deforestation processes and finally, the interruption of natural evolution, due to stand fragmentation. Selective use of the best mahogany trees, not only directly affects the resilience of the species in tropical ecosystems, it also presents a risk of genetic diversity erosion in natural stands, and a hazard for its natural regeneration, due to the elimination of the best seed producers.

Mahogany natural forest forms stands with low density of commercially exploitable trees per hectare. Mahogany stands are dispersed in groups of a few individuals, and in some regions, estimations show a tree density average of 0.7-2 individuals per hectare (Snook, 1993; Patiño, 1997). The low density of individuals directly affects the natural regeneration of the species, because seed production is not abundant, and the possibility of establishing seedlings, which facilitating the future presence of trees at different growth stages, is less likely to happen. During the rainy season abundant germination of small saplings may be observed close to mature trees, but once the dry season starts, a high percentage of them dies.

Mahogany grows in tropical ecosystems where a large number of tree species grow per hectare and species-level biodiversity is abundant. This high level of biodiversity is a huge challenge for the use and management of tree species with timber potential, mainly because these species grow in the same space and period of time as other species bearing different characteristics, such as wood hardness, colour and also heterogeneous growth rates. This circumstance is coupled by the fact that some species grow few individuals per hectare; besides, such species tend to be those with the highest commercial value, thus making their use and commercialization difficult. While this paper deals with some aspects of mahogany species in the context of its natural distribution range in the neotropical areas, it mainly tackles conditions and characteristics found in Mexico.

ECOLOGICAL ZONES AND MAHOGANY DISTRIBUTION

Ecological zones

Mahogany is mainly found in the tropical rain forest and tropical moist deciduous forest, in Central America, Mexico and South America (FAO, 2002). Rain forest is found in the coastal plains of the Gulf of Mexico region; in the Sierra Madre range in Chiapas, Mexico; and in the Caribbean coast along the Pacific coast in Central America. Tropical moist deciduous forest stretches along the lower region of the central mountain ranges of Central America, located towards the Pacific Ocean; the plains and hills of the Yucatan peninsula and the Gulf of Mexico. The tropical ecological zones in this region cover 134 million hectares (FAO, 2002).

These ecological zones are characterized by an average annual temperature ranging between 20 and 26º C, while rainfall in the tropical rain forest varies between 1 500 and 3 000 mm, and reaches up to 4 000 mm in certain areas, the Tropical moist deciduous forest receives between 1 000 and 1 500 mm average annual rainfall. The dry season in these areas lasts up to three months because they are not affected by the tropical depressions occurring between August and November, which produce rainfall during that season (FAO, 2002).

The forest in these areas is high and dense. The crown cover in the tropical rain forest reaches between 30 to 40 m heights, with emerging trees growing over 50 m, while trees in the Tropical moist deciduous forest reach a 30 m height. The lower storey of the crown cover is dense, with trees reaching between 5 and 25 m, strata in the under storey present a large variety of palms, ferns, climbers and grasses. Both ecological zones share many tree species and are characterized by a complex and varied flora, with approximately 5 000 vascular plant species and more than 60 different tree species per hectare.

The most common tree species associated to Mahogany in a natural distribution range are *Dialium guianense, Pimenta dioica, Brosimum alicastrum Ampelocera hottlei, Pseudolmedia cf spurea, Cordia alliodora, C. dodecandra, C. bicolor, Calophyllum brasiliense, Castilla elastica, Dendropanax arboreus, Tabebuia spp, Manilkara spp, Terminalia spp, Ochroma* spp, among others (FAO, 2002).

Natural distribution of mahogany

Swietenia macrophylla covers a wide natural distribution range stretching from 23º N latitude down to 18º S latitude, and covers a vast territory from the southern part of Tamaulipas, Mexico, along the Atlantic coast across Belice, Guatemala, Nicaragua, Costa Rica and Panama, and further down towards the north-eastern part of South America, along the border areas of the Amazon region down to Bolivia and the southern part of the Amazon in Brazil (Patiño, 1997; Bauer and Francis, 1998).

Mahogany is a native species in the following countries: Mexico, Belice, Guatemala, El Salvador, Honduras, Nicaragua, Costa Rica, Panama, Venezuela, Colombia, Ecuador, Peru, Bolivia and Brazil (Patiño, 1997; Bauer and Francis, 1998) and it is an introduced species in commercial plantations established in many regions in the world.

In Mexico, mahogany concentrates in the Mayan forest (Yucatan peninsula, Chiapas). In these areas, mahogany is relatively abundant; and presents a normal distribution of diametric classes, thus favouring the exploitation of tropical forest to concentrate in this species.

In Mexico, mahogany stands grow in different soils at intermediate depth, and stretch along the Gulf of Mexico, from the southern part of Tamaulipas to the northern part of Puebla and Veracruz, and to the Yucatan peninsula (Pennington and Sarukhan, 1968; Patiño, 1997). In Yucatan state, mahogany grows in the eastern and southern regions (Pennington and Sarukhán, 1968) and in stands with few individuals in the north-western part and in the Petenes area (Patiño, 1997). In Campeche and Quintana Roo mahogany is found in the central and southern areas, where it becomes the most frequently exploited species in the higher or middle subdeciduous stands of *Manilkara zapota* and in the north-eastern part of the petenes area, where it forms a continued vegetation joining with the northern stands of Yucatan (Patiño, 1997).

State of mahogany natural populations

Swietenia macrophylla trees are used along most of its natural distribution range, a trend confirmed by several authors (Rodan *et al.* 1992; Foster 1990; Lamb 1966; Smith 1965; Foster 1990; Navarro 1996; Patiño, 1997; Linares, 1996; Argüelles, 1999; Navarro, 1999). Detailed forest inventories that include information on this species are few and limited in scope and area, and vary in quality. Most of these inventories have been planned for forest use and concentrate only on high value species and trees with commercial dimensions, while they ignore trees of smaller dimensions that could favour

the natural regeneration of the species.

A recent assessment of mahogany stocks in Mesoamerica (Southern Mexico and Central America) (Calvo, 2000) estimates the natural distribution area of the species at 41 million hectares, while estimations show that the region (excluding Panama, due to the low density of trees, and Costa Rica and El Salvador, because commercial stands do not exist) boasts 14 million hectares of broadleaved forest with mahogany trees. Out of these, approximately 1.6 million hectares are located in protected areas, thus implying that there still are about 12.5 million hectares of forests with mahogany trees of some commercial potential. Nevertheless, not all the areas are suitable for forest management on have partially been exploited. According to the national assessments, the commercial volume of mahogany is not more than 5 percent of the total volume of the Mayan forest.

The historical trend of timber use of *Swietenia macrophylla* stands in the Yucatan peninsula in Mexico, Belice and Guatemala is characterized by the extraction of the best trees and a scarce application of silvicultural practices aimed at favouring natural regeneration of the species.

Mahogany use in the Mesoamerican region after the Maya era started in the 17th century. It initially involved populations located close to watercourses allowing the transportation of logs. It increasingly expanded to the interior, as new ways of access were opened for log extraction by means of animal traction, and later, railroad and trucks (Snook, 1993; Navarro, 1996; Argüelles, 1999; Navarro,1999).

FREQUENCY AND ABUNDANCE OF SPECIES IN TROPICAL REGION STANDS

Tree species

Tropical forests are reported to harbour the largest genetic and biological diversity of all terrestrial populations. The genetic quality of those resources is deteriorating across the world and this trend has grown at an accelerated rate in some regions. Populations and ecosystems are vastly disturbed, while scarce knowledge exists about the organization, dynamics, taxonomy, use and interaction among the components of tree populations within the tropical ecosystems.

Certain tree species growing in the forest stands occur at a very low density of individuals per hectare. On the contrary, other species, very few in general, grow in large numbers of individuals per hectare in those same stands. This is the result of the biodiversity existing in such ecosystems, where more than 60 tree species are often found per hectare.

According to Kageyama, Namkoong and Roberts (1991) a forest inventory carried out in the Atlantic forest of Brazil, showed that 30 percent of the tree species reported had only one tree or less per hectare. Besides, the authors report that three of the most abundant tree species reached 30 percent of all the trees per sampled hectare.

A forest inventory carried out in a 100 ha area of tropical forest (with a sampling intensity of 10 percent, where all trees with 10 cm diametre at DBH and above were measured) in Escárcega, Campeche State, Mexico, reported the existence of 94 tree species. Seven of these species had only one tree on the whole sampled area, while 15 species showed the largest number of individuals, with an average between 4.2 and 2 trees within the tropical ecosystem communities (Patiño, 1997).

The species with very low presence per hectare are designated as rare species according to Janzen (1970). The rarity of the species may be considered an event in the evolution of the tropics, where the interaction between plants and animals created a plant defence system against other organisms, being rarity, one of the strategies to protect the plants against animals and micro-organisms. Many trees considered as species of high commercial value, such as mahogany and cedar, both of which belong to the *Meliaceae* family, may be classified within the group of rare species.

Mahogany trees

Forest inventories carried out in the southern communities (Ejidos) of Quintana Roo state in Mexico indicated that mahogany trees per hectare were distributed as shown in the table below (Patiño, 1995). It must be stressed that trees with smaller diametre, or as large as 15 cm diametre, were more numerous, thus it was possible to incorporate more trees to the diametre class of commercial-sized trees and therefore ensure the permanence of the species without decreasing timber availability.

Forest inventories carried out in the central and southern regions of Campeche state in Mexico reported similar numbers, in terms of frequency and abundance of tree species, to those shown by populations in Quintana Roo and generally matched with the findings reported by Patiño, 1987 and Patiño, 1997, at an average of 0.7 trees of commercial dimensions per hectare. Also, according to FAO's (2001) quoting Snook

Location	DIAMETER BELOW OR EQUAL TO 15 CM.			COMMERCIAL SIZED DIAMETER ABOVE 50 CM.		
	Number of trees per ha	Basal Area m²/ha	Clean stem Vol. m³/ha	Number of trees per ha	Basal area m²/ha	Clean stem vol. m³/ha
Caobas	3.63	0.45	2.74	0.53	0.23	1.24
Plan de la Noria	5.78	0.33	2.34	0.08	0.03	0.16
Divorciados	11.11	1.13	7.95	1.07	0.38	2.28
Manuel Ávila Camacho	4.91	0.67	3.68	0.66	0.27	1.41
Petcacab	6.68	0.84	5.36	0.92	0.36	2.00
Nohbec	6.06	1.00	6.63	1.14	0.58	3.63
Tres Garantías	4.68	0.56	3.12	0.51	0.25	1.20
Botes	8.29	0.78	4.72	0.70	0.24	1.21

Mahogany (*Swietenia macrophylla* King) trees growing in permanent forest areas located in southern Quintana Roo, 1990

(1993, 1996) and Patiño (1997) mahogany grows in low densities in forest stands, at an average of one to two trees of commercial dimensions per hectare, mixed with 60 different tree species with lower or no commercial value.

Genetic diversity and conservation of resources

Genetic diversity in the tropical forests, where mahogany grows, is rapidly decreasing due, among other reasons, to deforestation processes and natural populations' fragmentation. The first phenomenon reduces the population size and natural communities, sometimes eliminating them completely. In the case of forest communities fragmentation, it makes the gene exchange difficult and may isolate continuous populations of a given species until its genetic diversity is lost, as a result of endogamy and genetic erosion. These phenomena highlight the enormous risks that tropical forest resources face, especially some species of commercial value such as mahogany, therefore justifying the urgent need to better understand genetic diversity at its different levels and use such knowledge

in the management, improvement and conservation practices of those important genetic forest resources.

Although *Swietenia macrophylla* is the most important tropical broadleaved species in neotropical regions, little attention has been given to the knowledge related with the genetic variation existing among the populations growing within their natural range of distribution. Newton *et al.* (1993b) carried out an exhaustive revision of the aspects concerning genetic variation, its capture and use for genetic improvement, and conservation of genetic resources, and found that few studies have been made on these issues. Probably one of the reasons of scarce genetic experimentation on mahogany is due to the difficulty of establishing plantations of this species, given its vulnerability to attacks and damages provoked by *Hypsipyla grandella* Zeller, a terminal shot borer, which is the main pest affecting the development of *Meliaceae*.

In Puerto Rico detailed studies have been made on the growth characteristics of *Swietenia* species, namely, *S. macrophylla*, *S. mahogani* and *S. humilis* and some natural and induced hybrids of these species (Weaver, 1987; Weaver and Bauer, 1986;

Newton *et al.*, 1993b), nevertheless, little information describing their different provenance has been published. Geary *et al.* (1973) report preliminary results of provenance tests of *Swietenia* species at the age of 4.4 years. In wet areas *S. macrophylla* was superior in height growth and survival rate, but in dry areas the performance of tree species was very similar (Geary *et al.*, 1973).

In Mexico Patiño (1997) reports provenance and progenies tests of *S. macrophylla* established in 1988; in this trial he reported an inheritance coefficient for growth in height, for progenies of provenances respectively from Escárcega at $h^2 = 0.038$, Cayal at $h^2 = 0.265$ and Zoh-laguna at $h^2 = 0.164$.

Navarro (1999) indicates that both progenies from Costa Rica and those from Central America and Mexico show high added genetic variation levels in terms of yield in height and diametre. As regards resistance to *Hypsipyla* no variation was found due to the presence or absence of that borer in any of the three tests.

It is important to notice that several tests have been carried out in several Latin American countries, among them those jointly carried out by the *Centro Agronómico Tropical de Investigación y Enseñanza* (CATIE) with the support of research and academic institutions from Central America and Mexico, such as *Instituto Nacional de Investigaciones Forestales, Agrícolas y Pecuarias* (INIFAP) on the behaviour and origin of mahogany progenies established in Costa Rica (Navarro, 1999).

Currently, several projects are being carried out by INIFAP in the Yucatan peninsula, in Mexico. The projects are aimed at obtaining information on genetic diversity of populations and at establishing provenance and progeny tests, as well as *ex situ* germplasm conservation banks for progenies and provenance of mahogany and *Cedrela odorata* of native populations from Mexico.

Population variability

Large morphological variability (leaves, fruits, wood properties) of *S. macrophylla* and other *Meliaceae* have been pointed out by several authors, suggesting that the species shows a wide genetic variability along its geographic range of distribution and also among many of the single populations, due to natural obstacles or fragmentation of populations caused by anthropogenic or natural causes.

This assumption suggests that each genetic population represents the subdivision of species in adapted biotypes that would correspond to different habitats. Trends seem to point out that tree morphology and resistance to *Hypsipyla grandella* borer are inherited characters.

This assumption underlines the importance of genetic conservation of populations and individuals with those characteristics, that may be lost if the current trends of fragmentation, population decrease and exploitation of the best phenotypes continue. Selective exploitation acts as a source of dysgenic selection, because the best individuals, in terms of growth and morphology, are being used in forestry activities. This practice leads to populations with decreasing genetic quality with poor phenotype crossings that produce this type of genetic erosion.

The best knowledge of genetic diversity in populations of *Swietenia* is key to define strategies of genetic improvement, management and genetic resources' conservation. Besides, the research of genetic resistance to *Hypsipyla* also requires better knowledge of populations in all the distribution range of species of these genera.

Initiatives are underway in almost all the countries in the region, yet, they only cover one part of the diversity contained in the distribution area. Thus, there is need to promote the development of cooperative studies between the different institutions in Mexico, Central America and the Caribbean, that allow to extend the scope of the research, and also promote the establishment of trials with similar characteristics that allow to analyze the variability in most of the distribution range of the species.

Genetic resources conservation

Genetic loss within and between populations, due to the exploitation of the species and the fragmentation and reduction of its populations, is a critical factor to value the conservation state of the populations. Genetic loss is a critical factor to assess the conservation state of *Meliaceae* populations in their natural distribution range in neotropical regions, including Mexico.

One of the main concerns arising in the forestry scientific community aims at preventing that certain species, such as *Swietenia macrophylla,* be affected by the genetic erosion suffered by other species like *S. mahogani.* Due to a series of causes, *S. mahogani* lost the best individuals in the past and now their descendants are bifurcated, deformed, with many branches and are very different from the original populations.

In spite of *in situ* and *ex situ Swietenia macrophylla* conservation projects undergoing in the Central American region and Mexico, there is no doubt that more efforts are needed to better understand what the distribution and dimension of the biological diversity are, in order to correctly plan, manage, use and encourage mahogany conservation.

It is important to gather information, especially on themes like: geographic distribution and size, damage and hazard level, number of individuals per hectare, dispersion and growth behaviour, phenology, reproductive biology, interaction with pollinators, seed characteristics, seeds and sapling damage, natural regeneration.

All these factors together, will allow to adjust the objectives and work goals, as well as to define the location and size of protection areas, and facilitate other necessary tasks for genetic improvement, seed collection, and species' propagation for plantation and orientation of activities towards the sustainable management of the species.

Management and use of mahogany populations

The use of mahogany within its natural distribution range started many centuries ago, for instance, the Mayans in Central America built large canoes to carry out trading activities in the region (Hammond, 1982). In more recent times, during the Spanish colonization, tropical forests were used to extract timber of different species, among them *Haematoxylon campechanum* and mahogany *Swietenia macrophylla,* and export it to Europe for cabinet-making and furniture, mahogany in the region was generally used for ship construction and repair, and for domestic use.

In Mexico, mahogany timber extraction began at large scale in the 20th century in many areas, mainly in the southern region of the country, including Oaxaca, Chiapas and the Yucatan peninsula. During this period, large transnational companies were established and were granted concessions by the government to exploit natural resources. The use of mahogany timber intensified during the Second World War,

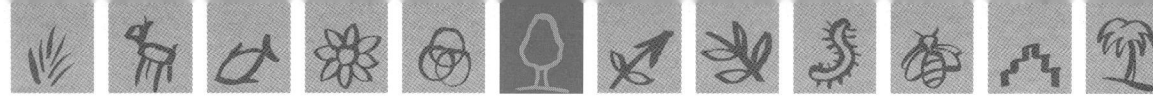

when the companies expanded extensive exploitation. These activities continued throughout the 70s.

Silvicultural aspects

Species of the *Meliaceae* family differ in terms of silvicultural characteristics. *Cedrela odorata* is considered a fast growing species, belonging to the more advanced stages of succession and a shade tolerant species.

One of the main problems arising from the management of mahogany populations is the scarce natural regeneration it shows in natural communities where the species grows. Besides the problem related to the decreasing number of mature individuals that could contribute to seed dispersion, an additional difficulty is due to the scarce quality of the remaining trees after the timber has been extracted, as well as the lack of information on the processes that regulate the natural regeneration.

Rodríguez, Chavelas and García (1994) carried out a study to assess the capacity for *S. macrophylla* seed dispersion and reported their findings regarding the successive establishment of saplings. Seed dispersion occurred from February to April, and a total of 6 861 seeds were found, that is to say 84 percent of the potential expected production of a tree. The maximum dispersion distance was 60 m from the mother tree. In August, when the first germination and survival appraisal was made, the authors found a total of 1 608 saplings. The conclusion was that 20 percent of the dispersed seeds were able to germinate and establish in the next two months.

Gullison and Hardner (1993) point out that selective use of tropical forests is damaging to the remaining trees in stands with large numbers of tree species with commercial value. The authors of this paper assessed the damage provoked by selective exploitation of *S. macrophylla,* a species with very low density, located in the Chimanes forest in Bolivia. Damages occurred mainly along the main roads and access tracks for extraction. Minor damage reached 4.4 percent of the total area of extraction assessed.

Nevertheless, the main obstacle to increase use practices through selective harvesting methods in Latin America the lack of international markets for many of the less known species, although in some cases local or national markets exist and could absorb the timber produced.

García, Negreros and Rodríguez (1993) carried out a study in subdeciduous forest in Quintana Roo, Mexico, in order to observe the effects of partial removal of the first storey on natural regeneration of mahogany (*Swietenia macrophylla* King). Five plots of 0.5 ha were established, where different intensity clearings were carried out, taking the original basal area of the population as a basis and removing respectively 0,8,28,45 and 55 percent of it. The work was assessed four years after the clearings were carried out. The authors observed marked differences in the average number of young mahogany individuals per hectare, varying from 500 in the reference sample (0 percent) to 2 100 in the 45 percent intensity clearing, obtaining 700 individuals in the 8 percent intensity clearing, and 1 400 in the 28 percent clearing and 900 in the 55 percent one.

The authors point out the outstanding effect on regenerating saplings in relation to the different clearing intensities. Although significant statistical differences were observed when comparing the 45 percent intensity clearing to the 0 percent clearing

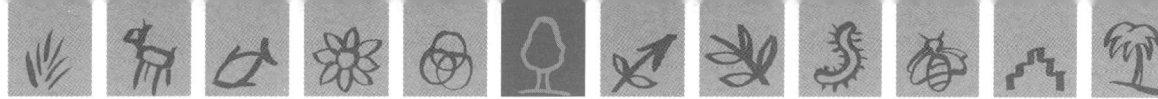

of reference, in absolute numbers, the former showed up to 4.2 times more individuals than the 0 clearing reference sample, while it was three times more numerous than the 8 percent clearing, 1.5 times more numerous than the 28 percent clearing and 2.3 times more than the 55 percent one. These figures suggest that a direct relation exists between the opening of the first *S. macrophylla* storey and the quantity of regenerating saplings produced in experimental plot conditions. Nevertheless, in the 55 percent clearing treatment, the number of regenerating saplings decreased, because the species requires shade during the first growth stages.

In terms of maximum height, observations show that the largest average tree dimensions were obtained in the demonstration plot where 45 percent of the existing basal area had been removed. These trees were 2.1 times higher than the reference sample, 2.4 times higher than the sample at 8 percent clearing, 1.9 times more than the sample at 28 percent clearing and 1.4 times more than at 55 percent clearing.

Negreros and Mize (1993) presented the following results: when the trees reached three years of age, their aim was to measure the effect of partial clearings (creation of multiple clearings) in the natural regeneration with special attention to tree species of commercial value, such as mahogany *(S. macrophylla)*. After three years, the population obtained by regeneration was similar in density and composition to the original population before the clearings were made. Regeneration of commercial tree species (with or without tolerance to shade), non-commercial species and those which are not tree species, were compared with the residual basal area and the percentage of basal area removed. The frequency of commercial, shade tolerant species were not affected by residual basal area or by the percentage of basal area removed.

Sustainable management perspectives

One of the main challenges that forest management programmes face in tropical regions consists in maintaining both timber productivity under long-term management, and conservation of biodiversity in the populations subject to management.

Today, an international trend associates management of forest resources with the sustainable development of communities, while conserving those communities and the genetic diversity within and between plant species' populations.

Sustainable management may be interpreted in different ways, and it may be considered simply as the management capable of maintaining timber productivity of the species in the management cycle. In a wider interpretation, sustainable management may also be considered as the management that preserves productivity, the forest structure, diversity and the basic ecological process of populations, communities and the ecosystem.

In many tropical regions a wide range of forest management methods may be identified, including those that carry out selective use of the best individuals of some species and those that aim at using the resources in the framework of sustainable management and encourage its natural regeneration.

One of the few examples of sustainable forest management is rubber tree *(Hevea brasiliensis)* latex extraction in the Brazilian Amazon region, by rubber collectors, who do not damage the trees and conserve the genetic resources of the species under management, while maintaining the diversity and structure of the natural forest (Kageyama, 1991). Another similar example would be latex extraction of "chicle" gum obtained from several species of Manilkara

spp. in the neotropical regions, where trees are maintained while carrying out a non-wood forest use, as well as timber use of other species.

Mahogany forests located in south-eastern Mexico are one example of sustainable forest management carried out in the Ejidos of Quintana Roo state, where the use of forests in the past was carried out through concessions that excluded the owners of the resources, a phenomenon, that according to Arguelles (1999), produced a social claim arising from the ejidos and communities to directly manage the forest themselves, as well as the commercialization of its products.

The result was a change in government policy, that granted licenses directly to forest owners, while a project known as the Forest Pilot Project was set up at later stage. These measures filled a gap in forest management left by the forestry concessions in southern Quintana Roo. The greatest success of this project is that 45 ejidos established 500 thousand hectares for permanent forest use, thus stopping uncontrolled clearings that had prevailed during decades.

The success of the Forest Pilot Plan encouraged researchers to replicate it in other regions such as Marqués de Comillas in Chiapas and the Calakmul region in Campeche; Nevertheless, these programmes did not have the expected success, because they were implemented under the influence of political considerations and not according to a productive organization of the users, in order to achieve a rural economic setting based on good management of forest resources.

The importance of a socially oriented initiative in terms of conservation efforts and development of forest populations, stems from the forest lands registration system in Mexico, where 80 percent of the forest land is managed by the communities, 15 percent by private enterprises and only five percent is still under the control of the Federal and State Governments. In terms of participation in productive activities, it is interesting to notice that today 40 percent of timber production is provided by communal forest enterprises. In 1975, only 2-3 percent of timber production was directly managed by the communities. In 1985, this figure had already grown to 17 percent.

CULTIVATION PROBLEMS

The most important risk factor in the establishment of mahogany plantations in neotropical regions is the attack of the terminal shoot borer (*Hypsipyla grandella* Zeller) (Arreola and Patiño, 1987; Newton *et al.*, 1993). This is an endemic problem in the tropics, especially when these species are planted in monoculture.

Two species of *Hypsipyla* exist on the American continent (*H. grandella* and *H. ferrealis*); the most damaging species for plantations is *Hypsipyla grandella*. This species is found across the tropical regions of Mexico, Central America and South America, excluding Chile. The species also occurs in the Caribbean islands and in the southern part of Florida, in the United States (Entwistle, 1967). The species' distribution is closely related to the distribution of *Meliaceae*, especially *Swietenia* and *Cedrela* genera. All the species of these genera are subject to the attack of *Hypsipyla*. This is an endemic problem in the tropics, especially when the species are planted in homogeneous populations.

The high value of mahogany, its current demand and the rapid decrease of its

natural populations require urgent measures to effectively control this pest. Several attempts have been made to control *Hypsipyla* through biological, chemical and silvicultural methods (Grijpma, 1973a and b, 1974; Whitmore, 1976a and 1976b) but these have not been effective in reducing the damage produced by the borer to acceptable levels of economic loss.

The borer usually attacks the trees during its larvae stadium, when it is easily able to penetrate the soft tissue of the shoot, and sometimes provokes its death. As a result of the death of the terminal shoot, the general growth of the plant is reduced and several new shoots may appear, producing bifurcated and deformed trees. (Rodríguez, 1981; Arreola y Patiño, 1987).

Borer attacks on young trees may start before they leave the nursery to plantation locality. Nevertheless, the most damaging attacks occur during the first four years of the plantation establishment (Grijpma, 1974; Arreola and Patiño, 1987). Young trees are more affected by the shoot borer, because they depend more on the apical meristem growth.

Currently, mahogany plantations in Mexico are being subject to intensive application of solutions containing some of the fungi belonging to *Beauveria* and *Metharricium* genera that affect the larvae of *Hypsipyla grandella* when these encounter the spores, which germinate and penetrate in the larvae's body reducing its activity and provoking its death. This control method alternated with the application of systemic insecticides during the higher rate of incidence of the pest. The insecticide must be sprayed every month or every 45 days in order to accurately protect the plantation.

Role of forest plantations in conservation

The natural forests considered available for wood production, including mahogany, are experiencing heavy pressures for harvesting, which contribute to continuing deforestation and which spill over to "presently unavailable" forests from the production point of view. The expected harvest from the natural forests is expected to decline.

On the other hand, global consumption of timber is projected to increase. Total roundwood consumption was projected as increasing at an annual rate of 1.12 percent between 1994-2010, from 3.21 billion m^3 to 3.84 billion m^3 (FAO, 1997a). Industrial roundwood was projected to increase 1.20 percent annually, from 1.47 billion m^3 to over 1.78 billion m^3. This increase of over 300 million m^3 by 2010 far exceeds the estimated net current growth from industrial plantations of 84 million m^3 (FAO, 2000).

An optimistic estimate of the permanent loss of existing natural forest production through further deforestation and degradation, based on recent deforestation rates, suggests that the annual decline in production from disturbed forests could be 6.2-7.0 million m^3 ha^{-1} yr^{-1}. Removing additional areas of natural forests through the establishment of new protected areas or expanded logging bans would further reduce harvests. Much of this protected area would likely be taken from available undisturbed forests that have higher current growth and productivity (approximately double) than the disturbed forests (FAO, 2000).

Without significant increases in both the available area and productivity of industrial plantations of commercial species the present net growth of 84 million m^3 annually would be overtaken by reductions due to deforestation alone within 12-14 years, a

shorter period than the growing cycle of most industrial plantations (FAO, 2000).

Forest plantations grown to supply raw material for industry and for other uses, such as fuelwood, also provide additional non-wood forest products and benefits from the trees planted or from other elements of the ecosystem that they help create. They contribute environmental, social, cultural and economic benefits. (Carle *et al.*, 2002).

The potential for forest plantations to partially meet the demand for wood and fiber for industrial uses is increasing. In several countries, a significant portion of the wood supply for industrial uses comes from plantations, rather than natural forest resources.

A large majority of timber production is harvested from currently available natural forest. Globally, an estimated 3 354 million m^3 was removed from the world's forests of which 56 percent is fuelwood. Fuelwood is most significant in Asia and Africa while industrial roundwood production is heavily concentrated in North America, Asia and Europe (FAO, 2000).

The global forest plantation estate has increased from 17.8 million hectares in 1980 and 43.6 million hectares in 1990 to 187 million hectares in 2000. Although in 2000, 26 percent of plantations continued to be for unspecified purposes, there was a significant increase in plantations for industrial purposes in the past decade: from 39 percent in 1980 and 36 percent in 1990 to 48 percent in 2000. There has been a corresponding decrease in forest plantations for non-industrial purposes (Carle *et al.*, 2002).

FAO (2001) reports that 326 007 hectares of mahogany plantations have been established in 18 countries in the world. The main plantation area occurs in Indonesia (187 500 ha), Fiji (42 000) Philippines (34 000 ha) and Mexico (21 400 ha), with 87.4 percent of the total. Sri Lanka, Bangladesh, Solomon Islands, Guadeloupe and Samoa have already been planted with 33 300 hectares (10.21 percent of total) and another nine countries have been planted with 7 807 hectares (2.4 percent).

Forest plantations make up only 3.5 percent of the global forest area and tropical plantations, including those of *Meliaceae* species, make up 45 percent of the estimated 187 million hectares of plantations.

The role of planted forests in increasing forest areas, including trees on farms and trees outside the forests, is very important in order to meet the rising demand for wood and non-wood products and ecological services, including carbon sequestration. In countries with low forest cover the planted forests are an option for rehabilitating degraded areas and where possible, the base for re-establishing natural forests.

The planted forests will be established taking in account all the practicable steps to avoid replacing natural communities and ecosystems of high ecological, economical, social and cultural value.

In the case of mahogany, the density of trees in nature is an average of one tree per hectare (Patiño, 1987; Snook, 1993; Snook, 1996; Patiño, 1997) and the final density of mahogany trees in Mexico in a 25 years rotation length, is on average 250 trees per hectare.

In that case, one hectare of plantation corresponds to 250 hectares of natural forests. This shows us the importance of promoting forest plantations in previously deforested lands to produce wood to meet rising demand and help the protection of natural forests.

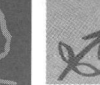

Initiatives to regulate the use and conservation of mahogany

For several decades FAO has made increasing efforts to link national, regional and international institutions in order to facilitate and promote better knowledge, management and use of mahogany genetic resources. Several years ago, the FAO Panel of Experts on Forest Genetic Resources included mahogany *(Swietenia macrophylla King)* improvement and genetic conservation as one of its priorities. FAO (1994, 1997, 2001) considers that *in situ* conservation is a high priority and it has suggested strengthening activities to carry out studies on genetic diversity and promote improvement and *ex situ* conservation.

In neotropical regions, where mahogany naturally occurs, national and regional institutions, as well as non-governmental organizations work to improve the knowledge, enhance management, use and conservation of mahogany genetic resources. These efforts are carried out in collaboration with international organizations and governmental institutions of all the countries.

Central American countries are working jointly to deal on common issues to strengthen regional integration in environmental matters, with the aim of promoting a regional approach oriented towards economic, social and ecological sustainability. One of these efforts is the *Comisión Centroamericana de Ambiente y Desarrollo* (CCAD) (Central American Commission for Environment and Development). The organizations acts on a regional basis and aims at harmonizing policies and management systems and promotes like-minded positions in all extra-regional and world fora.

One of the most important projects developed under the umbrella of CCAD are the studies carried out on the biological corridor in Mesoamerica, as well as issues related to species that have a conservation interest, such as mahogany. During its 33rd meeting held in July 2002, the Central American Commission on Environment and Development agreed, at the suggestion of Nicaragua, to request the Convention on International Trade in Endangered Species of Wild Fauna and Flora, (CITES) to include mahogany *(Swietenia macrophylla* King) in Appendix II of the Convention, during its next Conference of the Parties.

Several non-governmental organizations, both national and international are also present in the region. These organizations have, as part of their most important mandate, to obtain information on the status, management, use and commercialization of wildlife species, and to give more emphasis to those related to the CITES Convention. These organizations are the World Wide Fund for Nature (WWF) and the World Conservation Union (IUCN) which jointly manage the programme TRAFFIC, established to regulate the trade of plants and animals and thus monitor wildlife trade in the world. TRAFFIC works in cooperation with the CITES Secretariat.

Since the 1990s, attempts have been made to include mahogany in the Appendix lists of CITES. These attempts were based on studies carried out by several authors, who have analyzed the deforestation rates within the geographical distribution area in the neotropical regions, as well as the commercialization and general state of the species. In 1994 a proposal was presented to include *Swietenia macrophylla* King in Annex II of CITES, but it was rejected by CITES parties.

In 1997, the proposal made by Bolivia and the United States to include this species in Appendix II of CITES was rejected by few votes. Nevertheless, the most notorious agreement was reached in 1997 by a

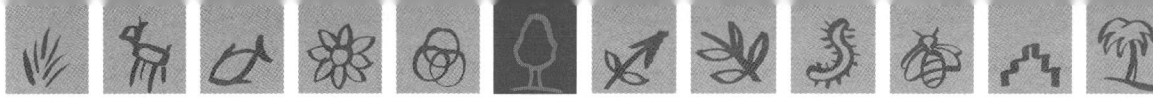

exporting countries like Brazil and Bolivia and the largest importer country, the United States, to create a working group in order to examine the state, management and trade of this species along all its geographic distribution areas.

Other countries belonging to the distribution area where mahogany grows, expressed their support for this initiative. Bolivia, Brazil and Mexico also engaged to include their mahogany logging practices in accordance with Appendix III of CITES.

Bolivia has included its mahogany population in Appendix III, as of 19 March 1998. Brazil as well, adopted this decision on 26 July 1998. Mexico included mahogany in Appendix III in November 1997. (Mahogany populations in Costa Rica are included in Appendix III since November 1995.)

The working group on mahogany - integrated by governmental organizations of countries belonging to the geographical area of mahogany distribution and importer countries, as well as non-governmental organizations working in the domain of conservation of natural resources - is currently working to develop text and content to present a renewed proposal to include mahogany in Appendix II of CITES, during its meeting in Chile in 2002.

CONCLUSIONS

In Mexico, mahogany (*Swietenia macrophylla* King) presents a natural distribution range that stretches from the Yucatan peninsula to the coastal plain in the southern Gulf, the warm regions in the states of Chiapas, Oaxaca and Puebla.

Apparently, topography is not a determining factor in the species' distribution, at least up to a limit of 800 m, below this altitude temperatures of more than +20º C, allow niche areas favourable to the species. Another key element is an annual rainfall intensity ranging from 1 000 to 3 500 mm.

In addition, soil does not appears to be a limiting factor for the development of this species, as mahogany grows in a large variety of soils, from clay superficial soils to alluvial deep soils. Nevertheless, it does not grow appropriately in terrains subject to floods, or soils with a high proportion of sand with high draining potential.

According to Arguelles (1999) forest systems that harbour mahogany populations, amount to an approximate area of 1 500 million hectares in Mexico. Out of these, 500 000 ha are located in Quintana Roo; 450 000 in Campeche; 300 000 in Chiapas; and 220 000 between Oaxaca and Veracruz and the rest of them stretch between Yucatan and Tabasco states.

From a historical point of view, mahogany has undergone many stages of depredation. This is due, on one side, to the advantages that the species has with regard to its wood, colour and homogeneous wood grain, and on the other, to its friendly processing properties that allow excellent finishings.

A large part of mahogany genetic resources, with few exceptions, is in danger. It is therefore urgent to make a joint effort to decrease deforestation as much as

possible, to include clear guidelines in forest management and use plans that allow concrete conservation of genetic resources.

The knowledge of genetic diversity and the factors linked to the reproduction of the species will allow to take some important measures to achieve its natural regeneration, among which: to conserve the number of individuals necessary to maintain appropriate levels of allogamy and crossing; to identify the most adequate distance between parent trees in order to effectively achieve those results; pollen and seed dispersion patterns and other aspects that allow to maintain genetic diversity and avoid endogamy in fragmented populations.

Both progenies from Costa Rica (Navarro, 1999) and progenies from Central American and Mexico show high levels of added genetic variation in terms of height and diametre. As regards resistance to *Hypsipyla* tests carried out so far do not show any relation to the presence and the number of attacks. The Mayans managed the forest by saving mahogany and other useful species for their community, while they would cut all the remaining trees to cultivate the land, this clearing allowed enough light to penetrate through the forest and space for the forest to successfully regenerate after the land was abandoned, or the people migrated to other locations.

There is no sustainable management of the species in harvesting areas in Mexico and Central America, the reproductive cycle is not taken into consideration and silvicultural treatments do not exist to help in promoting and encouraging regeneration in an efficient way.

The promotion and marketing of unknown species, which still do not have commercial value in the market, necessary. This is particularly important in mahogany forests where this is the only species with acceptable commercial value, thus leading to over-exploitation and a relative decrease in the number of *Swietenia macrophylla* trees, as compared with the other species growing in the forest.

Fostering the conservation of protected areas located within the distribution range of the species and increasing the knowledge of the genetic diversity they harbour is key. It is also important to study unprotected areas in order to determine if gene pools exist and how they should be maintained, in order to ensure the conservation of those sites.

The role of planted forests in increasing forest areas is very important in order to meet the rising demand for wood and non-wood products and ecological services. The planted forests constitute an excelent option for rehabilitating degraded areas and, where possible, to constitute the base for re-establishing natural forests.

The planted forests will be established taking in account all the practicable steps to avoid replacing natural communities and ecosystems of high ecological, economical, social and cultural value.

Finally, and following Kageyama *et al.*, (1992) it is important to point out that in order to achieve sustainable management of mahogany populations and other species of tropical forests, it is necessary to fill many of the gaps of scientific knowledge that still exist, such as following:

- Is sustainable management of timber producing species possible, especially if these are rare species growing in natural populations?
- May species with low density (rare), such as mahogany, be subject to regeneration control in order to replace harvested trees?
- Which of the two options has the highest

priority in sustainable management: biological and genetic diversity or the forest structure, or both? and

● How to achieve accurate population control of the species under management?

These are controversial questions that are the subject of intense debate and analysis in many countries, by different actors and institutions, while major research efforts are needed to clarify these aspects, especially in tropical ecosystems.

There is no doubt that the answer to these and other questions, that must be considered to achieve the sustainable management of tropical forest resources, lays within the domain of the scientific community, whose efforts focus on better knowing, managing, using and conserving these resources. The task is considerable and an answer may only be given, as soon as possible, by linking the efforts of countries, national and international institutions, as well as professionals who have a genuine interest in achieving the sustainable use of tropical forest resources.

FAO's mandate as the main United Nations agency working with forestry development, can play a catalytic role in joining efforts aimed at reaching the objectives pursued by all the actors: achieving a sustainable use of forest populations, communities and ecosystems, an effort that implies the use and conservation of genetic diversity of tropical forest resources.

REFERENCES

Arguelles, A,. 1999. Diagnóstico de la caoba en Mesoamérica: México. Centro Científico Tropical. PROARCA/CAPAS.

Arreola, V.M.C., Patiño, V.F., 1988. Influencia de factores climáticos en la incidencia de ataque de *Hypsipyla grandella* Zeller; Lep.: *Pyralidae* en caoba *Swietenia macrophylla* King y cedro Cedrela odorata L. INIFAP. Publicación Especial 59: IV Simposio Nacional sobre Parasitología Forestal.301-313.

Calvo, C.C.J., 2000. Diagnóstico de la caoba (*Swietenia macrophylla* King) en Mesoamérica, Visión General. Centro Científico Tropical. PROARCA/CAPAS.

Carle, J.; Vourinen, P.; Del Lungo, A., 2002, Status and trends in global forest plantations development. Forest Products Journal. 52(7), 2-13.

Bauer, G.P.; Francis, J.K., 1998. Swietenia macrophylla King, Honduras mahogany, caoba. SO-ITF-SM-81. New Orleans, LA, USA. Department of Agriculture, Forest Service, Southern Forest Experiment Station.

FAO, 1994. Informe: Cuadro de Expertos de la FAO en recursos genéticos forestales. Octava Reunión. FO:FGR/8/Rep. Roma, Italia, 28-30 de junio de 1993. 57p.

FAO, 1997. Informe: Cuadro de Expertos de la FAO en recursos genéticos forestales. Novena Reunión. FO:FGR/8/Rep. Roma, Italia, 28-30 de junio de 1996. 57p.

FAO, 1997a. FAO Yearbook: Forest Products 1995. Food and Agriculture Organization of the United Nations, Rome, 442 pp.

FAO, 2000. Global Forest Products Outlook Study: Thematic Study on Plantations by C. Brown. Working Paper No. GFPOS/WP/03. Food and Agriculture Organization of the United Nations, Rome, 129 pp.

FAO, 2001. Situación de los Bosques del Mundo 2001. Organización de las Naciones Unidas para la Agricultura y la Alimentación, Roma, Italia

FAO, 2002. Global Forest Resources Assessment 2000, Main Report, FAO Forestry Paper # 140, FAO of the United Nations, Rome, Italy

FOSTER, R.B., 1990. The floristic composition of the Río Manu floodplain forest. pp 99-111 in Gentry, A.H.and Guzmán Teare, M. eds. Four Neotropical Rainforests. Yale University Press, New Haven and London.

García-Cuevas, X., Negreros, C.P., Rodríguez-Santiago, B., 1993. Regeneración natural de caoba (*Swietenia macrophylla* King) bajo diferentes densidades de dosel. *Ciencia Forestal en México*, INIFAP, Coyoacán, 18 (74): 25 - 44.

Geary, T.F., Barnes, H., Barra-Coronada, R.Y., 1973. Seed source variation in Puerto Rico and Virgin Islands grown mahoganies. Res. Pap. ITF-17. Río Piedras, PR. US Department of Agriculture, Forest Service, Southern Forest Experiment Station.

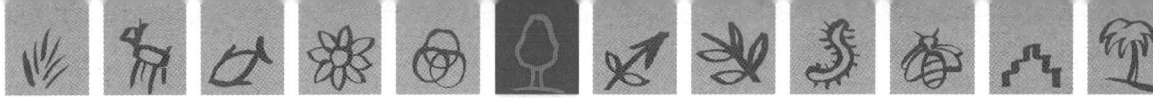

Grijpma, P., 1973a. Studies on the shootborer Hypsipyla grandella (Zell) (Lep. Pyralidae) XVIII. Records of two parasites new to Costa Rica. CATIE - IICA, Costa Rica. Turrialba 23 (2): 235 - 236.

Grijpma, P., (editor), 1973b. Proceedings of the first symposium on integrated control of Hypsipyla. Turrialba, Costa Rica, CATIE.

Gullison, R.E. y Hardner, J.J., 1993. The effects of road design and harvest intensity on forest damage caused by selective logging. Empirical results and a simulation model from The Bosque Chimanes, Bolivia. *Forest Ecology and Management* 59 (1-2): 1-14.

Janzen, D.H., 1970. Herbivores and the number of tree especies in tropical forests. *American Naturalist,* (104): 501-528.

Kageyama, P.Y., 1991. Extractive reserves in brasilian Amazonia and genetic resources conservation. Tenth World Forestry Congress. september.Paris

Kageyama, P.Y.; Namkoong, G. and Roberds, J., 1992. Genetic diversity in tropical forests in the state of *São* Paulo-Brazil.Piracicaba, ESALQ/USP (*Não* Publicado).

Lamb, F.B., 1966. Mahogany of tropical America: its ecology and management. Ann Arbor University of Michigan Press. 220 p.

Linares, B.C., 1996. Recursos genéticos de especies de la familia Meliaceae en los neotrópicos: prioridades para acción coordinada, Peru y norte de Sudamérica. Reporte No Publicado. FAO, Departamento de Montes.

Navarro, C., 1996. Recursos geneticos de la familia meliaceae en Centro America. prioridades para actividades coordinadas. Reporte No Publicado. FAO, Departamento de Montes.

Navarro C., Hernández M., 1998. Evaluación de la diversidad genética de especies tropicales de importancia económica y ecológica en Centro America y el Caribe, implicaciones para la conservación, la utilización sostenible y el manejo. *In Memoria I Congreso Latinoamericano de IUFRO. Valdivia, Chile. Noviembre 1998.*

Navarro, C., 1999 Diagnóstico de la caoba *(Swietenia macrophylla King)* en Mesoamérica: Silvicultura — Genética. Centro Científico Tropical. PROARCA/CAPAS.

Negreros Castillo, P., Mize, C., 1993. Effects of partial overstory removal on the natural regeneration of a tropical forest in Quintana Roo, México. Forest Ecology and Management. 58 (3-4): 259-272.

Newton, A.C., Baker, P., Ramnarine, S., Mesén F., y Leakey, R.R.B.,1993a. Mahogany shoot borer: prospects for control. Forest Ecology and Management

Newton, A.C., Leakey, R.R.B., y Mesén F., 1993b. Genetic variation in mahoganies: its importance, capture and utilization. Biodiversity and Conservation 2. 114-126.

Patiño,V.F., 1987. Los inventarios forestales y la conservación in- situ de los bosques tropicales. In: Lund,H.G. *et al.,* Evaluación de tierras y recursos para la planeación nacional en las zonas tropicales. Gen. Tech. Rep. WO-39. : 99-104, Washington, USDA, FS.

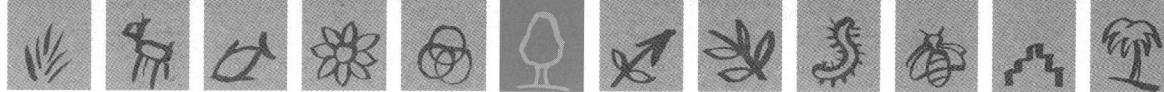

Patiño, V.F., 1995. Informe de la Región México. IX Reunión Cuadro de Expertos en Recursos Genéticos Forestales, FAO. Roma Italia. 9 p.

Patiño, F., 1997. Genetic resources of Swietenia and Cedrela in the neotropics: proposals for co-ordinated action. FAO of the United Nations, Rome, Italy.

Pennington, T.D., Sarukhan, J., 1968. Manual para la identificación de campo de los principales Arboles Tropicales de México. INIF - FAO.,413 p.

Rodan, B.D., Newton, A.C., Verissimo, A., 1992. Mahogany conservation: status and policy initiatives. Environmental Conservation. 19 (4): 331-338.

Rodríguez, S.B., Chavelas, P.J., García, C.X., 1994. Dispersión de semillas y establecimiento de caoba *(Swietenia macrophylla)* después de un tratamiento mecánico del sitio. In: Snook, K.L. y Barrera, A. (Editoras) Madera, Chicle, Caza y Milpa. Contribuciones al manejo integral de las selvas de Quintana Roo, México: 81 - 90.

Snook, L. K.,1993. Stand dynamics of mahogany (Swietenia macrophylla King) and associated species after fire and hurricane in the tropical forest of the Yucatan Peninsula, Mexico. Doctoral Dissertation. Yale School of Forestry and Environmental Studies. University Microfilms International # 9317535, Ann Arbor, MI, USA.

Snook, L.K., 1996. Catastrophic disturbance, logging and the ecology of mahogany (*Swietenia macrophylla* King): grounds for listing a major tropical timber species in CITES. Botanical J. Linnean Society 122:35-46.

Weaver, P.L. y Bauer, G.P., 1986. Growth, survival and shoot borer damage in mahogany plantings in the Luquillo Forest in Puerto Rico. Turrialba 36 (4): 509-522.

Weaver, P.L., 1987. Enrichment planting in tropical America. In: Figueroa, J.C.; Wadsworth, F.H. y Branham, S. editores Management of the forests of tropical America: Prospects and technologies pp 259-278. Puerto Rico, USDA - Forest Service, Institute of tropical forestrry.

Whitmore, J.L., 1976. Studies on the shootborer Hypsipyla grandella (Zeller) Lep. Pyralidae. CATIE Misc. Pub. 1 Turrialba, Costa Rica, Centro Agronómico Tropical de Investigación y Enseñanza.

BIODIVERSITY AND THE ECOSYSTEM APPROACH IN AGRICULTURE, FORESTRY AND FISHERIES   FAO INTER-DEPARTMENTAL WORKING GROUP ON BIOLOGICAL DIVERSITY FOR FOOD AND AGRICULTURE

RESPONSIBLE
TECHNICAL DIVISION

Forest Products Division
Wood and Non-Wood Products Utilization Branch

Sven Walter

IMPACT OF CULTIVATION AND GATHERING OF MEDICINAL PLANTS ON BIODIVERSITY:

GLOBAL TRENDS AND ISSUES

AUTHORS

Uwe Schippmann

Federal Agency for Nature Conservation
Bonn, Germany
uwe.schippmann@bfn.de

A. B. Cunningham

WWF/UNESCO/Kew People and
Plants Initiative
Fremantle, Australia
peopleplants@bigpond.com

Danna J. Leaman

IUCN Medicinal Plant Specialist
Group Ottawa, Canada
djl@green-world.org

CONTENTS

INTRODUCTION

Since time immemorial, people have gathered plant and animal resources for their needs. Examples include edible nuts, mushrooms, fruits, herbs, spices, gums, game, fodder, fibres used for construction of shelter and housing, clothing or utensils, and plant or animal products for medicinal, cosmetic or cultural uses. Even today, hundreds of millions of people, mostly in developing countries, derive a significant part of their subsistence needs and income from gathered plant and animal products (FAO, 1993; Walter, 2001). Gathering of high value products such as mushrooms (morels, matsutake, truffles) and medicinal plants (ginseng, black cohosh, goldenseal) also continues in developed countries for cultural and economic reasons (Jones *et al.*, 2002).

Among these uses, medicinal plants play a central role, not only as traditional medicines used in many cultures, but also as trade commodities which meet the demand of often distant markets. For the purpose of this paper the term "medicinal and aromatic plant" (MAP) is defined to cover the whole range of plants used not only medicinally *sensu strictu* but also in the neighbouring and often overlapping fields of condiments, food and cosmetics.

Demand for a wide variety of wild species is increasing with growth in human needs, numbers and commercial trade. With the increased realization that some wild species are being over-exploited, a number of agencies are recommending that wild species be brought into cultivation systems (BAH, 2002; Lambert *et al.*, 1997; WHO, IUCN and WWF, 1993). Cultivation can also have conservation impacts, however, and these need to be better understood. Medicinal plant production through cultivation, for example,

can reduce the extent to which wild populations are harvested, but may also lead to environmental degradation and loss of genetic diversity as well as loss of incentives to conserve wild populations (Anon., 2002b).

The relationship between *in situ* and *ex situ* conservation of species is an interesting topic with implications for local communities, public and private land owners and managers, entire industries and, of course, wild species. Identifying the conservation benefits and costs of the different production systems for MAP should help guide policies as to whether species' conservation should take place in nature or the nursery, or both (Bodeker *et al.*, 1997).

In this paper, we review global trends in the close relationship between cultivation and wild harvest of MAP species, then make recommendations on steps that should be taken to achieve a balance between consumption, conservation and cultivation.

CBD and the ecosystem approach

Since its adoption in 1992, the Convention on Biological Diversity (CBD) has strived to implement its three major goals: the conservation of biological diversity, the sustainable use of its components, and the fair and equitable sharing of the benefits from the use of genetic resources. Although MAP have not been explicitly on the agenda of the various CBD meetings, all three goals of the Convention are fully applicable to MAP resources.

In decision V/6, the Conference of the Parties of the CBD adopted the ecosystem approach as the primary framework for action under the Convention. It is a strategy for the integrated management of land, water and living resources that promotes

conservation and sustainable use in an equitable way. The ecosystem approach is based on the application of appropriate scientific methodologies focused on levels of biological organization which encompass the essential processes, functions and interactions among organisms and their environment. In April 2002, the CBD adopted the Global Strategy for Plant Conservation which provides a policy environment that is particularly well suited to addressing the conservation challenges for MAP in a coherent way (see Appendix 3).

Concept of sustainability

As a baseline element of the ecosystem approach it has to be recognized that humans, with their cultural diversity, are an integral component of ecosystems. In conceptual terms, the essence of sustainable development is expressed by the relationship between people and the ecosystem around it. This implies that ultimately one is entirely dependent upon the other. Human and ecosystem well-being need to be assessed together. A society is thought to be sustainable when both the human condition and the condition of the ecosystem are satisfactory or improving. The system improves only when both the condition of the ecosystem and the human condition improve (Prescott-Allen and Prescott-Allen, 1996).

SOME FIGURES TO START WITH...

How many MAP are used world-wide?

The number of plant species which have at one time or another been used in some culture for medicinal purposes can only be estimated. An enumeration of the WHO from the late 1970s listed 21 000 medicinal species (Penso 1980). However, in China alone 4 941 of 26 092 native species are used as drugs in Chinese traditional medicine (Duke and Ayensu 1985), anastonishing 18.9 percent. If this proportion is calculated for other well-known medicinal floras and then applied to the global total of 422 000 flowering plant species (Bramwell 2002; Govaert 2001), it can be estimated that the number of plant species used for medicinal purposes is more than 50 000 (Table 1).

We recognize, however, that certain plant families have higher proportions of medicinal plants than others. Good examples are the Apocynaceae, Araliaceae, Apiaceae, Asclepiadaceae, Canellaceae, Guttiferae and Menispermaceae. In addition, these families

TABLE 1

How many plants are used medicinally world-wide?

COUNTRY	PLANT SPECIES	MEDICAL PLANT SPECIES	%
China	26 092	4 941	18.9
India	15 000	3 000	20.0
Indonesia	22 500	1 000	4.4
Malaysia	15 500	1 200	7.7
Nepal	6 973	700	10.0
Pakistan	4 950	300	6.1
Philippines	8 931	850	9.5
Sri Lanka	3 314	550	16.6
Thailand	11 625	1 800	15.5
USA	21 641	2 564	11.8
Viet Nam	10 500	1 800	17.1
Average	13 366	1 700	12.5
World	422 000	52 885	

Sources: Duke and Ayensu (1985); Govaerts (2001); Groombridge and Jenkins (1994, 2002); Jain and DeFillipps (1991); Moerman (1996); Padua et al. (1999)

are not distributed uniformly across the world. As a consequence, not only do some floras have higher proportions of medicinal plants than others, but also have certain plant families a higher proportion of threatened species than others (Table 9 in Appendix 1).

How many MAP species are traded?

It is difficult to assess how many MAP are commercially traded, either on a national or even an international level. The bulk of the plant material is exported from developing countries while major markets are in the developed countries. An analysis of UNCTAD trade figures for 1981–1998 reflects this almost universal feature of MAP trade (Table 2). Adding the volumes for the five European countries in this list (94 300

tonnes) marks the dominance of Europe as an import region. Germany ranks fourth and third as importer and exporter, expressing the country's major role as a turntable for raw medicinal plants.

Iqbal (1993) estimates that about "4 000 to 6 000 botanicals are of commercial importance", another source refers to 5 000–6 000 "botanicals entering the world market" (SCBD, 2001). A thorough investigation of the German medicinal plant trade identified a total of 1 543 MAP being traded or offered on the German market (Lange and Schippmann, 1997). An extension of this survey to Europe as a whole arrived at 2 000 species in trade for medicinal purposes (Lange, 1998). Recognizing the role of Europe as a sink for MAP traded from all regions of the world, it is a qualified guess that the total number of MAP in international trade is around 2 500 species world-wide.

TABLE 2

The 12 leading countries of import and export of medicinal and aromatic plant

No. 7

COUNTRY OF IMPORT	VOLUME [TONNES]	VALUE [1 000 US$]	COUNTRY OF EXPORT	VOLUME [TONNES]	VALUE [1 000 US$]
Hong Kong	73 650	314 000	China	139 750	298 650
Japan	56 750	146 650	India	36 750	57 400
USA	56 000	133 350	Germany	15 050	72 400
Germany	45 850	113 900	USA	11 950	114 450
Rep. Korea	31 400	52 550	Chile	11 850	29 100
France	20 800	50 400	Egypt	11 350	13 700
China	12 400	41 750	Singapore	11 250	59 850
Italy	11 450	42 250	Mexico	10 600	10 050
Pakistan	11 350	11 850	Bulgaria	10 150	14 850
Spain	8 600	27 450	Pakistan	8 100	5 300
UK	7 600	25 550	Albania	7 350	14 050
Singapore	6 550	55 500	Marocco	7 250	13 200
Total	342 550	1 015 200	Total	281 550	643 200

Figures based on commodity group pharmaceutical plants (SITC.3: 292.4 = HS 1211).
Source: UNCTAD COMTRADE database, United Nations Statistics Division, New York (Lange, 2002).

How many MAP are threatened world-wide?

To satisfy the regional and international markets, the plant sources for expanding local, regional and international markets are harvested in increasing volumes and largely from wild populations (Kuipers, 1997; Lange, 1998). Supplies of wild plants in general are increasingly limited by deforestation from logging and conversion to plantations, pasture and agriculture (Ahmad, 1998; Cunningham, 1993).

In many cases, the impact through direct off-take goes hand-in-hand with decline owing to changes in land use. Species favoured by extensive agricultural management like *Arnica montana* in central Europe go into decline with changes in farming practices towards higher nutrient input on the meadows. This requires habitat management as the key factor in managing species's populations (Ellenberger, 1999).

One of the goals of the IUCN Medicinal Plant Specialist Group is to identify the species that have become threatened by non-sustainable harvest and other factors (see Appendix 4). The enormity of this task is illustrated by the following estimate: According to Walter and Gillett (1998), 34 000 species or eight percent of the world's flora are threatened with extinction. If this is applied to our earlier estimate that 52 000 plant species are used medicinally, it leads us to estimate that 4 160 MAP species are threatened (Table 3).

How many MAP are under cultivation?

Many medicinal plants, especially the aromatic herbs, are grown in home gardens, some are cultivated as field crops, either in sole cropping or in inter-cropping systems and rarely as plantation crops (Padua *et al.*, 1999).

In a survey carried out for the Rainforest Alliance, companies involved in trade and production of herbal remedies and other botanical products were asked what percentage of their material is from cultivated sources and what percentage from the wild. On average, companies reported that 60–90 percent of material was cultivated, with the remainder harvested wild. However, when asked about species' numbers rather than volume of material, the figures are generally inverted (Laird and Pierce, 2002). Lange and Schippmann (1997) state that of the 1 543 species traded in Germany, only 50–100 species (3–6 percent) are exclusively sourced from cultivation.

Of more than 400 plant species used for production of medicine by the Indian herbal industry, less than 20 species are currently under cultivation in different parts of the country (Uniyal *et al.*, 2000). In China, about 5 000 medicinal plants have been identified and about 1 000 are commonly used, but only 100–250 species are cultivated (Xiao Pei-Gen, 1991, He Shan-An and Ning Sheng, 1997). In Hungary, a country with a

TABLE 3

How many medicinal plant species are threatened?

Number of flowering plant species world-wide (Govaert, 2001)	422 000
12.5% of them are used medicinally	52 000
8% are threatened (Walter and Gillett, 1998)	4 160

long tradition of MAP cultivation, only 40 species are cultivated for commercial production (Bernáth, 1999; Palevitch, 1991). In Europe as a whole, only 130–140 MAP species are cultivated (Pank, 1998; Verlet and Leclercq, 1999).

Based on these figures, we assume that the number of MAP species currently in formal cultivation for commercial production does not exceed a few hundred world-wide. A global survey on the extent of MAP cultivation in terms of species, volumes and values would be highly desirable. On the other hand, however, we recognize that many more MAP species are cultivated on a small-scale in home gardens, either as home remedies or by herbalists or that cultivation by local people can take place as enrichment planting.

WILD OR CULTIVATED: WHAT DOES THE MARKET WANT?

Given the demand for a continuous and uniform supply of medicinal plants and the accelerating depletion of forest resources, increasing the number of medicinal plant species in cultivation would appear to be an important strategy for meeting a growing demand (Uniyal et al., 2000).

But why are so few species cultivated and why are some species cultivated and many others not?

One explanation may be found in the observation that cultivated plants are sometimes considered qualitatively inferior when compared with wild gathered specimens. For instance, wild ginseng roots are 5–10 times more valuable than roots produced by artificial propagation. The reason is primarily cultural, as the Chinese community, which is the largest consumer group of wild ginseng, believes that the similarity in appearance of gnarled wild roots to the human body symbolizes the vitality and potency of the root. Cultivated roots lack the characteristic shape of wild roots and are therefore not as highly coveted by consumers (Robbins, 1998). In Botswana, traditional medicinal practioners said that cultivated material was unacceptable, as cultivated plants did not have the power of the material collected from the wild (Cunningham, 1994).

Scientific studies partly support this. Medicinal properties in plants are mainly due to the presence of secondary metabolites which the plants need in their natural environments under particular conditions of stress and competition and which perhaps would not be expressed under monoculture conditions. Active ingredient levels can be much lower in fast growing cultivated stocks, whereas wild populations can be older due to slow growth rates and can have higher levels of active ingredients. While it can be presumed that cultivated plants are likely to be somewhat different in their properties from those gathered from their natural habitats, it is also clear that certain values in plants can be deliberately enhanced under controlled conditions of cultivation (Palevitch, 1991; Uniyal et al., 2000).

In general, in all countries, the trend is towards a greater proportion of cultivated material. The majority of companies, the mass-market, over-the-counter pharmaceutical companies, as well as the larger herb companies, prefer cultivated material, particularly since cultivated material can be certified as biodynamic or organic (Laird and Pierce, 2002).

From the perspective of the market, domestication and cultivation provide a number of advantages over wild harvest for production of plant-based medicines:

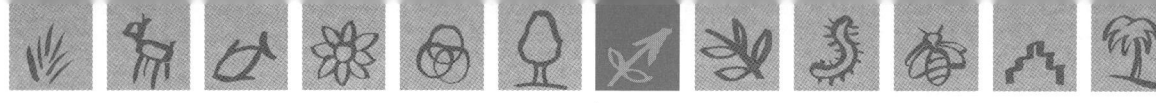

(i) While wild collection often offers material adulterated with other unwanted, sometimes harmful plant species, cultivation provides reliable botanical identification.

(ii) Wild harvest volumes are dependent on many factors that cannot be controlled and the irregularity of supply is a common feature. Cultivation guarantees a steady source of raw material.

(iii) Wholesalers and pharmaceutical companies can agree on volumes and prices over time with the grower.

(iv) The selection and development of genotypes with commercially desirable traits from the wild or managed populations may offer opportunities for the economic development of the medicinal plant species as a crop.

(v) Cultivation allows controlled post-harvest handling and therefore

(vi) Quality controls can be assured and

(vii) Product standards can be adjusted to regulations and consumer preferences.

(viii) Cultivated material can be easily certified as organic or biodynamic although certifiers are also presently developing wildcrafting standards (Leaman, 2002; Palevitch, 1991; Pierce, et al., 2002).

However, domestication of the resource through farming is not always technically possible. Many species are difficult to cultivate because of certain biological features or ecological requirements (slow growth rate, special soil requirements, low germination rates, susceptibility to pests, etc.).

Economical feasibility is the main rationale for a decision to bring a species in cultivation but is also a substantial limitation as long as sufficient volumes of material can still be obtained at a lower price from wild harvest. Cultivated material will be competing with material harvested from the wild that is supplied the market by commercial gatherers who have incurred no input costs for cultivation. Low prices, whether for local use or the international pharmaceutical trade, ensure that few species can be marketed at a high enough price to make cultivation profitable (Cunningham, 1994). Domestication of a previously wild collected species does not only require substantial investment of capital (up to US$ 200 000; Plescher in litt.) but also requires several years of investigations (e.g. 12 years for Alchemilla alpina; Schneider et al., 1999).

On a time scale of sometimes many decades, the transition from wild harvesting to possible cultivation goes through various phases (Figure 1):

(i) Discovery Phase: At this point the demand can be met by wild harvest. Extractivism is done for local use or for barter with others.

(ii) Expansion Phase: It is clear that the product is potentially useful and that demand is likely to increase. Harvest is done for local or regional sale and eventually for international markets. In general, species with naturally low densities are unlikely to become important sources of commercially large quantities.

(iii) Stabilization Phase: The species is unlikely to be attractive to growers unless prices are high enough and wild-harvested resources are scarce enough. However, desirable species may be grown on farm land and planted around settlements.

(iv) Decline Phase: Prices increase with scarcity due to transport costs, search time and long-distance trade. Wild populations will have to decline further before cultivation is a viable option. The trade is characterized by fluctuations in

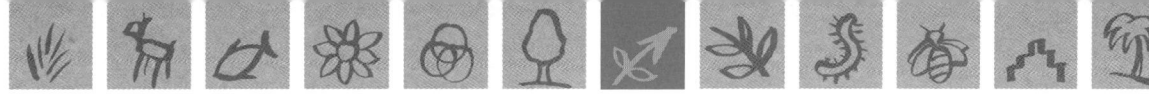

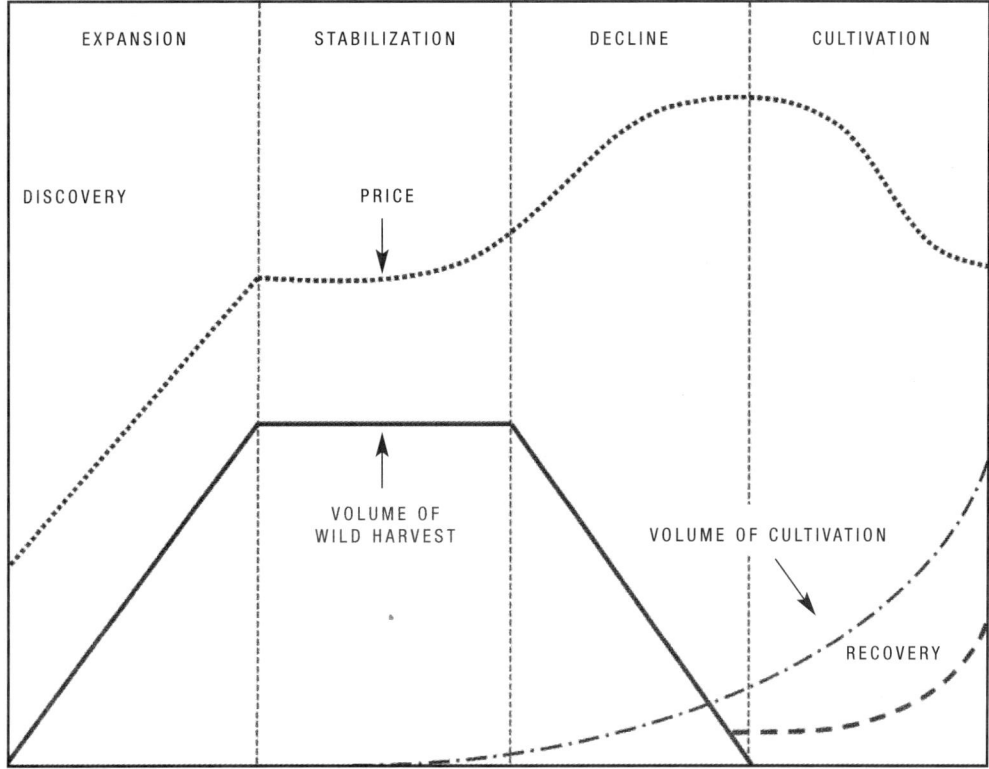

FIGURE 1

Transition phases from wild harvesting to cultivation: after wild resources decline with over-harvesting, raw material prices increase and cultivation becomes economically feasible; more resilient species can recover (after Homma, 1992 and Cunningham, 2001).

Figure labels: EXPANSION · STABILIZATION · DECLINE · CULTIVATION · DISCOVERY · PRICE · VOLUME OF WILD HARVEST · VOLUME OF CULTIVATION · RECOVERY

supplies, often to the extent of disrupting the trade balance. For slow growing species, if controls on collection are not strictly enforced, wild populations will be more seriously eroded before cultivated material is available (Cunningham, 1994; FAO, 1993).

(v) Cultivation Phase: Now, formal cultivation systems are developed and instituted. The plants are domesticated and incorporated in agroforestry systems sometimes for the benefit of small-scale farmers. If international market opportunities exist, commercial plantations are created with substantial investment and genetic selection, cloning, breeding and biotechnology may be applied. More resilient species may recover in their wild populations (Figure 1).

WILD OR CULTIVATED: WHAT DO PEOPLE NEED?

Health care needs

There is a world-wide trend of increasing demand for many popular, effective species in Europe, North America and Asia, growing between 8–15 percent per year (Grünwald and Büttel, 1996). Rapid urbanization and the importance of herbal medicines in African health care systems stimulated a growing national and regional trade in Africa (Cunningham, 1993). Demand for medicinal plants also reflects distinct cultural preferences. In the USA, for example, only three percent of people surveyed had used herbal medicine in the past year (Eisenberg *et al.*, 1993), whereas in Germany, with a strong tradition of medicinal plant use, 31 percent of

the over-the-counter products in pharmacies in 2001 were phyto-pharmaceutical preparations (BAH, 2002).

The level of herbal medicine use in most developing countries is much higher than this. While most traditional medicinal plants are gathered from the wild, they are not static health care systems, and introduced species are commonly adopted into the repertoire of plants used by African or South American herbalists. In many cases, herbal medicines can also be cheaper than western medicines, particularly where access to traditional healers is easier. Demand for traditional medicine continues in the urban environment even if western biomedicine is available (Anon, 2002b; Mander *et al.*, 1997).

Income generation

Wild harvesting of medicinal plants is a chance for the poorest to make at least some cash income. Especially those people who do not have access to farmland at all depend on gathering MAP to earn at least some money. However, local people generally get a low price for unprocessed plant material. Although income from *Prunus africana* bark sales is an important source of revenue to villagers in Madagascar, in some cases generating >30 percent of village revenue, the price paid to collectors is negligible compared to Madagascan middlemen (Walter and Rakotonirina, 1995). In Mexico, Hersch-Martinez (1995) found that medicinal plant collectors only received an average 6.17 percent of the medicinal plant's consumer price.

Whether fruits, roots, bark or whole plants are involved, the potential yield from wild stocks of many species is frequently over-estimated, particularly if the effects of stochastic events is taken into account. As a result, commercial harvesting ventures based

on wild populations can be characterized by a "boom and bust" situation where initial harvests are followed by declining resource availability.

Small-scale cultivation and home gardens

Small-scale cultivation, which requires low economic inputs, can be a response to declining local stocks, generating income and supplying regional markets. This can be a more secure income than from wild harvest which is notoriously inconsistent. For farmers who integrate MAP into agro-forestry or small-scale farming systems, these species can provide a diversified and additional source of income to the family. Home gardens are increasingly a focus of medicinal plant propagation and introduction programmes intended to encourage the use of traditional remedies for common ailments by making the plant sources more accessible (Agelet *et al.*, 2000).

Large-scale cultivation

As outlined by Leakey and Izac (1997), large-scale cultivation has a number of socio-economic impacts on rural people: "Commercialization is both necessary and potentially harmful to farmers. It is necessary in that without it the market for products is small and the opportunity does not exist for rural people to generate income. A degree of product domestication is therefore desirable. On the other hand, commercialization is potentially harmful to rural people if it expands to the point that outsiders with capital to invest come in and develop large-scale monocultural plantations for export markets. Rural people may benefit from plantations as a result of

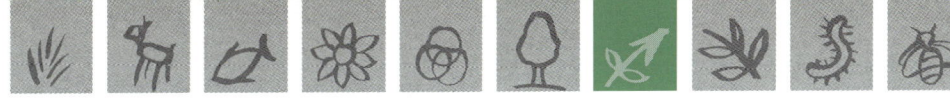

available employment and hence off-farm income [...]. However, plantations may also distort market forces to their advantage, for example, by imposing low wages which will restrict the social and economic development of local people. The major beneficiaries of large-scale exports will probably be the country's elite and, perhaps, the national economy".

Also, those socially disadvantaged groups who actually depend on gathering MAP for their survival and cash income, may not have access to farmland at all and are therefore not able to compete with large-scale production of MAP by well-established farmers (Vantomme in Anon., 2002a). Other limitations to the domestication approach include boom-bust and fickle markets that let farmers down when consumers turn their attention elsewhere (Laird and Pierce, 2002).

WILD OR CULTIVATED: WHAT DO THE SPECIES AND ECOSYSTEMS REQUIRE?

Cultivation of medicinal plants is widely viewed not only as a means for meeting current and future demands for large-volume production of plant-based drugs and herbal remedies, but also as a means for relieving harvest pressure on wild populations (FAO, 1995; Lambert *et al.*, 1997; Palevitch, 1991; de Silva, 1997; WHO, IUCN and WWF, 1993). In this chapter we want to assess

the benefits and risks associated with such recommendations.

Booming markets with rapidly rising demands often have devastating effects on wild collected species. A closer look reveals that not all species are affected in the same way by harvesting pressures. The seven forms of rarity described by Rabinovitz (1981) make clear that a species which (i) has a narrow geographic distribution, (ii) is habitat-specific, and (iii) has small population sizes everywhere, is more easily over-harvested than species of any other pattern (Table 4).

TABLE 4

Seven forms of rarity (after Rabinowitz, 1981)

GEOGRAPHIC DISTRIBUTION			
	HABITAT SPECIFICITY		
		LOCAL POPULATION SIZE	
wide	broad	somewhere large	least concern
		everywhere small	
	restricted	somewhere large	
		everywhere small	
narrow	broad	somewhere large	
		everywhere small	
	restricted	somewhere large	
		everywhere small	highly susceptible

Secondly, the **susceptibility or resilience** to collection pressure varies among species owing to biological characters such as different growth rates (slow-growing vs. fast-growing), reproductive systems (vegetative or generative propagation; germination rates; dormance; apomixis) and life forms (annual, perennial, tree).

Species can be distinguished quite well for their susceptibility to over-collection if their life form and the plant parts collected are viewed together (Table 5). Harvesting

	WOOD	BARK	ROOT	LEAF	FLOWER	FRUIT/ SEED
Annual	---	---	high	medium	medium	high
Bi-annual	---	---	high	medium	medium	high
Perennial	---	medium	high	low	low	low
Shrub	medium	medium ?	medium ?	low	low	low
Tree	medium	medium ?	medium ?	low	low	low

TABLE 5

Susceptibility of species to over-collection as a function of life form and plant parts used

fruits from a long-lived tree presents a far lower threat to the long-term survival of the species than does collecting seeds from an annual plant. In the latter case, if the seed is gone the plant is gone. In some cases the harvest impacts are more complex, e.g. with slow-growing trees which reproduce from seed but only produce few, large fruits (example: *Araucaria araucana,* monkey puzzle tree). This will increase their susceptibility to over-harvest from low to medium or even high. A thorough summary of predictors of resilience or vulnerability to harvesting wild populations is presented by Cunningham (2001).

In summary, we can state that species most susceptible to over-harvesting are habitat specific, slow-growing and destructively harvested for their bark, roots or the whole plant. These species suffer most from harvesting and many of them have been seriously depleted, for example *Prunus africana* in West Africa, *Warburgia salutaris* in southern Africa and *Saussurea costus* in the Himalayas.

For threatened medicinal plant species **cultivation is a conservation option** because the constant drain of material from their populations is much higher than the annual sustained yield. If the demand for these species can be met from cultivated sources the pressure on the wild populations will be relieved. In these cases, the need for strict conservation of remaining populations, improved security of germplasm *ex situ* and investment in selection and improvement programmes is extremely urgent as the example of Jaborandi *(Pilocarpus jaborandi)* in Brazil shows (Pinheiro, 1997).

However, among the species that can be marketed at a high enough price to make cultivation profitable, only a few are in the highest threat categories. Examples of threatened but cultivated species are *Garcinia afzelii, Panax quinquefolius, Saussurea*

costus and *Warburgia salutaris* (Cunningham, 1994). With respect to economic viability many highly endangered MAP do not qualify for cultivation. This group of plants will enter cultivation only with the help of public domestication programmes.

For all other harvested MAP species the **priority conservation option is sustainable harvest from wild populations,** for a variety of reasons.

Let's imagine that a valuable medicinal plant is exploited by local collectors. A pharmaceutical company has domesticated and begun to cultivate the plant on a commercial scale. When the company no longer needs the wild harvested material, local harvesters have to abandon the harvest and any **incentive** the local collectors might have had to protect the wild populations is gone. The domestication of MAP species has an environmental implication in the sense that it reduces the economic incentives for forest-dependent people to conserve the ecosystems in which the MAP species occur (Leaman *et al.,* 1997; Vantomme in Anon., 2002a).

If collectors and collecting communities can be involved in the development of propagation and management methods, the likelihood of their having an interest in protecting the wild populations from over-exploitation, particularly if these are understood to be the genetic resource "bank" for the domestic enterprises, will be greater.

Another aspect to consider is the **genetic diversity** of the species which is in demand. Long before non-sustainable harvest practices lead to extermination of a whole species, selection of favoured growth forms and concentration on certain harvesting areas which may hold certain ecotypes will lead to a degradation of genetic the diversity of the wild populations. The same is true under domestication: Industrial requirements for

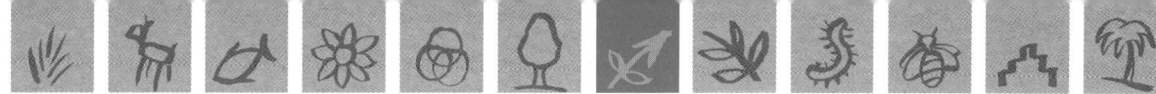

standardization encourage a narrow genetic range of material in cultivation. Domestication will not achieve conservation of genetic diversity because a narrow group of high-yielding individuals will be selected for planting.

As a summary of Chapters 3–5, Table 7 in Appendix 1 indicates the advantages and disadvantages of the three aspects distinguished: "species/ecosytems", "market" and "people".

CHALLENGES OF HARVESTING SUSTAINABLY FROM THE WILD

Sustainable harvesting is increasingly seen as the most important conservation strategy for most wild harvested species and their habitats, given their current and potential contributions to local economies and their greater value to harvesters over the long term. The basic idea is that non-destructive harvests and local benefits will maintain population, species and ecosystem diversity.

Besides poverty and the break-down of traditional controls, the major challenges for sustainable wild collection include: lack of knowledge about sustainable harvest rates and practices, undefined land use rights and lack of legislative and policy guidance.

Lack of information on the wild resource

"The most important ingredient required to achieve a truly sustainable form of resource use is information" (Peters, 1994). In reality, resource managers are always confronted with the lack of adequate information about the plants used, their distribution, the genetic diversity of wild

populations and relatives and, above all, the annual sustained yield that can be harvested without damaging the populations (Iqbal, 1993). Research on the conservation and sustainable use of medicinal plants and their habitats has fallen far behind the demand for this globally important resource. Each species has unique ecological, socio-economic, health and cultural associations that must be understood. Model research approaches are feasible, model solutions are not.

Problems of open access

In many cases, access to the resource is open to everybody, rather than limited or privately owned. To make a living, commercial medicinal plant gatherers therefore "mine" rather than manage these resources (Cunningham, 1994). Open access schemes to harvestable plant population prevent rational and cautious use and make it difficult to adhere to quotas and closed seasons.

Lack of legislative and policy support for wild harvesting schemes

Information on trade in MAP is scarce and data are rarely collected or published at a national level. Much production and consumption are at subsistence level and as a consequence the economic importance of these activities is largely under-estimated in government decision making regarding rural development, natural resource management planning and in government budget allocations (Vantomme in Anon., 2002a). Therefore, national legislation and policies mostly fail to provide frameworks for a rational and sustainable use of a wild resource.

Opportunities for governments to develop legislation to control and monitor harvest and trade of medicinal plant species and to consider conservation and sustainable use of medicinal plants as a priority in establishing protected areas have been greatly enhanced by two recent developments in international legislation: the addition of medicinal plant species to the Convention on International Trade of Endangered Species (CITES) and the entry into force of the CBD (see Appendix 2).

FUTURE TRENDS AND SOLUTIONS

How will the market demand develop in the future? People in developing countries are already dependent and will increasingly depend on medicinal plants as sources for their primary health care. An estimate by the World Health Organization (Bannerman, 1982) that more than 80 percent of the world's population relies solely or largely on traditional remedies for health care, is frequently cited.

Also in the northern countries, use of medicinal plants is expected to rise globally, both in allopathic and herbal medicine (WHO, 2002). This upward trend is predicted not only because of population explosion, but also due to the increasing popularity of natural-based, environmentally friendly products.

In general, the demand for medicinal plants and herbal remedies and especially its renaissance in the developed countries is driven by the following factors (Iqbal, 1993; Leaman, 2002):

- increasing costs of institutional, pharmaceutical-based health care;
- interest of individuals, communities and national governments in greater self-reliance in health care;
- interest of communities and national governments in small-and large-scale industrial development based on local/national biodiversity resources;
- increasing success in validating the safety and efficacy of herbal remedies;
- legislation improving the status of herbal medicine industry;
- renewed interest of companies in isolating useful compounds from plants;
- search for new drugs and treatments of serious and drug-resistant diseases;
- marketing strategies by the companies dealing in herbal medicine.

Most MAP species will continue to be harvested wild

The limitations of cultivation as an alternative to wild harvest have been examined by Sheldon *et al.* (1997) in several case studies. We share their conclusion that, notwithstanding the level of interest in cultivation as a means for enhanced production and in a few cases as an effort to contribute to conservation of the resource, most medicinal and aromatic plant species will continue to be harvested wild to some extent. There is therefore a need to recognize and strengthen the role of local people in forest inventory, monitoring and impact assessment processes and to integrate non-timber product uses into forest management.

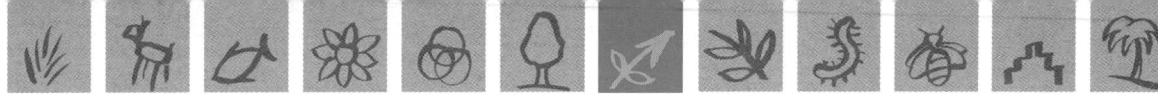

Need for implementation of management plans

Limiting the harvest to a sustainable level requires an effective management system and sound scientific information (Table 6). The management system must include annual harvest quotas, consider seasonal or geographical restrictions and restriction of harvest to particular plant parts or size classes. In addition, clarification of the access and user rights to the resources providing MAP is part of the essential baseline information. Continuous monitoring and evaluation of success is necessary to adapt the management strategy (FAO, 1995; Leaman *et al.*, 1997; Prescott-Allen and Prescott-Allen, 1996; Schippmann, 1997; WHO, IUCN and WWF, 1993).

In many cases harvesting techniques need to be improved as the extraction of the roots or bark is often negatively affecting the recovery of the species or may even kill it. Collection methods are often crude and wasteful, resulting in loss of quality and reduction in price (Iqbal, 1993; Vantomme in Anon,. 2002a).

Field-based methods have already been developed for sustainable harvest assessment and monitoring of non-wood forest products, resulting in the publication of research guidelines and predictive models (Cunningham, 2001; FAO, 1995; Nantel *et al.*, 1996; Peters, 1994).

Eco-labelling and certification

Given that sustainable harvesting from the wild is difficult to achieve, certification standards can play a role to assure that a product meets certain standards of sustainability. Certification programmes

● resource inventory of population abundance and distribution
● assessment of regional and global threat based on all available knowledge and expertise
● biological studies (growth and regeneration rates, pollination system, seed dispersal, potential for confusion with similar species, etc.) and assessment of harvest impact on viability of individuals
● assessment of annual sustained yield
● review of local knowledge and harvest practices
● review of harvest and trade levels in the past and evaluation of market trends
● revision of national regulations for the utilization in source country
● assessment of tenure and access
● design and implementation of management scheme: annual harvesting quota, seasonal or regional restriction and on certain plant parts or size classes, domestication programme
● installation of continuing monitoring and re-evalutaion (adaptive management)

TABLE 6

Steps and standard elements of a management plan for MAP utilization (Schippmann 1997)

related to natural resource use have mainly been developed for timber and agricultural products but they are presently being adapted for wild harvest of non-timber plants. Various schemes focus on different areas along the supply chain: production, processing, trade, manufacturing, marketing. Four categories of certification schemes have been identified to be of relevance for MAP products (Walter, 2002): (i) forest management certification (e.g. Forest Stewardship Council FSC), (ii) social certification (e.g. Fair Trade Federation FTF), (iii) organic certification (e.g. International Federation of Organic Agriculture IFOAM), and (iv) product quality certification.

The latter includes parameters such as product identity, purity, safety and efficacy. The Good Harvesting Practices (GHP) developed for medicinal plants cover to some degree ecological aspects (Harnischfeger, 2000) but need to be more clearly focused on this aspect before they can make a meaningful contribution to ensuring

sustainability. Dürbeck (1999), Walter (2002) and most comprehensively Pierce *et al.* (2002) present overviews of certification programmes and their activities.

RECOMMENDATIONS

(1) To overcome significant knowledge deficits, a global MAP **cultivation survey** should be commissioned by an international organization. Aims are to identify species cultivated, in which countries they are grown, volumes produced and their market values. This survey should also assess public domestication programmes, as well as *in situ* and *ex situ* conservation efforts of wild populations of species in cultivation (e.g. in protected areas, in genebanks and botanic gardens).

(2) Wild harvesting of MAP will continue to prevail due to the economic reasons outlined above. **Sustainable wild harvest management schemes need to be supported** by governments and authorities. Management plans need to be established as a standard prerequisite for any such harvesting in the wild. There is a need to monitor and audit the harvesting process to determine whether it is sustainable.

(3) Primary producers need help to improve returns from sustainable harvesting of MAP. Community-based **small-scale cultivation** enterprises need to be strengthened to enable them to compete with large-scale high-tech cultivation.

(4) Secure *ex situ* field **genebanks** need to be developed, particularly for habitat-specific, slow-growing species with high susceptibility of being over-harvested.

(5) Medicinal plant **domestication programmes** need to be expanded, taking fuller advantage of the genetic and chemical diversity within species over wide geographical areas.

(6) Capacity to assess and monitor the conservation status of MAP and to manage harvesting within the limits of sustainability is extremely limited world-wide and needs to be developed through training courses and curriculum development in ethnobotany and applied ecology. **Research** to investigate the sustainability of production systems is lacking and needs to be stimulated for a better understanding of the biological dynamics of the resource in the wild and in domestication.

(7) Management planning has to take the diversity of tenure systems which apply to medicinal plants into account to a far greater degree. Clarification of **user rights** over the resource and access to it, particularly where it is considered common property, needs to be recognized as a crucial factor enabling or preventing a sustainable harvest from wild populations.

(8) Eco-labelling and other social and economic incentives to strengthen market credibility and competitiveness of biodiversity-friendly products need to be promoted. The efforts of certifiers to develop **certification standards** for wild harvested plant material need to be supported as well as the approaches of industry to set up self-binding product quality standards. The private sector should be encouraged to consider local livelihoods and biodiversity when setting up ethical and environmental standards.

(9) Conservation of medicinal plants currently lacks priority in policy and law. There are opportunities to change this with the implementation of legal instruments such as the CBD and CITES. Government **policies and legislation** need to be adapted and implemented to recognize the value of and need for sustainable wild harvesting management regimes, to implement national and/or regional permit systems and make medicinal plant conservation a priority for national health and economic policy.

(10) The **Global Environment Facility** (GEF) needs to consider medicinal plant conservation as a

programme priority worthy of funding.

(11) Medicinal plants warrant priority in national efforts to implement the **Global Strategy for Plant Conservation** of the CBD.

(12) Local communities can take more responsibility for sustainable harvesting of medicinal plants only if they have the choices afforded by adequate income, control over the resource and the knowledge and skills required. On the issue of **intellectual property rights** it needs to be elaborated how the country, the local user or other entity can be adequately compensated for use of the resource by outsiders.

No. 7

REFERENCES

Agelet, A., M.A. Bonet & J. Valles, 2000. Homegardens and their role as a main source of medicinal plants in mountain regions of Catalonia (Iberian Peninsula). − *Economic Botany* 54: 295−309.

Ahmad, B., 1998. Plant exploration and documentation in view of land clearing in Sabah. In Nair, M.N.B. & N. Ganapathi, eds., *Medicinal Plants. Cure for the 21st Century. Biodiversity Conservation and Utilization of Medicinal Plants.* Proceedings of a seminar, 15−16 October 1998. − pp. 161−162. Serdang, Malaysia, Faculty of Forestry, Universiti Putra Malaysia.

Anon., 2002a. *Conservation impacts of commercial captive breeding workshop.* Briefing notes II. 7−9.12.2001, Jacksonville. − Cambridge, UK, IUCN/SSC Wildlife Trade Programme.

Anon., 2002b. *Assessing the impacts of commercial captive breeding and artificial propagation on wild species conservation. IUCN/SSC Workshop. 7−9.12.2001, Jacksonville. Draft workshop report.* − Cambridge, IUCN/SSC Wildlife Trade Programme. (Unpublished report)

BAH, 2002. *Pflanzliche Arzneimittel heute. Wissenschaftliche Erkenntnisse und arzneirechtliche Rahmenbedingungen. Bestandsaufnahme und Perspektiven.* 3rd edition. − Bonn, Bundesfachverband der Arzneimittelhersteller.

Bannerman, R.H., 1982. Traditional medicine in modern health care. − *World Health Forum* 3(1): 8−13.

Bernáth, J., 1999. Biological and economical aspects of utilization and exploitation of wild growing medicinal plants in middle and south Europe. *In* Caffini, N., J. Bernath, L. Craker, A. Jatisatienr & G. Giberti, eds., *Proceedings of the Second World Congress on Medicinal and Aromatic Plants for Human Welfare. WOCMAP II. Biological resources, sustainable use, conservation and ethnobotany.* − pp. 31−41, Leuven, Netherlands, ISHS (Acta Horticulturae 500).

Bodeker, G., K.K.S. Bhat, J. Burley & P. Vantomme, Eds., 1997. *Medicinal plants for forest conservation and health care.* − Rome, FAO (Non-wood Forest Products 11).

Bramwell, D., 2002. How many plant species are there? − *Plant Talk* 28: 32−34.

Cunningham, A.B., 1993. *African medicinal plants. Setting priorities at the interface between conservation and primary healthcare.* − Paris, UNESCO (People and Plant Working Paper 1).

Cunningham, A.B., 1994. Management of medicinal plant resources. In Seyani, J.H. & A.C. Chikuni, eds., *Proceedings of the 13th Plenary Meeting of AETFAT, Zomba, Malawi,* 2−11 April, 1991. Vol. 1. − pp. 173−189, Limbe, Cameroon, Montfort.

Cunningham, A.B., 2001. *Applied ethnobotany.* People, wild plant use and conservation. − London, Earthscan (People and Plants Conservation Manuals).

Dürbeck, K., 1999. Green trade organizations. Striving for fair benefits from trade in non-wood forest products. − *Unasylva* 198. Retrieved from FAO website, www.fao.org/docrep/x2450e/x2450e04.htm (viewed 28.9.2002)

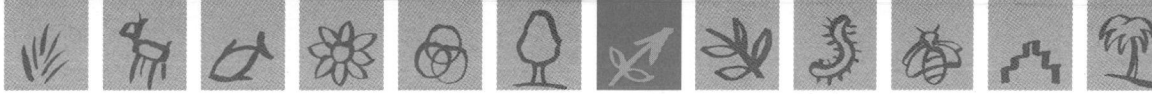

Duke, J.A. & E.S. Ayensu, 1985. *Medicinal plants of China*. Vol. 1 & 2. – Algonac, USA, Reference Publications (Medicinal Plants of the World 4).

Eisenberg, D.M., R.C. Kessler, C. Foster, F.E. Norlock, D.R. Calkins & T.L. Delbanco, 1993. Unconventional medicines in the United States. – *The New England Journal of Medicine* 328: 246–252.

Ellenberger, A., 1999. Assuming responsibility for a protected plant. Weleda's endeavour to secure the firm's supply of *Arnica montana*. In TRAFFIC Europe, ed., *Medicinal plant trade in Europe. Proceedings of the first symposium on the conservation of medicinal plants in trade in Europe, 22–23.6.1998, Kew.* – pp. 127–130, Brussels, Belgium, TRAFFIC Europe.

FAO, 2001. NWFP in Africa: A regional and national overview; PFNL en Afrique: Un aperçu regional et national. by S. Walter. FAO Working Paper FOPW/01/1. Rome. Also available at www.fao.org/DOCREP/003/y1515b/y1515b00.HTM.

FAO, 1993. International trade in non-wood forest products: an overview. by Iqbal, M. Working Paper Misc/93/11. Rome. 100 pp.. Also available at http://www.fao.org/docrep/x5326e/x5326e00.htm

FAO, 1995. *Non-wood forest products for rural income and sustainable development*. – Rome (Non-Wood Forest Products 7).

Farnsworth, N.R. & D.D. Soejarto, 1991. Global importance of medicinal plants. In Akerele, O., V. Heywood & H. Synge, eds., *Conservation of medicinal plants*. – pp. 25–51, Cambridge, UK, University Press.

Gaston, K.J. 1994., *Rarity*. – London, Chapman, London (Population and Community Biology Series 13).

Govaerts, R., 2001. How many species of seed plants are there? – Taxon 50: 1085–1090.

Groombridge, B. & M. Jenkins, 1994. *Biodiversity data sourcebook*. – Cambridge, UK, World Conservation Press (WCMC Biodiversity Series 1).

Groombridge, B. & M.D. Jenkins, 2002. World atlas of biodiversity. Earth's living resources in the 21st century. – Berkeley, USA, University of California Press.

Grünwald, J. & K. Büttel, 1996. The European phytotherapeutics market. – *Drugs* Made In Germany 39: 6-11.

Harnischfeger, G., 2000. Proposed guidelines for commercial collection of medicinal plant material. – *Journal of Herbs, Spices and Medicinal Plants* 7(1): 43–50.

Hersch-Martinez, P., 1995. Commercialization of wild medicinal plants from southwest Puebla, Mexico. – *Economic Botany* 49: 197-206

He Shan-An & Ning Sheng, 1997. Utilization and conservation of medicinal plants in China with special reference to Atractylodes lancea. In Bodeker, G., K.K.S. Bhat, J. Burley & P. Vantomme, eds., *Medicinal plants for forest conservation and health care*. – pp. 109–115, Rome, FAO (Non-wood Forest Products 11).

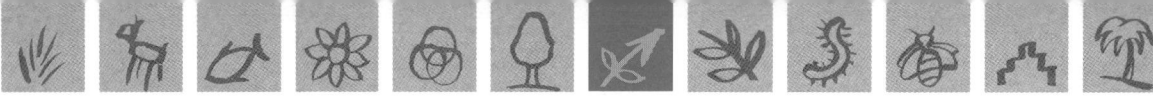

Homma, A.K.G., 1992. The dynamics of extraction in Amazonia. A historical perspective. – *Advances in Economic Botany* 9: 23–31.

Jain, S.K. & R.A. DeFillipps, 1991. *Medicinal plants of India.* Vol. 1 & 2. – Algonac, USA, Reference Publications (Medicinal Plants of the World 5).

Jones, E.T., R.J. McLain & J. Weigand, 2002. *Nontimber forest products in the United States.* – Lawrence, USA, University Press of Kansas.

Kuipers, S.E., 1997. Trade in medicinal plants. In Bodeker, G., K.K.S. Bhat, J. Burley & P. Vantomme, eds., *Medicinal plants for forest conservation and health care.* – pp. 45–59, Rome, FAO (Non-wood Forest Products 11).

Laird, S.A. & A.R. Pierce, 2002. *Promoting sustainable and ethical botanicals. Strategies to improve commercial raw material sourcing. Results from the sustainable botanicals pilot project. Industry surveys, case studies and standards collection.* – New York, Rainforest Alliance (www.rainforest-alliance.org/news/archives/news/news44.html, viewed 27.9.2002).

Lambert, J., J. Srivastava & N. Vietmeyer, 1997. *Medicinal plants. Rescuing a global heritage.* – Washington DC, World Bank (World Bank Technical Paper 355).

Lange, D., 1998. *Europe's medicinal and aromatic plants. Their use, trade and conservation.* – Cambridge, UK, TRAFFIC International.

Lange, D., 2002. *The role of east and southeast Europe in the medicinal and aromatic plants' trade.* – *Medicinal Plant Conservation* 8: 14–18.

Lange, D. & U. Schippmann, 1997. *Trade survey of medicinal plants in Germany.* – Bonn, Germany, Bundesamt für Naturschutz.

Leakey, R.R.B. & A-M.N. Izac, 1997. Linkages between domestication and commercialization of non-timber forest products. Implications for agroforestry. *In* Leakey, R.R.B., A.B. Temu, M. Melnyk & P. Vantomme, eds., Domestication and commercialization of non-timber forest products in agroforestry systems. – pp. 1–7, Rome, FAO (Non-wood Forest Products 9).

Leaman, D.J., U. Schippmann & L. Glowka, 1997. Environmental protection concerns of prospecting and producing plant-based drugs. In Wozniak, D.A., S. Yuen, M. Garrett & T.M. Schuman, eds., *International symposium on herbal medicine. A holistic approach. Documents, proceedings and recommendations.* 1–4 June 1997. Honolulu. – pp. 352–378, San Diego, USA, International Institute Human Resources Development.

Leaman, D., 2002. *Medicinal plants. Briefing notes on the impacts of domestication/cultivation on conservation. Paper for the "Commercial captive propagation and wild species conservation" workshop, 7–9.12.2001, Jacksonville.* (Unpublished report)

Mander, M., J. Mander & C. Breen, 1997. Promoting the cultivation of indigenous plants for markets. Experiences from KwaZulu-Natal, South Africa. In Leakey, R.R.B., A.B. Temu, M. Melnyk & P. Vantomme, eds., *Domestication and commercialization of non-timber forest products in agroforestry systems.* – pp. 104–109, Rome, FAO (Non-wood Forest Products 9).

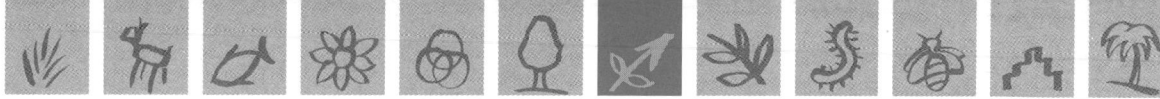

Moerman, D.E., 1996. An analysis of the food plants and drug plants of native North America. – *Journal of Ethnopharmacology* 52: 1–22.

Nantel, P., D. Gagnon & A. Nault, 1996. Population viability analysis of American Ginseng and Wild Leek harvested in stochastic environments. – *Conservation Biology* 10: 608–621.

Padua, L.S. de, N. Bunyapraphatsara & R.H.M.J. Lemmens, 1999. *Medicinal and poisonous plants.* Vol. 1. – Leiden, Netherlands, Backhuys (Plants Resources of South-East Asia 12/1).

Palevitch, D., 1991. Agronomy applied to medicinal plant conservation. In Akerele, O., V. Heywood & H. Synge, eds., *Conservation of medicinal plants*. – pp. 168–178, Cambridge, UK, University Press.

Pank, F., 1998. Der Arznei- und Gewürzpflanzenmarkt in der EU. – *Zeitschrift für Arznei- und Gewürzpflanzen* 3: 77–81.

Penso, G., 1980. *WHO inventory of medicinal plants used in different countries.* – Geneva, Switzerland, WHO.

Peters, C.M., 1994. *Sustainable harvest of non-timber plant resources in tropical moist forest. An ecological primer.* – Washington DC, Biodiversity Support Program.

Pierce, A., S. Laird & R. Malleson, 2002. *Annotated collection of guidelines, standards and regulations for trade in non-tim,ber forest products* (NTFPs) and botanicals. Version 1.0. – New York, Rainforest Alliance. Retrieved from www.rainforest-alliance.org/news/archives/news/news44.html (viewed 27.9.2002).

Pinheiro, C.U.B., 1997. Jaborandi (*Pilocarpus* sp., *Rutaceae*). A wild species and its rapid transformation into a crop. – *Economic Botany* 51: 49–58.

Prescott-Allen, R. & C. Prescott-Allen, 1996. *Assessing the sustainability of uses of wild species. Case studies and initial assessment procedure.* – Gland & Cambridge, IUCN (Occasional Paper of the IUCN Species Survival Commission 12).

Rabinowitz, D., 1981. Seven forms of rarity. *In* Synge, H., ed., *The biological aspects of rare plant conservation.* – pp. 205–217, John Wiley & Sons.

Robbins, C.S., 1998. American ginseng. The root of North America's medicinal herb trade. – Washington DC, TRAFFIC USA.

Schippmann, U., 1997. Plant uses and species risk. From horticultural to medicinal plant trade. In Newton, J., ed., Planta Europaea. *Proceedings of the first European Conference on the conservation of wild plants*, Hyères, France, 2–8 September 1995. – pp. 161–165, London, Plantlife.

Schneider, E., G. Stekly & P. Brunner, 1999. Domestikation von Bergfrauenmantel *(Alchemilla alpina* agg.). – *Zeitschrift für Arznei- und Gewürzpflanzen* 4: 134–140.

SCBD, 2001. *Sustainable management of non-timber forest resources.* – Montreal, Canada, Secretariat of the Convention on Biological Diversity (CBD Technical Series 6).

No. 7

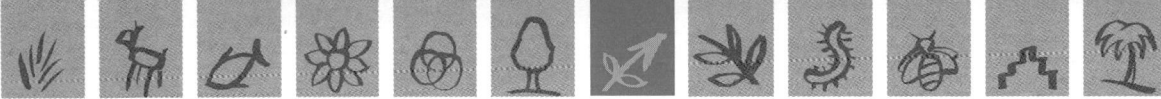

Sheldon, J.W., M.J. Balick & S. Laird, 1996. *Medicinal plants. Can utilization and conservation coexist?* – New York, New York Botanical Garden (Advances in Economic Botany 12).

Silva, T. de, 1997. Industrial utilization of medicinal plants in developing countries. In Bodeker, G., K.K.S. Bhat, J. Burley & P. Vantomme, eds., *Medicinal plants for forest conservation and health care.* – pp. 34–44, Rome, FAO (Non-wood Forest Products 11).

Uniyal, R.C., M.R. Uniyal & P. Jain, 2000. *Cultivation of medicinal plants in India. A reference book.* – New Delhi, India, TRAFFIC India & WWF India.

Verlet, N. & G. Leclercq, 1999. The production of aromatic and medicinal plants in the European Union. An economic database for a development strategy. In TRAFFIC Europe, ed., *Medicinal plant trade in Europe. Proceedings of the first symposium on the conservation of medicinal plants in trade in Europe, 22–23.6.1998, Kew.* – pp. 121–126, Brussels, Belgium, TRAFFIC Europe.

Walter, K.S. & H.J. Gillett, 1998. 1997 *IUCN Red List of threatened plants.* – Gland, Switzerland, IUCN.

Walter, S., 2002. Certification and benefit-sharing mechanisms in the field of non-wood forest products. An overview. – *Medicinal Plant Conservation* 8: 3–9.

Walter, S. & J.C.R. Rakotonirina, 1995. *L'exploitation de Prunus africanum à Madagascar.* – Antananarivo, Madagascar. (Unpublished report)

WHO, IUCN & WWF, 1993. Guidelines on the conservation of medicinal plants. – Gland & Geneva, Switzerland.

WHO, 2002. *Traditional medicine strategy 2002–2005.* – Retrieved from WHO website, www.who.int/medicines/library/trm/trm_strat_eng.pdf (viewed 30.9.2002)

Xiao, Pei-Gen, 1991. The Chinese approach to medicinal plants. Their utilization and conservation. In Akerele, O., V. Heywood & H. Synge, eds., *Conservation of medicinal plants.* – pp. 305–313, Cambridge, UK, University Press.

APPENDIX 1

TABLE 7

Wild harvesting
versus cultivation of
medicinal and
aromatic plants:
A summary of

▲ advantages

▼ disadvantages

FOR **SPECIES AND ECOSYSTEMS** IT IS BETTER TO...

WILD HARVEST BECAUSE...

▲ it puts wild plant populations in the continuing interest of local people

▲ it provides an incentive to protect and maintain wild populations and their habitats and the genetic diversity of MAP populations

but...

▼ uncontrolled harvesting may lead to the extinction of ecotypes and even species

▼ common access to the resource makes it difficult to adhere to quotas and the precautionary principle

▼ in most cases knowledge about the biology of the resource is poor and the annual sustained yields are not known

▼ in most cases resource inventories and accompanying management plans do not exist

CULTIVATE BECAUSE...

▲ it relieves harvesting pressure on very rare and slow-growing species which are most susceptible to threat

but...

▼ devaluates wild plant resources and their habitats economically and reduces incentive to conserve ecosystems

▼ narrows genetic diversity of gene pool of the resource because wild relatives of cultivated species become neglected

▼ it may lead to conversion of habitat for cultivation

▼ cultivated species may become invasive and have negative impacts on ecosystem

▼ reintroducing plants can lead to genetic pollution of wild populations

THE **MARKET** DEMANDS...

WILD HARVESTED PLANTS BECAUSE...

▲ it is cheaper since it does not require infrastructure and investment

▲ many species are only required in small quantities that do not make cultivation economically viable

▲ for some plant parts extra-large cultivation areas are required (e.g. Arnica production for flowers)

▲ successful cultivation techniques do not exist, e.g. for slow-growing, habitat-specific taxa

▲ no pesticides are used

▲ it is often believed that wild plants are more powerful

but...

▼ there is a risk of adulterations

▼ there is a risk of contaminations through non-hygienic harvest or post-harvest conditions

CULTIVATED MATERIAL BECAUSE...

▲ it guarantees continuing supply of raw material

▲ it makes reliable botanical identification possible

▲ genotypes can be standardized or improved

▲ quality standards are easy to maintain

▲ controlled post-harvest handling is possible

▲ production volume and price can be agreed for longer periods

▲ resource price is relatively stable over time

▲ certification as organic production is possible

but...

▼ it is more expensive than wild harvest

▼ it needs substantial investment before and during production

FROM A PERSPECTIVE OF THE **PEOPLE** IT IS BETTER TO...

WILD HARVEST BECAUSE...

▲ it provides access to cash income without prior investment

▲ it provides herbal medicines for health care needs

▲ it maintains the resources for rural populations on a long-term basis (if done sustainably)

but ...

▼ unclear land rights create ownership problems

▼ this income and health care resource is becoming scarce through over-harvesting

CULTIVATE BECAUSE...

▲ it secures steady supply of herbal medicines (home gardens)

▲ it provides in-country value-adding

but ...

▼ capital investment for small farmers is high

▼ competition from large-scale production puts pressure on small farmers and on wild harvesters

▼ benefits are made elsewhere and traditional resource users have no benefit return (IPR)

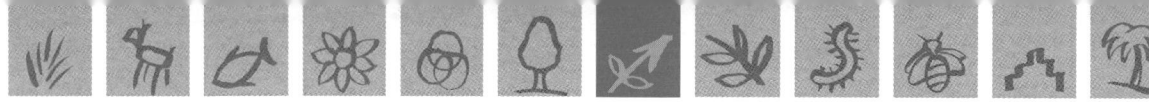

TABLE 8

Selected plant families characterized by high numbers of species used for medicinal purposes showing the number and proportion of threatened species. Data on proportion of threatened species per family according to the IUCN criteria from Walter and Gillett (1998)

FAMILY	NO. GENERA	NO. SPECIES	% OF TOTAL SPECIES THREATENED	MAIN USES	EXAMPLES OF OVERHARVESTED SPECIES
Stangeriaceae	1	1	**100**	Traditional medicine, symbolic	*Stangeria eriopus*
Zamiaceae	8	144	**90.3**	Horticultural collection & traditional medicine	*Encephalartos* species (56 listed as threatened)
Canellaceae	6	20	**35**	Traditional medicine, Molluscides	*Warburgia elongata. W. salutaris*
Leguminosae	590	12 000 - 14 200	**18**	Multiple uses: timber, medicinal, forage & food	*Dalbergia odorifera, D. tonkinensis; Afzelia* species
Araliaceae	47–70	700	**16.3**	Medicinal & carving medicinals	*Panax* species
Rosaceae	100	3 000	**14.4**	Stone fruit crops & medicines	*Prunus africana*
Guttiferae	50	1 200	**13.3**	Dyes, medicines, fruits, chewing sticks	West African *Garcinia* species over-exploited for chewing sticks
Lauraceae	35–50	2 000	**13**	Timber, medicines, cinnamon	*Ocotea bullata*
Menispermaceae	70	400	**9.5**	Medicines, dyes	*Stephania* (several species in SE Asia)
Apocynaceae	168-200	2 000	**7.5**	Medicines	*Holarrhena floribunda*

APPENDIX 2

How to implement sustainability: The role of CITES

The principal tool for monitoring or restricting trade in species threatened by over-exploitation is the **Convention on International Trade of Endangered Species of Wild Fauna and Flora,** or CITES, which entered into force in 1975. The 158 national governments that currently have signed CITES are obliged to monitor and control international trade in the plants and animals listed in its two main Appendices.

Appendix I prohibits trade in wild specimens, except for reasons such as scientific research. Appendix II requires parties to issue export permits that confirm non-detrimental harvest of listed species, and requires importing countries that are parties to CITES to check and monitor permits on incoming material. It is important to note that for Appendix II species it is solely the country of export that decides whether to issue a permit or not.

Having become parties to CITES, national governments are required to establish or designate scientific authorities to conduct non-detriment studies for listed species, and management authorities to issue permits and certificates.

Species can be added, removed, or shifted between Appendices through proposals passed at biennial meetings of the signatories, or Conferences of the Parties. Seventeen species have been added to CITES Appendices because of their exploitation as medicinal plants, the majority of them in the 1990s (Table 9).

TABLE 9

List of plant species which have been included in the CITES Appendices because of trade for medicinal purposes

SPECIES	FAMILY	DATE OF INCLUSION IN *CITES*	APPENDIX
Adonis vernalis	Ranunculaceae	16.08.2000	II
Aquilaria malaccensis	Thymelaeaceae	16.02.1995	II
Cistanche deserticola	Orobanchaceae	16.08.2000	II
Dioscorea deltoidea	Dioscoreaceae	01.07.1975	II
Guaiacum officinale	Zygophyllaceae	11.06.1992	II
Guaiacum sanctum	Zygophyllaceae	01.07.1975	II
Hydrastis canadensis	Ranunculaceae	18.09.1997	II
Nardostachys grandiflora	Valerianaceae	18.09.1997	II
Panax ginseng only populations of the Russian Federation	Araliaceae	16.8.2000	II
Panax quinquefolius	Araliaceae	01.07.1975	II
Picrorhiza kurrooa	Scrophulariaceae	18.09.1997	II
Podophyllum hexandrum	Berberidaceae	18.01.1990	II
Prunus africana	Rosaceae	16.02.1995	II
Pterocarpus santalinus	Leguminosae	16.02.1995	II
Rauvolfia serpentina	Apocynaceae	18.01.1990	II
Saussurea costus	Asteraceae (Compositae)	01.07.1975 App. II 01.08.1985 App. I	I
Taxus wallichiana	Taxaceae	16.02.1995	II

No. 7

For medicinal plant conservation, CITES accomplishes a number of tasks very well, including: (i) monitoring trade at the species level; (ii) focusing attention on high use, high priority species with global value; and (iii) calling international attention to threatened medicinal plant species.

As a conservation tool, CITES also has a number of limitations: (i) Many countries are reluctant to support inclusion of important commercial species in CITES, even when there is justification for restricting or monitoring trade, for fear of losing needed international exchange. (ii) Internal trade is not monitored. (iii) A substantial amount of international trade is not monitored because exported material, such as dried bark and extracts, can be difficult to tie to particular species. (iv) CITES focuses on species that are already threatened, rather than preventing the threat.

APPENDIX 3

Global Plant Conservation Strategy of the Convention on Biological Diversity

In April 2002, the CBD adopted the Global Strategy for Plant Conservation, including 14 outcome-oriented global targets for 2010. The ultimate and long-term objective of the strategy is to halt the current and continuing loss of plant diversity.

Policies relevant to the conservation challenges that arise from the increasing global demand for wild harvesting and cultivation of medicinal plants have been scattered over many different areas: forestry, health, agriculture, indigenous knowledge, access and benefit-sharing, and sustainable livelihoods. The Global Strategy for Plant Conservation provides a policy environment that is particularly well suited to addressing these challenges in a coherent way for medicinal and aromatic plant species.

Targets for the year 2010

Understanding and Documenting Plant Diversity

1. A widely accessible working list of known plant species, as a step towards a complete world flora;
2. An assessment of the conservation status of all known plant species;
3. An understanding of basic conservation needs for threatened plant species, with conservation protocols developed for 50 percent of such species;

Conserving Plant Diversity

4. 10 percent of each of the world's ecological regions and 50 percent of the world's threatened species effectively conserved *in situ*;
5. 90 percent of threatened plant species in accessible *ex situ* collections, and 20 percent of them included in recovery programmes;
6. 30 percent of production lands managed consistent with the conservation of plant diversity;
7. 70 percent of the genetic diversity of crops and other major socio-economically valuable plant species conserved;
8. Threats to plant diversity from invasive alien species tackled;

Using Plant Diversity Sustainably

9. No species of wild flora subject to unsustainable exploitation because of international trade;
10. 30 percent of plant-based products derived

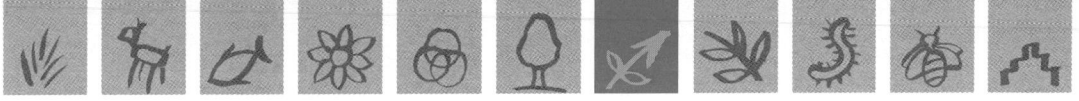

from sources that are sustainably managed;

11. The decline of plant resources that support sustainable livelihoods, local food security and health care, reversed;

Promoting Education and Awareness about Plant Diversity

12. Every child aware of the importance of plant diversity and the need for its conservation;

Building Capacity for the Conservation of Plant Diversity

13. The number of trained people working with adequate facilities in plant conservation and related activities doubled;

14. Networks for plant conservation activities established or strengthened at international, regional and national levels;

APPENDIX 4

Role of the IUCN/SSC Medicinal Plant Specialist Group

The Medicinal Plant Specialist Group (MPSG) is a global voluntary network of experts contributing within their own institutions and in their own regions to the conservation and sustainable use of medicinal plants. The MPSG was founded in 1994, under the auspices of the Species Survival Commission of the IUCN – the World Conservation Union-to increase global awareness of conservation threats to medicinal plants and to promote conservation action. Its

membership is made up of scientists, field researchers, government officials and conservation leaders.

Goal and objectives

The overall aim is to support and promote efforts leading to medicinal plant conservation and rational, sustainable use. The approach is to provide information, tools and strategy coordination that builds on the efforts of local, national, regional, and global partners to conserve and use medicinal plants sustainably, focusing particularly on actions that reduce threats to endangered species and habitats.

The programme has, among others, the following objectives:

- To identify priority medicinal plant taxa and habitats threatened by unsustainable harvesting, high levels of trade, environmental degradation, and other factors contributing to loss of species and genetic diversity;

- To work with local, regional, national, and global partners to design and implement conservation action plans for priority medicinal plant taxa and habitats;

- To support and encourage the sharing of information and collaboration among all stakeholders in finding common solutions to the sustainable use and conservation of medicinal plants;

- To provide opportunities for consumers, industry and other beneficiaries to understand and participate more directly in conservation and sustainable use of medicinal plants and their habitats.

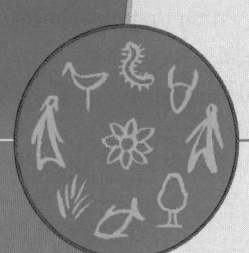

RESPONSIBLE
TECHNICAL DIVISION

Regional Office for Europe
Sustainable Development Department Group

Rainer Krell

IMPACT OF CULTIVATION AND GATHERING OF MEDICINAL PLANTS ON BIODIVERSITY:

CASE STUDIES FROM INDIA

8

IMPACT OF CULTIVATION AND GATHERING OF MEDICINAL PLANTS ON BIODIVERSITY:
CASE STUDIES FROM INDIA

AUTHORS

Kamalappa Ramakrishnappa

Biotechnology Centre Bangalore, India

jdhhulimavu@vsnl.net

CONTENTS

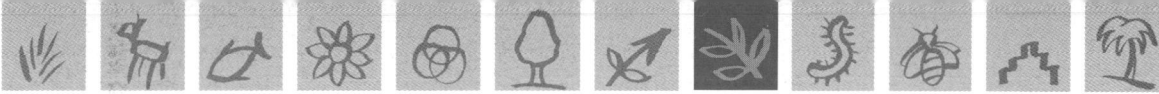

INTRODUCTION

Biodiversity encompasses all biological entities occurring as an interacting system in a habitat or ecosystem and plants constitute a very important segment of such biological systems. Biodiversity of plants collectively known as "plant genetic resources" is a key component of any agricultural production system, indeed, of any ecosystem, without which natural evolutionary adjustment of the system to the changing environmental and biotic conditions would be impossible. Plant biodiversity is an irreplaceable resource, providing raw materials for introduction, domestication as well as improvement programmes in agriculture and forestry. Conservation and use of genetic diversity for sustainable ecosystem or agro-ecosystem should be continuous to meet food, clothing, shelter and health requirements of India's growing population.

Indian biodiversity

India is a treasure chest of biodiversity which hosts a large variety of plants and has been identified as one of the eight important "Vavilorian" centres of origin and crop diversity. Although its total land area is only 2.4 percent of the total geographical area of the world, the country accounts for eight percent of the total global biodiversity with an estimated 49 000 species of plants of which 4 900 are endemic (Kumar and Asija, 2000). The ecosystems of the Himalayas, the Khasi and Mizo hills of northeastern India, the Vindhya and Satpura ranges of northern peninsular India, and the Western Ghats contain nearly 90 percent of the country's higher plant species and are

therefore of special importance to traditional medicine. Although, a good proportion of species of Medicinal Plants (MP) do occur throughout the country, peninsular Indian forests and the Western Ghats are highly significant with respect to varietal richness (Parrota, 2001).

Peninsular India extending downwards from Gujarath, Madhya Pradesh and Southern Bihar was once dominated by a continuum of tropical forests, namely: thorn forests, dry deciduous forests, moist deciduous forests, dry evergreen forests, wet evergreen forests and semi-evergreen forests. The complexity with respect to soils, topography and climate has created an exceptional variety of bio-mass and specialized habitats within this region. The ecosystems of southern peninsular India including the southern Western Ghats contain more than 6 000 species of higher plants including an estimated 2 000 endemic species. Of these, 2 500 species representing over 1 000 genera and 250 families have been used in Indian systems of medicine (Jain, 1991).

Medicinal plants

Medicinal plants which constitute a segment of the flora provide raw material for use in all the indigenous systems of medicine in India namely Ayurveda, Unani, Siddha and Tibetan Medicine. According to the World Health Organization (WHO), 80 percent of the population in developing countries relies on traditional medicine, mostly in the form of plant drugs for their health care needs. Additionally, modern medicines contain plant derivatives to the extent of about 25 percent.

FIGURE 1

*The vegetation map
of India indicating
study sites*

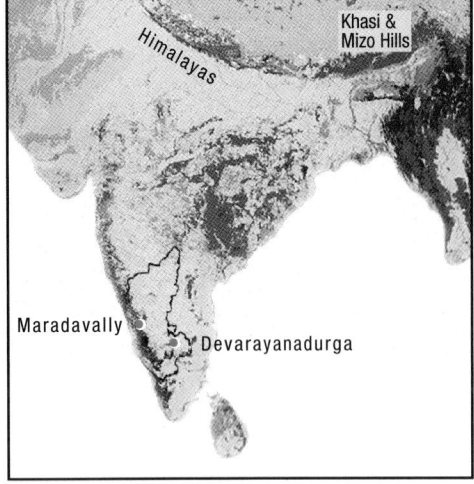

On account of the fact that the derivatives of medicinal plants are non-narcotic having no side-effects, the demand for these plants is on the increase in both developing and developed countries. There are estimated to be around 25 000 effective plant based formulations available in Indian medicine. Over 1.5 million practitioners of the Indian system of medicine in the oral and codified streams use medicinal plants in preventive, promotional and curative applications. It is estimated that there are over 7 800 medicinal drug manufacturing units in India, which consume about 2 000 tonnes of herbs annually (Singh, 2001). According to Exim Bank, the international market for medicinal plant-related trade is to the tune of US$ 60 billion having a growth rate of seven percent per annum. The annual export of medicinal plants from India is valued at Rs. 1 200 million.

Collection

Currently more than 75 percent of the herbal requirement is met through wild collections. While the demand for medicinal plants is increasing, their survival in their natural habitat is under growing threat. Species like *Rauvolfia serpentina, Terminalia*

chebula, Sapindus laurifolius, Jatropha curcas are becoming uncommon in the Western Ghat forests (Anonymous, 2001). Collection of herbs from the wild by destructive harvesting followed by unscientific handling have resulted in poor quality products.

Cultivation

Cultivation of medicinal plants in a grower's field is a recent phenomenon. Industry prefers raw material from cultivated sources because of authentication, reliability and continuity. Nonavailability of quality planting material coupled with poor development and extension support in the cultivation and processing and also unorganized markets are the major constraints coming in the way of commercialization of cultivation. Therefore, concentrated efforts are required, both in collection and cultivation of medicinal plants, in order to ensure sustainability of the industry.

To elicit the impact of cultivation and gathering on the biodiversity of medicinal plants, two case studies, one at Maradavally Forest Range in the semievergreen forest of Western Ghats and the other at Devarayanadurga state forest in the dry deciduous forest of Deccan Plateau were conducted.

CASE STUDY: MARADAVALLY STATE FOREST

Profile

Much information was elicited from the members of "Saravathy Valayabhivrudhi Sanga" of Maradavally who were involved in group discussions and field visits. The profile of the area is as follows:

VILLAGE	MARADAVALLI
Location	25 km from Sagar Town, Shimoga District in Western Ghats
Geographical Area	38.6 sq. km
Population	466
Elevation	600-700 m MSL
Rock type	Metamorphic
Soil types	Red sandy & lateritic
Rain fall	250-300 cm
Forest area	23.2 sq. km
Forest type	Semi-evergreen (intermediate between evergreen & moist deciduous)

The village is in the catchment of River Saravathy, which has been dammed for the Linganamakki Reservoir. Plantation crops dominate the agricultural land area of 76 ha. The village has a forest area of 232 888 ha which is rich in natural resources and a high hill towards the north from which the backwaters of Linganamakki Reservoir, Sagar town and Kodachadri hill range are visible as shown in Figure 2.

FIGURE 2

Semi-evergreen forest landscape of Maradavally forest range

Principal observations

These are:

- The forest area on the northern part of the village was covered by dense forests a decade ago, but now due to rapid destruction, only thin forest exists.
- The water requirements of the village and cattle are met by three lakes in the village. They also serve as the water source for plantations and other agricultural fields in the village. There is depletion of water in the lakes which could largely be influenced by the reduction of forested areas.
- Collection of MP is largely through the deployment of landless labourers by traditional middlemen who either serve as contractors to the large pharmaceutical companies or who run small tightly knit family enterprises on their own. This is augmented by collections from local traditional herbal practitioners.
- Cultivation of MPs is only through the government's joint forest management plan and to the extent of 25 hectares. This is a seed initiative which after three years of government. management was recently handed over to the "Saravathy Valayabhivrudhi Sanga". This organization will allow the community to share in the benefits though not in the ownership of the forest area.

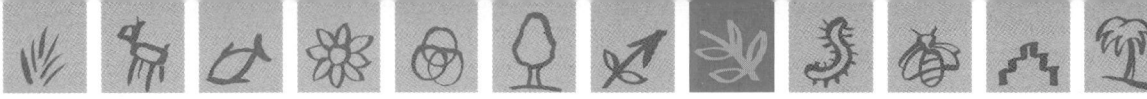

Socio-economic conditions

The village has a population of 310 adults and 166 children with 64 being under the age of six. The overall literacy of the village is 60 percent and female literacy is low. The major traditional community groups own agricultural lands and plantations. Carpenters and agricultural labourers belong to backward communities and are often landless.

As landless labourers are principally deployed by contractors for collection of MP, there is relatively low awareness of the role of biodiversity conservation.

Medicinal plant wealth

The villagers have listed 147 species of medicinal plants in Maradavally State Forest. Of these, 14 are listed as endangered. The medicinal plant wealth includes tall trees, shrubs and climbers. The important medicinal trees, shrubs and climbers and their usage as listed by the members of the Saravathi Valayabhivrudhi Sanga are given in Tables 1, 2, and 3 respectively.

CASE STUDY: DEVARAYANADURGA

Profile

The main source of local information accrued from field visits and discussions with the members of the Local Traditional Herbal Practitioners Association. The profile of the study area is as follows:

VILLAGE	URUDUGERE
Location	14 Km from Tumkur, Deccan plateau
Geographical area	61.5 sq. km
Population	1200
Elevation	1266 m MSL
Rock type	Metamorphic, Presence of granite
Soil types	Sandy and red
Rainfall	80-85 cm
Forest Area	42.27 sq. km
Forest type	Dry deciduous & thorny

Principal observations

These are:

- Devrayanadurga is in the midst of a chain of hills running across the eastern part of Tumkur district in Karnataka.
- This forest is a place of origin of many small streams and tributaries. The prominent one is Jayamangali.
- The climate is pleasant for most part of the year, and therefore attracts tourists and naturalists throughout the year.
- Two historically famous temples at the top of the hill attract many devotees across the state round the year.
- There is no apparent cultivation of MP

FIGURE 3

Dry deciduos forest landscape of Devrayanadurga

in this region but there is a recent initiative from the Forest Department to create a Medicinal Plant Conservatory in an area of 20 hectares. At this early stage, there is no involvement of the local community in the management of this conservatory.

● The labour force deployed for collection in this study area tends to be labour employed from outside the area by contractors. Hence, the issue of preservation of biodiversity is confounded by the absence of vested interest.

Socio-economic conditions

Little is known about the village population in this area. Population density is relatively low and it is possible that literacy rates are lower too. Very little data is available in terms of socio-economic conditions.

Medicinal plant wealth

Of the total of 307 species of plants reported from Devarayanadurga forest, 167 plants (54 percent) are found to have medicinal importance and are used locally by the people (Bhat, 2000). Herbs, shrubs and climbers constitute the major category of medicinal plants, with more than 60 percent coming under this group. The rest are trees with varying growth habits. The majority of medicinal plants of this habitat are higher flowering plants with wide medicinal properties. Herbs are abundant in the foothills while climbers, shrubs and middle - size trees are in the lower range valley.

HABIT-WISE DISTRIBUTION OF MEDICINAL PLANTS

● Analysis of habit-wise distribution of medicinal plants in Maradavally and Devarayanadurga forests indicates that the trees constitute nearly 50 percent of the plant population. Herbs form 11 percent, shrubs 30 percent and climbers 10 percent of the population.

● One impact of the distribution is that trees are most vulnerable due to their multiple uses. In the process of harvesting a tree for any one use, e.g. extraction of bark, the remaining value of the tree is eliminated.

● Shrubs, herbs and climbers are probably better suited to cultivation even in small holdings. This phenomenon does not appear to occur in these study areas. The habit-wise distribution of medicinal plants in Maradavally and Devarayanadurga forests is shown in (Figure 4).

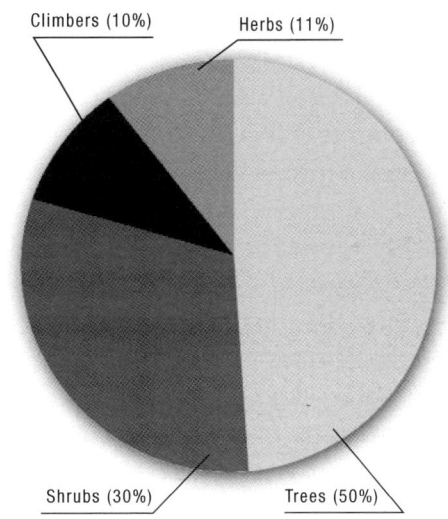

Climbers (10%)   Herbs (11%)
Shrubs (30%)   Trees (50%)

FIGURE 4

Habit-wise distribution of medicinal Plants in Maradavally and Devarayanadurga forests

No. 8

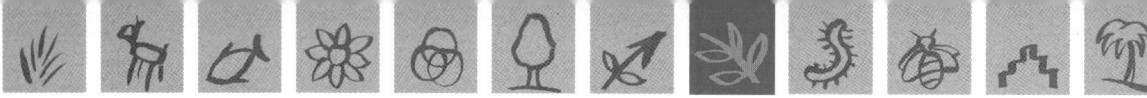

COMPARATIVE ANALYSIS OF CASE STUDY AREAS

Maradavally is a relatively high rainfall area with a generally more protected environment. It also enjoys lower development pressures and lower population pressure. Both areas suffer from a decreasing biodiversity, with generally low levels of awareness of and activism against biodiversity depletion and its consequences, although both have small community groups which seek to ameliorate the situation. These groups collaborated in this study.

The primary differences between the two areas is in the context of the beneficiaries of the collection process. In Maradavally, this benefit, small as it is, accrues to landless labourers. In Devrayanadurga, the local community is not benefited in any apparent way.

In both places, the primary beneficiaries are small contractors who either act as middlemen to large manufacturers or are manufacturers themselves. These groups are possibly conscious of the effect of their actions but tend to be tightly knit and generally inaccessible organizations motivated by profit without being adequately sensitive to the long-term impact of their actions, in their totality. Creating awareness in this segment about the need for efficient rather than destructive harvesting would be an essential first step.

It appears that providing opportunities to produce MPs through cultivation is not necessarily a complete solution. In Maradavally, the MPs resulting from the cultivation process initiated by the Forest Department are sold by auction. However, at least one forest contractor obtains more than 40 product types from the cultivated area although only five products are auctioned by the Forest Department.

Better management of the collection process would require a concerted effort to map the depletion and its effects and awaken local communities to the consequences of such loss. Much of this would be facilitated by the existence of an infrastructure for education and for basic services in the areas themselves so that the young and the fit can be encouraged to develop, study and work within their own communities rather than choose to migrate to urban centres. The quantification of biodiversity benefits and of the effects of their decrease would constitute appropriate first aid. Groups (NGOs) such as those which assisted in these case studies could assist in this effort.

Finally, the development of cultivation as a source for MP would potentially have the effect of ameliorating some of the current habitat destruction and, at the same time, serve to create awareness of the role of MPs in the community. This also requires the development of alternative marketing mechanisms, so that a higher percentage of profits would flow to members of the local community and thus encourage small landowners to become MP suppliers. There is a need for a coherent, transparent and equitable process of generation, processing, marketing and revenue sharing.

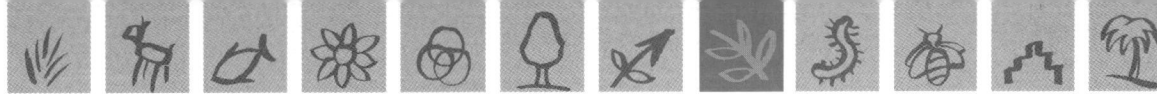

LOSS OF BIODIVERSITY OF MEDICINAL PLANTS

The medicinal plant wealth of both Maradavally State and Devrayanadurga forests is declining constantly over the years. Of the 147 species of medicinal plants reported from Maradavally range forest, fourteen species are listed as endangered of which *Catuneregam spinnosa* is at the verge of extinction. The other endangered species are: *Garcinia cambogea, Acacia pinnata, Ficus benghalensis, Zanthoxyllum rhesta, Hemidesmus indicus, Terminalia chebula, Wrightia zeylanica, Cinnamomum verum, Bombax ceiba, Sapindus laurifolius, Alangium salvifolium* and *Calophyllum inophyllum.*

In Devarayanadurga forest, out of 307 species reported, *Abrus precatorius, Adenanthera paronina, Aegle marmelos, Caesalpinia bonducella, Cardiospermum halicacabum, Corallocarpus epigaeus, Gloriosa superba, Andrographis paniculata* have become uncommon and *Jatropha curcas* is endangered.

Causal factors

The major factors threatening the species and genetic diversity of medicinal plants in Maradavally and Devarayanadurga are similar to those operating elsewhere in India. Many factors both natural and man-made have been responsible for limiting the distribution of medicinal plant species and are causing them to become rare or even extinct.

Environmental factors
Rainfall:

In the forest areas the annual rainfall has decreased to the extent of 40 percent in Maradavally forest and 50 percent in Devarayanadurga forest resulting in the death of many herbaceous species during the summer months.

Deforestation:

Deforestation to an extent of 10 percent in Maradavally and 25-30 percent in Devarayanadurga forests has been reported over the last two decades. The spread of agriculture, logging, firewood collection, heavy grazing, etc. are the main reasons for reduction in area under valuable forest.

Siltation of water bodies:

In Maradavally forests, four tanks have silted up to 30 percent of their capacity and in Devarayanadurga forest, siltation is up to 40 percent. Siltation of water bodies in both the forests has resulted in the reduction of water-holding capacity leading to depletion of underground water (Figure 5).

Lack of pollinators:

Honey bee colonies have declined in numbers to the extent of 60 percent in Maradavally and 90 percent in Devarayanadurga forests. Loss of pollinators has resulted in reduced seed set and dispersal of seeds.

Developmental influences
Submersion:

The Maradavally forest is the catchment of Linganamakki Dam, the main reservoir of Karnataka for irrigation and power generation. Submersion of nearly 10 sq. km of forest area during monsoons has resulted in loss of valuable medicinal plant species.

FIGURE 5

A silted water body in Devarayanadurga state forest

No. 8

Infrastructure:

Expansion of roads of about 14 kms in Devarayanadurga and installation of power lines in Maradavally forests have caused extensive damage to forests and medicinal plants.

Agriculture and forestry methods

Monoculture:

There has been a progressive increase in monoculture plantations of economically important indigenous as well as exotic species in both Maradavally and Devarayanadurga forests. Nearly 5-10 percent in Maradavally and 10-15 percent in Devarayanadurga forests have been planted with Eucalyptus and Acacia species. Monoculture plantation totally affects the organic productivity and reduces the natural stability and complexity resulting in loss of medicinal plants (Figure 7).

Encroachments:

Encroachments over forest lands in Devarayanadurga forests have assumed alarming levels. Apart from felling of trees and clearing vegetation, the cultivation practices followed on high slopy lands have caused soil erosion and decline of medicinal plant wealth. Encroachment is minimal in Maradavally forest (Figure 8).

Over-exploitation:

Gathering of medicinal plants from both the forests of Maradavally and Devarayanadurga is rampant. About 20-25 collectors engaged by a licensed contractor, gather nearly 40 types of medicinal plant products from the forests of Maradavally between November and April, whereas the forest record shows only five types. The approximate quantity of material collected every year would be 20-25 tonnes. In Devarayanadurga forests, collection of medicinal plants by outsiders was severe a few years ago. Every year about 5-10 tonnes of material were gathered by outsiders. The collection was by unorganized forest collectors, who in turn sold the product to a contractor at the price fixed by the latter. But now, due to the awareness created by the members of the "Local Traditional Medicinal Practitioners' Association", illegal gathering is controlled to a certain extent.

FIGURE 6

Destruction of forest due to installation of power lines in Maradavally forest

FIGURE 7

Monoculture of Eucalyptus leads to destruction of biodiversity

FIGURE 8

Encroachment is a common feature in Devrayanadurga forest

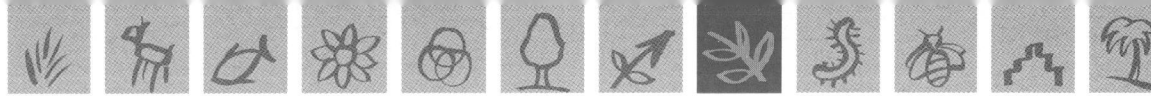

COMPLEXITY OF COLLECTION

Destructive harvesting

It is reported that more than 800 species of plants are being used currently by the industries for large-scale production of herbal products of which less than 20 species are cultivated commercially, that is, more than 95 percent of the medicinal plants used by the Indian industry are collected from the wild (Anonymous, 2001). Mr. Ramesh, head of "Banajalaya", a voluntary organization in Kodluthota in the Maradavally State Forest, established to educate the local people on various aspects of forestry management, reveals that more than 70 percent of the collections from the wild involve destructive harvesting of roots, bark, wood, stems and whole plants (Figure 9). This poses a serious threat to the genetic stock and to the diversity of medicinal plants.

Lack of awareness

Medicinal plants are gathered from the wild by collectors through the tribals, forest dwellers and other local people. The collected material is passed on to the traders in towns and cities. Each one of the major traders has one or more traditional drug manufacturers and private pharmacies as their customers purchasing raw plant material. Generally, as the price paid to the gatherers is very low, the gatherers often mine the plants excessively to generate more income. Greed to earn more, coupled with the ignorance of the collector about plant biology and selective harvesting leads to the whole plant being destroyed. For example, the indiscriminate

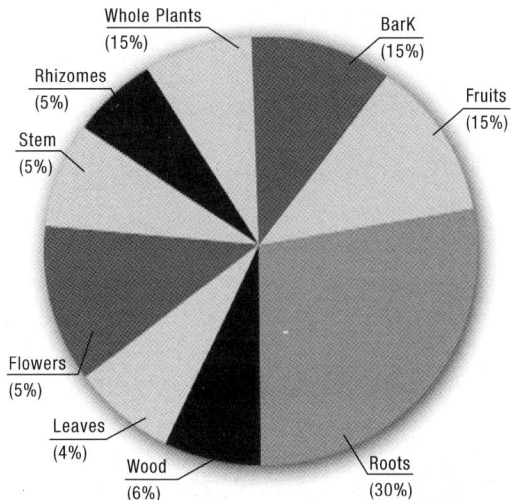

FIGURE 9

Medicinal plants break-up by parts used

removal of bark from *Diospyros montana* in the forests near Kodluthota village in Maradavally SF has resulted in the death of several trees.

Inefficiency of handling

The collected materials are mostly dumped with the traders, who with their limited knowledge sort out the saleable ingredients in a crude manner, resulting in contamination with other materials leading to poor quality. Many of the medicinal plants are sensitive to climatic conditions, requiring proper drying and storage under specific temperatures and humidity. This aspect is largely neglected by the collectors, growers and traders leading to deterioration and rejection of the produce particularly for the international market. For example: the forest collectors gather 10-15 tonnes of *Sapindus laurifolius* fruits every year in Maradavally SF between October and December. The limited sunshine during these months results in improperly dried fruits, leading to inferior quality and rejection, and as a consequence a loss of income and wasted resources.

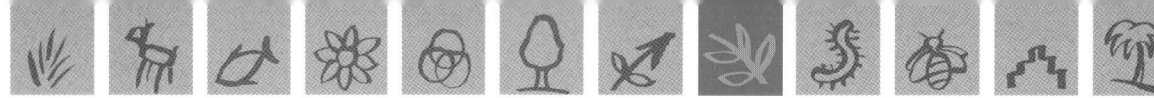

FIGURE 10

*Indiscriminate
removal of bark
resulted in the death
of a valuable tree*

Lack of traceability and certification

In an unorganized collection system the collector who gathers the material from the wild will have no idea about its destination or usage. Similarly, the companies which receive the material also will have no knowledge of its source of origin. There is an example of a company in Bangalore which receives several tonnes of raw materials of *Cassia fistula, Acacia pinnata, Emblica officinalis, Coleus forskohlli, Piper longum,* etc., through their contractors every month. These products are either gathered from the wild or cultivated. No documents are maintained either by the contractors or the company about the source of the material, methods of handling and cultivation practices. For instance, many farmers use copper fungicides indiscriminately against bacterial wilt in cultivation of *Coleus forskohlli.* This is not

known to the company. The complexity of collection is such that it would be very difficult to adopt any strategy of certification on the basis of sources of origin or product quality. Social certification and management certification of the medicinal plant products collected from the wild are also complex and demand a variety of conditions which require additional knowledge and resources and create extra costs.

Transparency and accountability

In the absence of any regulatory mechanism, trade in medicinal plants is very secretive. Stakeholders like collectors, contractors, traders, wholesale dealers and companies involved in collection and processing of medicinal plants do not understand nor trust each other. Similarly in the absence of a proper management system, neither the collector nor the contractor who extracts medicinal plant products from the wild will have any responsibility of replenishing the natural resources through re-planting. For this purpose, the role of local non-political voluntary organizations becomes significant and allows them to participate in collection and maintenance of natural resources. In Maradavally SF, a forest contractor residing in Sagar town collects minor forestry products including medicinal plants through local agents. The contractor obtains a collection contract through bidding in annual auctions for minor forestry products by the Forest Department. The Forest Department, on record, auctions only five minor products, whereas the contractor collects more than 40 types from the forest.

IMPACT ON ECO- AND AGRO-ECOSYSTEMS

Different landforms which exist in an ecosystem support different and specific vegetation. Ecosystem diversity is difficult to measure since the boundaries of the communities, which constitute the various sub-ecosystems, are elusive. Forests perform important ecological functions such as maintaining delicate ecological and hydro-biological balances, conserving soil, controlling floods, drought and pollution. Forests also provide habitats for innumerable plants, animals and micro-organisms. Agro-ecosystems tend to become poorer with more intensive cultivation, but could be a valuable local genetic resource with proper management.

Ecosystem impact

There are examples where the depletion of a few species within a forest has caused a deleterious impact on the whole eco- and agro-ecosystems. *Aloe vera* and *Asparagus racemosus* species, in addition to their medicinal properties, are also good soil binders. Removal of these plants for their underground parts has caused large-scale soil erosion in Maradavally forests.

Bombax ceiba commonly known as "Mahamara" in Maradavally forest area, has been the main shelter tree for honey bee colonies. Over-exploitation of the species for its latex, has resulted in reduction of honey bee populations which are the main pollinating agents in forests as well as in cultivated fields.

Agro-ecosystem impact

Paddy growers of Maradavally village have been using *Hasiosiphon eriocephalus* as a green manure crop which was once available in plenty in the forest area. This species, in addition to its nutritional value, also possesses the insecticidal properties believed to protect rice fields from the attack of insects and pests. Large-scale exploitation of the species by the pesticide manufacturing companies during the last few years has deprived local people of its use.

FIGURE 11

Asparagus is also a good soil binder

IMPACT ON PROVIDERS OF MEDICINAL PLANTS

Biodiversity is inextricably woven into the social fabric of the local people besides providing the sources of livelihood for them. Different kinds of plants form essential requirements of ceremonies and festivals. Loss of biodiversity would lead to fragmentation of the society and the, decline of social and religious practices. The people depending on forest resources for their livelihood will have to find alternative ways of living, resulting in rural poverty, occupational hazards and sometimes migration to industrial areas.

Primary health care

Several traditional herbal practitioners living in rural areas have been serving local people by giving medicines collected from the surrounding forests. This is a family tradition for many practitioners and is traditionally rendered gratis. For example, Narayanamurthy, a traditional medicinal practitioner and an honorary member of "Saravathy Valayabhivrudhi Sanga" in Maradavally forest range provides herbal medicines to a large population of patients every week. Similarly, hundreds of local villagers receive medicine from Mr. Thirumalaiah, a folk-medicine practitioner in Urudugere Village. A majority of the patients who receive herbal medicine from these practitioners are poor villagers who have no financial resources to go to towns and cities for their health care. Moreover, villagers tend to have a strong belief in traditional herbal medicine, and trust in local practitioners. Depletion of medicinal plants in the natural habitat, is liable to destroy this facility and could lead to socio-

economic problems. Urban dwellers who lean towards this system of medicine in the hope of a better cure will also be affected.

Cultural impact

There is a very interesting relationship between biodiversity and cultural diversity in the area of medicinal plants. This relationship is being lost because of the loss of cultural diversity associated with the natural habitat and the various pressures that generally operate on biodiversity.

Medicinal plants have the potential to make the rural people self-reliant in primary health care. In Maradavally village, every family knows the uses of at least five medicinal plants for their immediate health care. More than 50 percent of the families have the knowledge of more than 25 species and their availability in surrounding areas. The depletion of certain species of plants has eliminated certain cultural traditions of the community, for e.g. the leaves of *Cariota urens* were invariably used to decorate the marriage pendals and the juice of *Garcinia cambogia* was a sacred drink in all religious and ceremonial gatherings in the village. Presently both these traditions are no longer found because of the unavailability of these species in the surrounding areas.

Growth of urban demand

As urban demand in India begins to lean towards traditional herbal medicines, possibly partly due to the preference for herbal cosmetic products and the increase in prices of allopathic drugs following World Trade Organization reforms, there is a concomitant need for planned development in the source regions for MP. The current

REFERENCES

Anonymous 1996. Karnataka at a glance. Bureau of Economics and Statistics. Technical report, Government of Karnataka, Bangalore.

Anonymous, 2001. The key role of forestry sector in conserving India's medicinal plants. Technical report, Foundation for Revitalization of Local Health Traditions, Bangalore.

Bhat, H.R., 2000. *Medicinal plants of Devarayanadurga Forests.* Deputy Conservator of Forests, Government of Karnataka.

Jain, S.K., 1991. *Dictionary of Indian Folk Medicine and Ethnobotany.* Deep Publication, New Delhi.

Kumar, V. and Asija, 2000. *Biodiversity Conservation in: Biodiversity-Principles and Conservation.* Agrobiosis (India), Jodhpur.

Parrota, J.A., 2001. *Healing Plants of Peninsular India.* CABI, New York.

Singh, H.P., 2001. National perspective on development of medicinal and aromatic plants. Technical report, Agri Watch.

No. 8

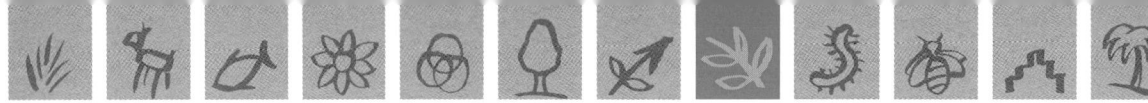

TABLE 1

A list of important medicinal trees and their uses in Maradavally Forest

S.N.	COMMON NAME	SCIENTIFIC NAME	PART USED	MEDICINAL USES
1	Alexandrian laurel (Sura honne)	Calophyllum inophullum	Bark	Juice, ulcers, inflammation, internal hemorrhage, purgative, sore eyes
2	Torchwood tree (Arishina gurige)	Cochlospermum religiosum	Gum	Stomachic, sedative gonorrhea, syphilis, asthma
3	Spanish chesy (Bakula, Ranjalu)	Mimusops elengi	Fruit pulp and Seeds	Chronic dysentery, purgative, chronic constipation
4	Garcinia (Uppage)	Garcinia cambogia	Fruit	Antiseptic, rheumatism
5	Grewia (Dhadasalu/ Bhootale)	Grewia tyliaefolia	Bark	Dysentery, diarrhea cowitch
6	Wild Jack (Hebbalsu)	Artocarpus hisutus	Leaves & Juice	Inflammatory swellings
7	Cinnamom Cinnamomum (Nishini/Dalchini)	Cinnamomum verum	Bark	Stimulant, expectorant, carminative, gastric irritations, nausea
8	Alangium (Ankole)	Alangium salvifolium	Root bark	Antipyretic, diaphoretic, purgative, emetic,snake bites, diarrhea, worms, syphilis, skin diseases
9	Indian Trumphet Flower (Anangi)	Orexyllum indicum	All the parts	Dashamoola, astringent, carminative, diuretic, aphrodisiac, fevers, cough, respiratory disorders
10	Emetic nut (Aramadalu)	Catunaregam spinnosa	Fruit	Nausea, expectorant, diaphoretic, hysteria
11	Black Myrobolan Tree (Alale)	Terminalia chebula	Fruits	Laxative, carminative, digestion, expectorant, antihelmentic, alternative purpose, healing of wounds, ulcers, swellings, diabetes, anemia, cardiac disorders.
12	Arjun tree (Matti)	Terminalia arjuna	Bark	Alexiteric, styptic, antidysentric, cardiac diseases, blood diseases, fever, fractures, obesity, skin diseases
13	Bellaric Myrobolan (Thare)	Terminalia bellarica	Fruits	Laxative, antihelmentic, bronchitis, sore throat, asthma, opthalmia.
14	Matchstick tree (Maddale)	Alstonia scholaris	Bark	Anti choleric, vulnerary properties, ailments to relieve sprains, bruises and dislocated joints
15	Kokam butler tree (Murugana/Dhupada mara)	Garcinia indica	Fruits	Anthelmentic, cardiotonic, piles, dysentery, tumors, heart ailments
16	Silk cotton tree (Bhuruga)	Bombax Ceiba	Flowers, Leaves, Fruits	Astringent, cooling, relieves swellings, skin troubles. expectorant, stimulant, diuretic, ulcer of kidneys.
17	Indian laburnum (Golden shower/kakke)	Cassia fistula	Dried Pods, pulp Bark	Laxative, antiviral, purgative, disorders of liver and biliousness
18	Indian Sago Palm (Bhagini)	Caryota urens	Nuts	Hemicrania, fatigue, laxative.
19	Fruit Root Bark Oil	Zanthoxyllum rhesta	Fruit Root Bark Oil	Astringent, stomachic, dyspepsia, rheumatism, diarrhea purgative, of kidneys.
20	Indian Ell Jungle (Thapasi)	Holoptelea integrifolia	Bark leaves	Edema, diabetes, leprosy, skin diseases,intestinal disorders, piles.

S.N.	COMMON NAME	SCIENTIFIC NAME	PART USED	MEDICINAL USES
1	Indian Gooseberry (Nelli)	*Phyllanthus embelica*	Fruit	Acidic, acrid, astringent diuretic, laxative,rich in vitamin'c', leprosy, piles, anemia, triphala
2	Wild Jasmine (Kare)	*Canthium parvifolium*	Leaves, Fruit	Astringent against cough and indigestion,anti-spasmodic
3	Solid Bamboo (Bidiru)	*Dendrocalamus strictus*	Siliceous matter near joints, Juice	Cooling, astringent, healing of cuts, eardrops
4	Embelia (Vayuvilanga)	*Embelia ribes*	Fruit, Seeds	Vermicide, antispasmodic, carminative, anthelmenthic, stomachic, skin diseases, oedema, rheumatism
5	Camel's Foot Climber (Parige)	*Banhinia Vahlii*	Seeds Leaves	Aphrodisiac, demulcent, mucilageneous
6	Elephant Apple (Ganagalu)	*Delania indica*	Fruit	Tonic, mild laxative, abdominal pains, fever, cough mixture
7	Christ's Thorn (Kavali)	*Carrissa Carandus*	Roots, Leaves, Fruits	Stomachic, antihelmentic, cardiotonic activity.remittent fevers,antiscorbutic, pickles and beverages
8	Queen of Flowers (Holedasavala)	*Laegestroemia speciosa*	Leaves, Roots, Bark	Purgative, deobstruent, diuretic astringent, stimulant febrifuge
9	White Emetic Nut (Bikke)	*Gardenia resinifera*	Leaf, buds, Young shoots	Cumbi gum-antispasmodic, expectorant, diaphoretic, carminative, antihelmentic.relieves constipation, pain, treats worms
10	Gantubharangi	*Clerodendrum serrata*	Roots	Respiratory diseases, antispasmodic, carminative, expectorant, epilepsy, intermittent fevers, asthma, dropsy, mental

TABLE 2

A list of important medicinal shrubs and their uses in Maradavally Forest

No. 8

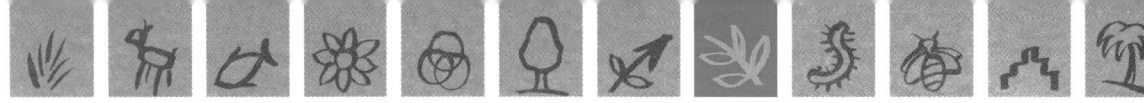

TABLE 3

This table identifies important medicinal climbers and their uses in Maradavally Forest

S.N.	COMMON NAME	SCIENTIFIC NAME	PART USED	MEDICINAL USES
1	Amrutaballi	*Tinospora cordifolia*	Mature stem	Tonic and stomachic, fever, jaundice, burning sensations, diabetes, piles, skin ailments, respiratory disorders, neural diseases, urinary diseases
2	Chariot tree (Hotta balli)	*Desmodium latifolium*	Roots	Infant's digestion, diarrhea, cardiotonic, fever, dysentery
3	Purple Convolvuluso (Kalluballi)	*Argyreia cuneata*	Root	Arthritis
4	Vishnukanti	*Evolvulus alcinoides*	Whole plant	Alterative, antiflogistic, brain tonic, nervous debility, memory loss, fever, diarrhea and indigestion
5	Indian liqourice (Gulagangiballi)	*Abrus precatorius*	Roots, Leaves, Seeds	Cough, catarrhal affections, gonohorrea, jaundice, haemoglobnuric bile, peptic ulcers
6	Australian cow (Madhunashini)	*Gymnema sylvestre*	Whole Plant	Antiperiodic, diuretic, stomachic, headache, diabetes, leprosy, pruritis, polyurea
7	Indian birthwort (Eshwariballi)	*Aristolochia indica*	Dried roots, Leaves	Snakebite, gastric stimulant dyspepsia, nausea, bowel troubles
8	Shathavari	*Asperagus racemosus*	Roots	Dysentery, tumors, inflammation, blood diseases, biliousness, rheumatism, neural disease, kidney & liver disorders
9	Emetic Swallow Wort (Adumuttadaballi)	*Tylophora indica*	Roots	Expectorant, chronic bronchitis, whooping cough, stimulant, antirheumatic, alterative, jaundice
10	Black Oil plant (Jyothiskmathi)	*Celastrus paniculatus*	Seeds	Emetic, diaphoretic, nervous and febrifugal for memory loss, cures sores, ulcers, rheumatism and gout
11	Indian Sarasaparilla (Sogadeberu)	*Hemidesmus indicus*	Dried Roots	Alterative, demulcent diaphoretic, diuretic, and blood purifier, bowel complaints, elephantiasis

S.N.	COMMON NAME	SCIENTIFIC NAME	PART USED	MEDICINAL USES
1	Yellow thistle (Daturi)	*Argemone mexicana*	Milky juice of plant, Side shoots	Diuretic, hypnotic, anodyne, malarial fevers, jaundice, leprosy, skin diseases, laxative, emetic, demulcent
2	Hogweed (Balavadike)	*Boerhaavia diffusa*	Whole plant	Stomachic, laxative, diuretic, emetic, edema, anemia, heart diseases, kidney stones, rheumatism
3	Balloom vine (Agni Balli)	*Cardiospermum halicacabum*	Roots, Leaves, Seeds	Rheumatism, nervous diseases, piles, bronchitis, snake bite
4	(Bitharrige)	*Ceropegia tuberosa*	Roots	Tonic, increasing digestive power
5	Velvet leaf (Padvali)	*Cissampelos pereria*	Root, Bark, Leaves	Fever, diarrhea, dysentery, dropsy, nepheritis
6	Kuntigenagida	*Daemia extensa*	Leaves, Roots, Root Bark.	Anthelmenthic, asthma, snake bite, rheumatic swelling
7	Figi Yam (Kadukumbala)	*Diascoria pentaphylla*	Tubers	Swellings, rheumatism
8	Emilia	*Emilia sonchifolia*	Whole plant	Sudorific, to relieve cuts and wounds, sore eyes, sore ears, headache
9	Indian Sarasaparilla (Sogadeberu)	*Hemidesmus indicus*	Dried roots, Root bark, Juice	Diuretic, demulcent, alterative, dyspepsia, loss of appetite, skin diseases, rheumatism, leucorrhea
10	Phyllanthus (Kiru nelli)	*Phyllanthus amaras*	Whole plant	Astringent, doobstruent, stomachic, carminative, febrifugal, antiseptic, jaundice, gastro intestinal, ailments

TABLE 4

List of important herbs and their medicinal uses of Devarayanadurga Range Forest

No. 8

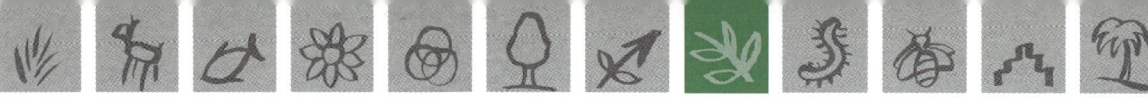

List of important
shrubs and their
medicinal uses of
Devarayanadurga
Range Forest

S.N.	COMMON NAME	SCIENTIFIC NAME	PART USED	MEDICINAL USES
1	Indian Jamaica (Gulaganji)	*Abrus precatorius*	Root, Seeds, Leaves	Antiphlogistic, aphrodisiac, antiopthalmic, painful swellings and paralysis
2	King of Bitters (Kalmegh)	*Andrographis paniculata*	Whole plant	Tonic, syphilitic, ulcers, intermittent fevers, stomachic, antipyretic, antihelmentic, properties
3	(Indian birthwort) Eshwariballi	*Aristolochia indica*	Root, Stem, Leaves	Antidote to snake-bite and poisonous insects, leprosy, dropsy, bowel complaints
4	Cambiresin (Bikke)	*Gardenia gummifera*	Exudes of fruits	Fever, flatulence, dyspepsia, nervous disorders, toothache, ulcers, roundworm infections
5	East Indian Screw tree (Bhootha karalu)	*Helicteres isora*	Fruits, Root, Bark, Juice, Seeds	Intestinal problems like colic, flatulence diarrhea, dysentery, diabetes, snakebite
6	Ceylon leadwort (Chithramoola)	*Plumbago zeylanica*	Root	Leprosy, skin diseases, scabies, ulcers, piles, leucoderrma, intermittent fevers

T A B L E   6

List of important
climbers and their
medicinal uses of
Devarayanadurga
Range Forest

S.N.	COMMON NAME	SCIENTIFIC NAME	PART USED	MEDICINAL USES
1	Molucca Bean (Gajjuga)	*Caesalpinia bonducella*	Seeds, Root, Bark, Leaves	Intermittent, fevers, asthma, gumboils, hydrocele, swellings, leprosy and rheumatism, antiperiodic, antispasmodic, bitter tonic and antihelminthic properties
1	Bittarige	*Ceropegia tuberosa*	Root	Increases digestive power, used as tonic
3	Velvet leaf (Paduvali)	*Cessampelos periria*		Fever, diarrhea, dysentery, dropsy, nepheritis

S.N.	COMMON NAME	SCIENTIFIC NAME	PART USED	MEDICINAL USES
1	Haldu (Aarishinatega)	Adina cordifolia	Stem bark	Febrifuge, antiseptic properties, malarial fever, stomach disorders
2	Bael fruit (Bilpathre)	Aegle marmelos	Fruit, Root Bark, Leaves, Flowers	Constipation, chronic, dysentery, dyspepsia inflammation, vomiting, gonorrhea
3	(Tree of heaven) (Hiremara/ Doddabevu)	Ailanthus exelsa	Bark, Leaves	Bronchitis, asthma, dyspepsia
4	East Indian Walnut (Doddabage)	Albizzia lebbeck	Root, Leaves, Bark	Snake-bite, scorpion sting, skin diseases, leucoderma, asthma, blood diseases
5	Button tree (Bejjalu/ Dindiga)	Anogeissus latifolia	Gum Bark	Astringent, scorpion sting, snake-bites, chronic disrrohea, leprosy, anemia, piles, polyuria
6	Indian laburnum (Kakke)	Cassia fistula	Pulp, Root Bark, Flowers, Pods, Leaves	Laxative, Gout, Rheumatism, snake-bite, fever, heart diseases, purgative febrifuge, ring worm
7	Indian Gooseberry (Amla/ Bettadanelli)	Embelica officinalis	Fruits, Leaves, Roots, Bark, Flowers	Hemorrhage, diarrhea, dysentery, stomach ache, vermifuge, painful respiration
8	Wood apple (Bheladahannu)	Ferronia elephantum	Fruit, Gum, Leaves, Bark, Pulp	Sore throat, hiccups, dyspepsia, diarrhea, dysentery
9	White teak (Shivani/ Bachanigemara)	Gmelinia arborea	Root, Bark, Fruit, Leaves	Demulcent, stomachic, laxative, snakebites, scorpion stings, gonorrhea fever, indigestion, headache
10	Karihuli	Kirganelia reticulata	Leaves	Skin diseases, rashes, diuretic, diseases of blood, syphilitic source
11	Maddimara	Morinda tomentosa	Root, Leaves, Fruits	Astringent, cathartic, diarrhea, dysentery wounds and ulcers, spongy gums
12	Soapnut tree Antvala	Sapindus laurifolius	Fruits, Seeds, Leaves	Purgative, colic, snake-bite, gout diarrhea, cholera, epilepsy
13	Belleric myrobolan Thare	Terminalia bellarica	Fruits	Astringent, tonic, laxative, cough, eye diseases, scorpion sting, sore throat, piles, fever, diarrhea
14	Sweet indrajao (Aate/Beppale)	Wrightia tinctoria	Leaves, Bark, Seeds, Fruit	Astringent, stomachic, tonic, febrifuge, stomach pain, bowel complaints, fever

TABLE 7

List of important trees and their medicinal uses of Devarayanadurga Range Forest

No. 8

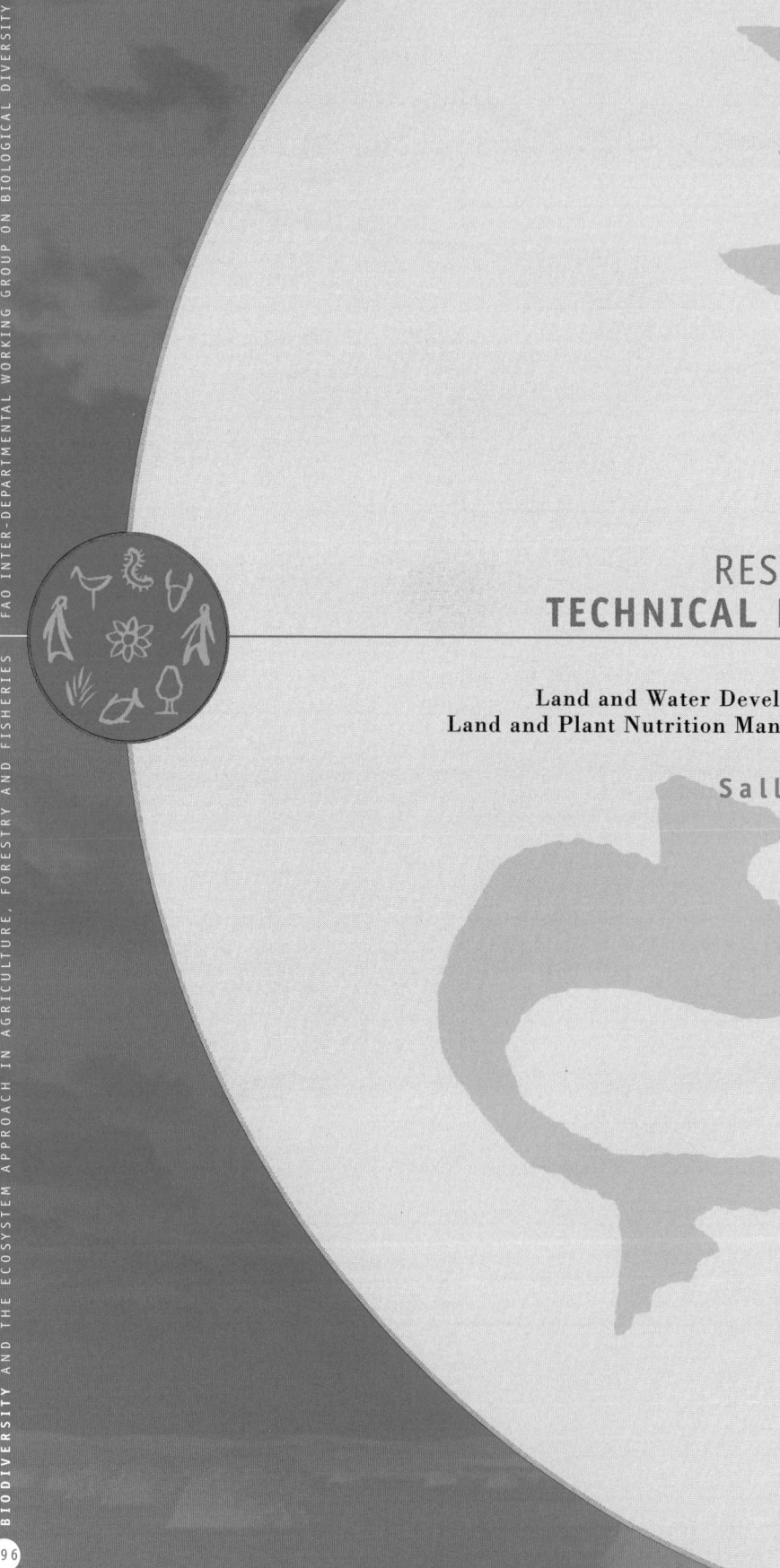

RESPONSIBLE
TECHNICAL DIVISION

Land and Water Development Division
Land and Plant Nutrition Management Service

Sally Bunning

SOIL BIODIVERSITY MANAGEMENT FOR SUSTAINABLE AND PRODUCTIVE AGRICULTURE:

LESSONS FROM CASE STUDIES

9

SOIL BIODIVERSITY MANAGEMENT FOR SUSTAINABLE AND PRODUCTIVE AGRICULTURE:
LESSONS FROM CASE STUDIES

AUTHORS

Dan Bennack

Instituto de Ecología, Xalapa, Mexico
bennack@ecologia.edu.mx

George Brown

Embrapa-Soja, Londrina – PR, Brazil
browng@cnpso.embrapa.br

Sally Bunning

FAO Land and
Water Development Division
sally.bunning@fao.org

Mariangela Hungria da Cunha

Embrapa-Soja, Londrina – PR, Brazil
hungria@cnpso.embrapa.br

ACKNOWLEDGEMENTS

Particular acknowledgements
are gratefully expressed to all
the original authors as indicated
in each case study.

CONTENTS

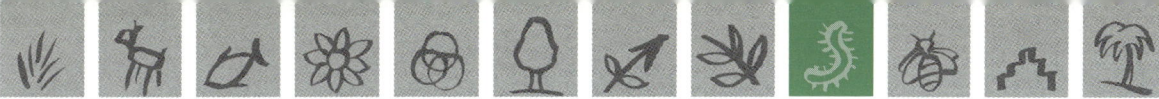

SOIL BIOLOGICAL MANAGEMENT FOR SUSTAINABLE AND PRODUCTIVE AGRICULTURE

International recognition

The third session of the Conference of the Parties (COP) to the Convention on Biological Diversity (CBD), in its development of the programme of work on Agricultural Biodiversity, identified the study of soil micro-organisms as a gap requiring attention (COP decision III/11). Subsequent compilation of case studies and recognition by the Subsidiary Body for Scientific, Technical and Technological Advise (SBSTTA) of the importance of soil biodiversity in the functioning of agricultural ecosystems led to the following decision:

"to establish an International Initiative for the Conservation and Sustainable Use of Soil Biodiversity as a cross-cutting initiative within the programme of work on agricultural biodiversity, taking into account case studies which may cover the full range of ecosystem services provided by soil biodiversity and associated socio-economic factors and, inviting FAO, and other relevant organizations, to facilitate and co-ordinate this initiative" (COP decision VI/5, paragraph 13, Nairobi April 2002).

In particular, FAO recognizes the importance of soil health and improved soil biological management for promoting sustainable agricultural systems and for the restoration of degraded lands. Soil organic matter management and enhanced biological nitrogen fixation (BNF) are already well-known practices in the agricultural and environmental sectors. However, the capacity to enhance soil biological functions through a better understanding of soil biodiversity processes and mechanisms and improved land use systems and practices have been seriously neglected.

Today's knowledge in this area is, however, fragmented and remains largely in the research domain with limited practical application by farmers. Various reasons include difficulty of observation and limited local understanding of below-ground interactions and processes, specialized research focus and lack of holistic or integrated solutions for specific farming systems, and lack of or inadequate institutional capacity or support services for a concerted resource management approach.

FAO is taking an active role in following up on the above decisions through networking with partners and institutions, collecting and initiating case studies and identifying priorities requiring attention. An important step in this process was the "international technical workshop on biological management of soil ecosystems for sustainable agriculture" organized jointly by EMBRAPA-Soya and FAO in Londrina, Brazil, in June 2002. The review and analysis of case studies from different countries and agro-ecological zones is found to be a useful means of sharing experiences and encouraging collaborative actions. Capacity building is also required, in particular, in the areas of assessment and monitoring and adaptive management for specific agro-ecological and socio-economic contexts, with a view of achieving food security and environmental benefits.

It is well known that farmers' management practices and land use decisions influence ecological processes and soil-water-plant interactions. However, farmers' decisions are often made to achieve short-term goals rather than long-term

No. 9

management of soil productivity and health. Unsustainable land use practices and agricultural intensification are significant causes of soil biodiversity loss and related impacts on ecosystem function and resilience. A better understanding of the linkages among soil life and ecosystem function and the impact of human intervention will allow us, not only to reduce the negative impacts, but also to more effectively capture the benefits of soil biological activity for sustainable and productive agriculture.

Given escalating population growth, land degradation and increasing demands for food, achieving sustainable agriculture and viable agricultural systems is critical to food security and poverty alleviation. Soil health and soil quality are fundamental to the sustained productivity and viability of agricultural systems world-wide. Improvement in agricultural sustainability and productivity requires, alongside effective water and crop management, the optimal use and management of soil fertility and soil physical properties, which rely on soil biological processes and soil biodiversity.

The soil is a very complex and multi-faceted environment providing the habitat for a diverse array of soil organisms. The activities of this wide range of soil biota contribute to many critical ecosystem services, including: soil formation; organic matter decomposition, and thereby nutrient availability and carbon sequestration (and conversely greenhouse gas emissions); nitrogen fixation and plant nutrient uptake; suppression or induction of plant diseases and pests; and bio-remediation of degraded and contaminated soils (through detoxification of contaminants and restoration of soil physical, chemical and biological properties and processes). The effects of soil organisms also influence water infiltration and run-off and moisture retention, through effects on soil structure and composition and

indirectly on plant growth and soil cover. These services are not only critical to the functioning of natural ecosystems but constitute an important resource for sustainable agricultural production.

Lessons from case studies

The CBD Secretariat has made a call for case studies as a follow-up to decisions on agricultural biodiversity and FAO is assisting in compiling and assessing experiences and lessons learnt. In this process the following six case studies on soil biodiversity/ecosystem management have been selected and reviewed on the basis of their potential to catalyse further work on enhancing the beneficial functions of soil biodiversity for sustainable and productive agriculture and application of the ecosystem approach as adopted by the Convention on Biological Diversity .

The road toward sustainability is not an easy one to follow as short-term economic goals are often perceived to be more desirable by decision makers than the longer-term process of developing socially and technologically acceptable solutions that are also economically viable. In particular, technical assessments, participatory processes of testing and adaptation of improved management practices by farmers/land managers and successful wider application of soil biodiversity management for sustainable and productive agriculture will require adherence to the *ecosystem approach*.

Application of the guiding principles of the ecosystem approach (as demonstrated through the following case studies) should provide a better understanding of the biological, physical, and human interactions associated with sustainable agro-ecosystems and the ways and means to better manage those interactions with a view to effectively contributing to food

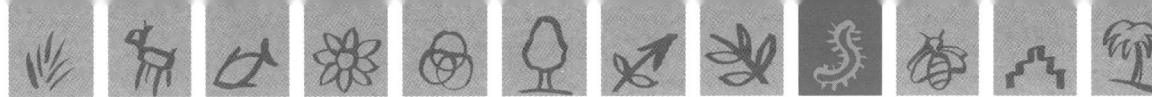

security and the well-being of rural populations. Extensive documentation and analysis of such case studies on soil biodiversity management for sustainable and viable production systems (including cropping, pastoral, forestry and mixed systems) will be part of that process. Such case studies should demonstrate the importance of integrated approaches that address and manage interactions between soil and other components of the agro-ecosystem (soil–plant–water–pest-predator interactions in the rhizosphere; soil-plant-livestock-atmospheric interactions through organic matter and nutrient management; and so forth).

The sharing of information, research and development experiences is expected to lead to raised awareness and understanding and wider application of improved soil biological and agro-ecological management approaches that will help ensure environmentally-friendly, productive and sustainable agricultural systems. This will also require policy and institutional support to provide an enabling enviornment for the adoption of such agro-ecological approaches.

It is hoped that this small selection of cases will encourage a greater compilation and dissemination of similar examples, in accordance with the call for case studies on soil biodiversity by the Conference of the Parties to the CBD and with FAOs mandate for assisting member countries in improving food security and sustainable agriculture.

INTRODUCTION

Agricultural studies of soil systems have historically been directed toward the biophysical and chemical aspects of crop production. The ecological dimensions of soil's systems have been considered less important. Currently, there is a need to develop greater knowledge of soil ecosystems, and their biological diversity and ecological functions, in order to build a broad basis for sustainable agricultural development. To this end, *an ecosystems management approach* (see below) is being advocated in many quarters to help carry forward the sustainable agriculture agenda.

This paper represents a selected review of case studies on the management of soil biological diversity for agricultural purposes. Of particular interest is the relevance of each study to the 12 guiding principles of the Convention on Biological Diversity (CBD) Ecosystem Approach[1], and the four thematic areas of the CBD/Food and Agriculture Organization (FAO), Programme of Work on Agricultural Biological Diversity[2].

THE CASE STUDIES PRESENTED HEREIN INCLUDE

- **CASE 1:** Successful farmer-to-farmer promotion of sustainable crop and soil management practices in the central highlands of Mexico.
- **CASE 2:** Managing termites and organic resources to improve soil productivity in the Sahel, Africa.
- **CASE 3:** Restoring soil fertility and enhancing productivity in Indian tea plantations with earthworms and organic fertilizers.
- **CASE 4:** Symbiotic nitrogen fixation in the common bean in Brazil.
- **CASE 5:** No-tillage agriculture in southern Brazil benefits soil macrofauna and their role in soil function.
- **CASE 6:** Management practices to improve soil health and reduce the effects of detrimental soil biota associated with yield decline of sugarcane in Queensland, Australia.

---

1 See Convention on Biological Diversity (CBD), Decision V/6 of the Conference of the Parties (COP), the Appendix on the Ecosystem Approach.

2 See in particular CBD COP Decisions III/11, V/5 and VI/5.

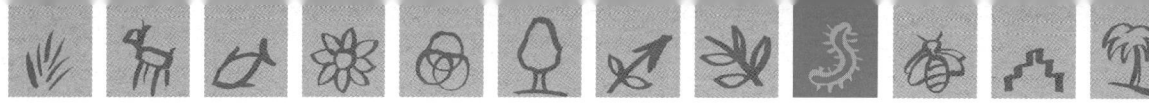

The overall scheme for the case study presentations is an outline of the *problem to be solved, objectives* for the study and *actors and actions* involved. Results of each study are then discussed and analyzed in the context of the CBD *ecosystem approach* and the joint CBD/FAO *Programme of Work on Agricultural Biological Diversity*. Finally, the major *outcomes and lessons learnt* from each study are summarized.

CASE 1

Successful farmer-to-farmer promotion of sustainable crop and soil management practices in the central highlands of Mexico

A Case Study from
North America - Tlaxcala, Mexico[3]

Problem statement
To motivate and empower peasant farming communities in the Central Highlands of Mexico to address the deterioration of soil quality, quantity and biological diversity using sustainable agricultural practices that restore ecosystem functions and meet livelihood needs.

The Central Highlands of Mexico have been under cultivation for thousands of years. Nevertheless, centuries of deforestation occurring since the fall of the Aztec empire at the hands of the Spaniards, plus recent intensive farming practices to feed the burgeoning population of Mexico City, have left soils in these agriculturally critical regions severely eroded and degraded. Deep gullies scour portions of the landscape, affecting water catchment and recharge capacity and reducing the productive potential of natural and agricultural systems. Severely eroded areas (known as *tepetates*) are characterized by hard, exposed subsoils, virtually no topsoil, and very little below ground life. The deterioration in soil quality, quantity and biodiversity has greatly challenged the capacity of Mexican peasant farmers *(campesinos)* to maintain even a subsistence living from the land.

Additional constraints to achieving sustainable agriculture, soil biological diversity and ecosystem functioning have been:

1) the immoderate application of agro-chemicals;
2) excessive conversion of vegetatively diverse lands to mono-cultures;
3) loss of traditional inter-cropping systems, especially the corn (maize)-bean-squash mixture;
4) lack of soil and water conservation measures;
5) scant knowledge of sustainable agro-ecological techniques, such as composting, cover crops and green manure;
6) inadequate access to credit;
7) low, guaranteed prices for basic grains;
8) high costs of agricultural inputs; and
9) little opportunity for capacity building among local farmers who spend most of their time meeting survival needs.

Objectives
In the western portion of the state of Tlaxcala, Mexico, also part of the Central Highlands, peasant farmers forming the Vicente Guerrero Group have experimented for more than 20 years with *integrated agro-ecological approaches to crop and soil management*. Their purpose has been to generate, share, and promote such approaches in order to improve the local quality of life, while respecting and caring for the fragile lands upon which they live.

---

3   Adapted from F. J. Ramos S. (1998) Grupo Vicente Guerrero de Españita, Tlaxcala. Dos decadas de promoción de campesino a campesino. Red de Gestión de Recursos Naturales and Rockefeller Foundation, Mexico City, Mexico

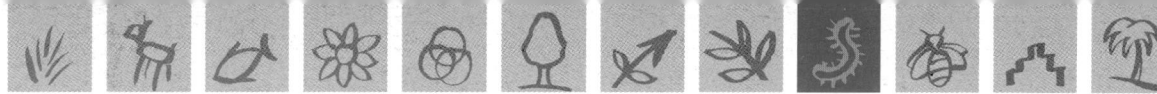

Actors and activities

The Vicente Guerrero Group (VGG; now a legally registered non-governmental organization in Mexico) is comprised of men and women from Españita, Tlaxcala, who have been acting as agricultural trainers since 1978. Their *farmer-to-farmer approaches and rural participatory* processes have led to notable successes in the adoption of integrated crop, water and land management practices.

Application of the ecosystem approach to soil biodiversity management

VGG advocates and teaches *adaptive management* in the maintenance of crop and soil resources, including soil biodiversity. This learn-by-doing approach is consistent with **Principle 9 of the ecosystem approach,** as adopted by the Convention on Biological Diversity (CBD) though decision V/6 of the Conference of the Parties (COP). These methods are also consistent with **Principle 5,** which advocates the **conservation of ecosystem structures and functions.** Adaptive management and conservation methods characterize various Vicente Guerrero programmes, such as:

The production of basic grains using techniques that enhance soil biodiversity functions, including the use of:

- crop rotations, leguminous cover crops, improved local seed varieties and diversified crop associations to broaden agro-ecosystem resilience and improve yields;
- low-impact tillage methods to reduce disturbances to soil structure and soil biota;
- stubble, harvest residues, livestock manure, and green manure to produce organic fertilizers;
- conservation measures to maintain soil structure and moisture content.

Land management that favours plant and animal diversity and its association with soil biological activity. This includes:

- mosaics of different crops and land uses;
- the capture and conservation of rainwater for plants, animals and people;
- the incorporation of backyard animals (native races of chickens, turkeys and rabbits, whose manure also provides soil organic matter for home gardens);
- the restoration of agricultural biodiversity by planting native crops, medicinal plants and tree species.

Participatory methods and various tools, including:

- visits to farmer fields;
- field demonstrations of crop and soil management techniques;
- on-farm experimentation;
- rapid participatory diagnostics;
- workshops, talks, courses, didactic games and community theatre.

The philosophical mainstay of the Vicente Guerrero Group is consistent with **Principle 1 of the ecosystem approach,** *that resource management is a matter of societal choice and that benefits should be shared in fair and equitable ways.*

To this end; promoters should be morally committed to their work. Promoters affirm their obligation to share all techniques and knowledge that they have acquired with other peasant farmers. This is characteristic of the farmer-to-farmer approach in which the promoter becomes aware of the wider social impact of his or her knowledge.

Principle 2 is also an important feature of VGG efforts because promoters teach that *management should be decentralized to the*

No. 9

lowest appropriate level to encourage greater efficiency, effectiveness and equity.

Accordingly, promoters and farmer clients should continue to cultivate their own lands. Neither the promoter nor the client should lose his or her identity as a farmer. Instead, they should remain connected to the livelihood practices of the rural community and aware of the local needs for assistance. The promoter is considered an example for other peasant farmers and should be visible in this capacity as a role model.

Relevance to the Programme of Work on Agricultural Biological Diversity

Besides *adaptive management,* the principal strength of VGG with respect to the FAO/CBD collaborative programme is *capacity building.* The latter includes strengthening the ability to manage biological diversity (including soil biodiversity) and promoting responsibility. To this end VGG members have elaborated mechanisms to promote awareness and maintain continuity in their actions over time.

Promoters must work as unpaid volunteers in order to demonstrate their community commitment. They work several days a week for one or two years, receiving only travel expenses. During this period, they are evaluated according to their management of a specialty area, their willingness to participate responsibly in group endeavours and their ability to work as part of a team. If a promoter is subsequently asked to stay on with the group, he or she will receive a small monetary compensation for his/her participation.

Outcomes

VGG promoters have trained more than 2 000 peasant farmers in Mexico and elsewhere in Latin America in integrated crop and soil biological management, and soil and water conservation practices, during the past two decades. Members of the group also count the following as some of their principal successes:

- An increase in local agricultural productivity;
- Significant reduction in agro-chemicals use by farmers who initially resisted natural or organic alternatives;
- Greater incorporation of stubble and crop residues into the soil;
- Increased adoption of soil and water conservation measures and soil fertility restoration efforts;
- Increased capacity to organize and attract outside funding.

Lessons learnt

The successes of the Vicente Guerrero Group highlight the importance of farmer-to-farmer approaches in achieving sustainable crop production, soil conservation and soil biological management on marginal and degraded lands. Furthermore, they suggest that intangible factors are as important as technical capacity.

These include:

- a profound respect for the environment, evidenced by an evolving, integrated and evermore sustainable use of local agroe-cological resources;
- the firm conviction that sharing knowledge with other farmers is an undeniable, and even moral obligation resting upon members of the group.

CASE 2

Managing termites and organic resources to improve soil productivity in the Sahel

A Case Study from
Africa - Bam Province, Burkina Faso[4]

Problem statement

To restore the productive capacity of crusted soils in the Sahel region of Africa in order to extend arable lands and sustain viable, agricultural livelihood systems.

Soil degradation, particularly crusting, is a major agricultural problem in the Sahelian region of Africa. The combined effects of extreme and difficult climatic conditions, over-grazing and trampling by cattle, continuous cultivation and other unsustainable management practices have resulted in the pervasive spreading of bare, infertile soils. Degraded soil structures and sealed crusts impede water infiltration and root growth, thus delaying soil regeneration and seriously limiting the use of local lands for crop and animal production. Because degraded soils constitute a serious threat to Sahelian agriculture, active restoration efforts must be undertaken if rural livelihoods are to be maintained.

Termites are widespread and abundant in tropical dry areas such as the Sahel. Although they are major agricultural pests, termites can also play an important role in recovering degraded ecosystems. Specifically, the burrowing and feeding activities of termites can be utilized to break up crusted soils and thereby counteract land degradation.

Objectives

The main purpose of this work was to evaluate the capacity of termites to improve the structure of crusted soils, including their ability to reduce soil compaction, increase soil porosity and improve the water infiltration and retention capabilities of soils. Such conditions encourage root penetration, vegetative diversity, and the restoration of primary productivity, all prerequisites for food and livelihood security in the Sahel.

Actors and activities

Organic mulch (cow manure or straw) was applied to soil surfaces during a three years study in order to trigger termite activity. It was assumed that termite-mediated processes (see above) would promote the recovery and rehabilitation of degraded soils. Dr. Abdoulaye Mando (Institut de l'Environnement et des Recherches Agricoles, Burkina Faso) completed this work as part of his PhD requirements under Drs. Leo Stroosnijder and Lijbert Brussaard (Wageningen Agricultural University, Holland).

Application of the ecosystem approach to soil biodiversity management

Three important principles of the ecosystem approach (CBD COP Decision VI/5) are evident in this case study. First, **Principle 10** advises *flexibility in balancing the goals of conservation, sustainable use, and agricultural production.* Flexibility, in this case, allowed the negative impacts of termites to be turned into positive benefits for soil productivity because termites, widely considered to be pests, eventually enhanced agricultural production:

● Flexible attitudes allowed investigators to see that conserving termite populations (instead of eradicating

No. 9

4 Adapted from Abdoulaye Mando, 1997. The role of termites and mulch in the rehabilitation of crusted Sahelian soils. Tropical Resources Management Paper 16. Wageningen Agricultural University. Wageningen, the Netherlands, 101 pp.

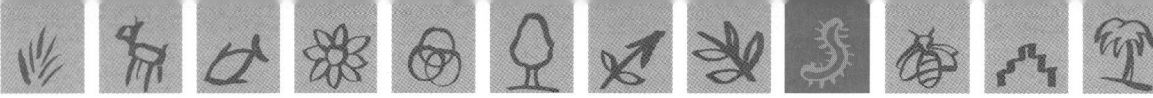

them) and stimulating their soil mixing capacities, would improve crusted soils.

Principle 9 recognizes that cyclical or successional change is normative, dynamically stable, and ultimately resilient. *Thus, ecological change should be incorporated into adaptive management plans.* Adaptively working with ecological change was important to this study because:

Termite disturbance (seasonal or successional burrowing, excavation, and foraging) turned out to be a viable management option. Specifically, investigators found that:

- termites feeding upon or transporting surface-applied mulch improved soil structure and water infiltration, and thereby enhanced nutrient release into the soil from mulch;
- native plant diversity, as well as cover, biomass and rainfall use, were all greater on mulched plots with termite activity than on plots without termite activity;
- growth and yield of cowpeas were far better on plots with termites than on no-termite plots. In particular, yields reached one ton per hectare where cow manure had been added and termites were present.

A third component of the ecosystem approach also characterized this work. **Principle 6** states that ecosystems should be managed within their functional limits; or more specifically, *that management objectives should be conceived within the environmental and biological conditions that limit natural productivity.* Such considerations were evident in this study, as investigators realized that:

Termite activity and weather can impose functional limitations on nutrient cycling in semi-arid, Sahelian conditions. For these reasons, it was acknowledged that:

- mulch application should be timed to optimally coincide with termite foraging periods; and
- mulch application should anticipate seasonal rainfall events, thereby allowing nutrient release to be better synchronized with plant growth demands.

Relevance to the Programme of Work on Agricultural Biological Diversity

In addition to *adaptive management,* the principal strength of this case study is its *technical assessment* of the goods and services, positive and negative impacts, and management options associated with termites as soil mixers, nutrient transporters and productivity enhancers (see above).

Outcomes

The principal outcome of this study was that termites successfully restored crusted Sahelian soils when their soil mixing and decomposing activities were properly managed by careful organic matter additions.

Lessons learnt

Perhaps the most important lessons learned from this study in Burkina Faso are that:

- significant soil degradation (compaction and crusting) results from eradicating native termite pests, but that ironically;
- the judicious application of surface organic matter to feed termites promotes their capacity to regenerate crusted soils.

Furthermore, this case illustrates a potentially extremely important, practical and cost-effective way of using biological activity

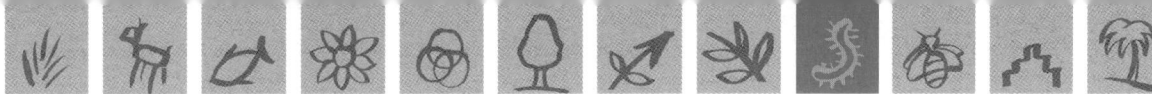

to restore seriously degraded and unproductive lands. Its practical application should be seriously investigated through participatory processes including on-farm experimentation and pilot project development.

CASE 3

Restoring soil fertility and enhancing productivity in Indian tea plantations with earthworms and organic fertilizers

A Case Study from
Asia - Tamil Nadu, India[5]

Problem Statement

To restore soil fertility and increase tea yields on intensively cultivated tea plantations in southern India.

Between the 1950s and the 1980s, fertilizer and pesticide use in India increased tea production from 1 000 to 1 800 kg ha^{-1}. Currently, national yields have stagnated as decades of intensive cultivation have left soil fertility greatly depleted. On some tea plantations, not even the use of external inputs and plant growth hormones has overcome 100 or more years of intensive exploitation.

Soil degradation on tea plantations is seen in: 1) the loss of soil biota (losses as high as 70 percent), 2) decreased organic matter, 3) lower cation exchange capacity, 4) reduced water retention, 5) soil compaction, 6) soil erosion, 7) nutrient leaching, 8) aluminium toxicity, 9) accumulated toxins (polyphenols) from tea leaves, and 10) acidification (pH levels as low as 3.8).

Objectives

The purpose of this study was to restore soil fertility and improve tea production on six private teas estates in Tamil Nadu, India, using organic matter and earthworms.

Actors and activities

Prof. Patrick Lavelle (Institut de Recherche pour le Développement; IRD) and Dr. Bikram K. Senapati (Sambalpur University, India) worked in collaboration with plantation managers from Parry Agro-Industries Ltd. In this effort, tea prunings, high quality organic matter, and vermicultured earthworms were applied in trenches between tea rows in order to evaluate effects on tea yields. Improvements in structural and biological properties of soils were expected to produce higher tea yields.

Application of the ecosystem approach to soil biodiversity management

Trenching is an old practice that has been mostly abandoned on plantations because of high human labour costs and substitutions by other techniques. In this study, it was thought that trenches would minimize soil loss and improve moisture and aeration conditions, so that nutrient cycling processes would be enhanced. The choice of trenching, an unconventional technique, illustrates **Principle 11** of the ecosystem approach; that *all forms of relevant knowledge (including traditional, indigenous and scientific) should be considered in developing management strategies.* Interestingly, trenching, prunings, organic material and earthworms between tea rows (Bio-Organic Fertilization, or FBO[6]) dramatically increased yields and profits:

5 Bikram Keshari Senapati, School of Life Sciences, Sambalpur University, Orissa State, India; Patrick Lavelle, Institut de Recherche pour le Développement, Bondy, France; Pradeep K. Panigrahi, Parry Agro-industries, Ltd., Tamil Nadu; Sohan Giri, State Pollution Control Board, Bhubaneswar; and George Brown, Embrapa Soja, Londrina PR Brazil.

6 Fertilisation Bio-Organique dans les Plantations Arborées, Patent ref. No PCT/FR 97/01363.

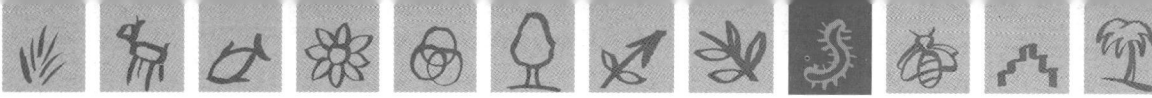

- FBO methods increased tea yields by 76 to 239 percent, compared to conventional inorganic fertilization.
- Profits from FBO were up to US$ 5 500 per hectare compared to conventional practices.

The surprising results of FBO soil biological management are a reminder that *agro-ecosystems should be managed within their unique biological and economic contexts;* a point clearly made by Principle 4 of the ecosystem approach. Principle 4 emphasizes that economic policies tend to undervalue sustainable agricultural practices, favouring instead a series of high-input/high-energy technologies suited to less diverse production systems. Nevertheless, results achieved with FBO demonstrate that even intensive agriculture, such as that practised on commercial tea plantations, can benefit from *adaptively managing soil biodiversity* **(Principle 9)**.

Principle 8 addresses one of the interesting features of FBO soil biological management encountered in this study. More specifically, it addresses the fact that yields and profits with FBO varied considerably between the six different tea plantations. In particular:

- The Sheikalmudi Estate was most responsive to FBO treatments and showed the greatest first-year gains.
- Yields and profits were variable and lower at the other five plantations.
- Low responses at the other plantations were due to site-specific conditions, including delays in soil recovery that were proportional to the degree of soil degradation.

These findings illustrate that *varying temporal scales and lag-effects characterize* *agro-ecosystems and favour long-term, rather than short-term management solutions,* a point aptly made by **Principle 8.**

In this study, project participants understood that variable yields indicated different degrees of land degradation and that long-term management commitments would be required to remedy these conditions. In particular, it was acknowledged that trenches at FBO sites would have to be opened and re-inoculated every three to four years in order to maintain high-level benefits.

Relevance to Programme of Work on Agricultural Biological Diversity

This case study addresses the need to identify *technical assessments* and *adaptive management methodologies.* In this respect, it determined that Bio-Organic Fertilization (FBO) is an affordable tool, adaptable to situational needs and appropriate to commercial management scales from small farms to plantations. The major components of this technological package include:

- large-scale vermiculture production;
- adaptable management practices;
- rearing different functional types of earthworms for inoculation;
- selecting and placing organic matter by quality and quantity criteria.

In addition, mainstreaming concerns are addressed in this study, as FBO has received patent protection and is now being disseminated within the agricultural sector. Mainstreaming activities include:

- The extension of FBO on more than 200 ha of Indian tea plantations and a large contract for extensive work in Chinese plantations (thousands of hectares).
- The patent holders' intention to transfer

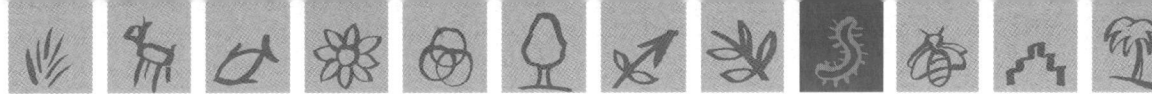

FBO technology to Sri Lanka and Australia for large-scale implementation.

- The anticipated inclusion of additional tree- or agroforestry-based cropping systems, such as coffee, citrus, banana, coconut, oil palm, eucalyptus and pine, into the FBO management portfolio.

Outcomes

The principal outcomes of this case study were:

- The development of a practical and conservation-oriented solution to soil degradation using earthworms and local organic material.
- The patenting of this technological package.
- The extension of this technology within India and to other countries.

Lessons learnt

The salient lessons learned during this project were:

- FBO is not a formula approach; it must be tailored to each site.
- FBO requires constant interventions by trained personnel to determine organic matter combinations and placements, employ vermiculture methods, and monitor soil faunal populations during soil rehabilitation.
- FBO is labour-intensive, being at present economically viable only in countries with an abundant and inexpensive human labour source.
- Farmers and agricultural agents tied to conventional methods continue to resist the FBO approach.

CASE 4

Symbiotic nitrogen fixation in the common bean

*A Case Study from
South America – Paraná, Brazil[7]*

Problem to be solved

To improve yields of the common bean on nitrogen-poor, tropical soils in Brazil.

Brazil is currently the second largest producer of the common bean *(Phaseolus vulgaris L.)* in the world. However, inadequate inoculation technology and low soil fertility (especially in relation to nitrogen content) are limiting national yields. In 2001, 4.3 million hectares were sown with this leguminous crop, yet an average yield of only 640 kg ha^{-1} was harvested nation-wide; an amount considered poor by experts.

Nitrogen fixing rhizobia (a bacterial soil component associated with legume roots) might cost-effectively provide enough nitrogen to increase bean yields. However, inoculations with commercially available rhizobial strains have shown poor results and raised doubts about the viability of this approach as a management option. Poor inoculation results may be attributable:

- to failure of commercial rhizobia to compete with abundant native rhizobia;
- to the high temperatures, acidity, and dryness sometimes associated with tropical soils under cultivation.

Objectives

The main emphasis of this study was to improve root nodulation and nitrogen fixation in the common bean through inoculations with soil bacteria. The strategy chosen was to isolate and select efficient native rhizobia from local bean production sites.

7 Mariangela Hungria, Embrapa Soja, Londrina, PR, Brazil; in collaboration with Rubens J. Campo, Júlio Cezar Franchini and Lígia M. O. Chueire, Embrapa Soja, Londrina, PR, Brazil; Ieda C. Mendes, Embrapa Cerrados, Planaltina, DF; Brazil; Diva S. Andrade, Arnaldo Colozzi-Filho and Élcio L. Balota, IAPAR, Londrina, PR, Brazil; Maria de Fátima Loureiro, UFMT, Cuiabá, MT, Brazil.

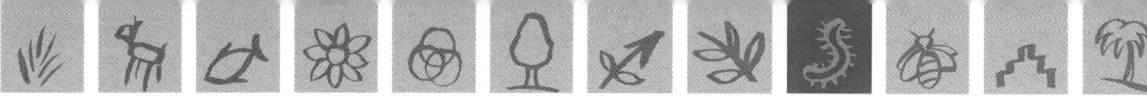

Actors and activities

Mariangela Hungria (Embrapa-Soya, Brazil), Diva de Souza Andrade (Instituto Agronômico do Paraná, Brazil) and Iêda Carvalho Mendes (Embrapa Cerrados, Brazil) coordinated efforts to collect local rhizobia in the state of Paraná during 1992 and 1993. More than 400 isolates were tested for nitrogen-fixing capacity, competitive ability, and other characteristics appropriate for performance in the tropics.

Application of the ecosystem approach to soil biodiversity management

It was thought that efficient native strains might perform better as inoculants than commercial strains. In particular, some indigenous strains might remain genetically stable in stressful tropical environments, when compared to commercial rhizobia, and thus better able to fix nitrogen. Also, superior local rhizobia should competitively exclude symbiotically inferior local strains for nodulation sites.

Principle 6 of the ecosystem approach indicates that *agro-ecosystems should be managed within the local environmental conditions that limit productivity*. In this study, stressful conditions of cultivated tropical soils (e.g. high temperature, acidity and dryness) were major deterrents to root nodulation and nitrogen fixation, and had to be overcome in the development of an effective inoculant. Eventually, three strains (PRF 35, PRF 54 and PRF 81) functioned well under tropical conditions. These strains were characterized by high nitrogen fixation rates, good competitive ability and tolerance to high temperatures.

Principle 4 addresses the need to *manage ecosystems in an economic context*. In this case, the directive was to improve bean yields using native rhizobial inoculants

because the alternative technologies were considered to be economically less viable. Specifically, commercial strains were known to be metabolically ineffective, resulting in low yields; and nitrogen fertilizers, the other option, contaminated ground water (at substantial cost because of impacts on biological diversity and human health).

After field testing, the superior performance and economic advantage of the PRF 81 inoculant was confirmed.

- Bean yields increased up to 906 kg ha^{-1} compared to non-inoculated controls.
- Total yields (1 571 kg ha^{-1} to 3 425 kg ha^{-1}) were significantly higher than the national average.

It was also determined that re-inoculation the following year improved the establishment of PFF 81. This practice was accordingly incorporated into management recommendations.

Another strain selection programme was started in 1998 for the Brazilian Cerrados region, an edaphic type of savannah. After three years of experiments, the H 12 and H 20 strains were selected as superior.

- The H 12 and H 20 strains showed yield increases of 437 and 465 kg of grains ha^{-1}, respectively, over controls composed of indigenous rhizobial populations.
- Total yields were approximately 2 500 kg ha^{-1}, confirming that economically important increases in bean yields can also be obtained in the Cerrados region as a result of inoculations with superior, native rhizobial strains.

Another brief, but important result of this study relates to the ecosystem approach. **Principle 3** urges agro-ecosystem managers to *consider the impacts of* management interventions both within and beyond the boundaries of the managed system. It is noteworthy that:

The selection of superior native strains from local bacterial diversity did not require genetic modification or the introduction of exotic species. The actual and potential risks of genetically altered micro-organisms and non-native strains in field situations have been hotly debated for decades. However, the use of locally selected rhizobia avoids these risks altogether, while assuring higher bean yields and associated economic gains.

Relevance to the Programme of Work on Agricultural Biological Diversity

The thematic focus of this case study with respect to the joint programme falls within *technical assessments*. In particular, this effort represents an innovative approach to improving inoculation responses that seldom considers the isolation and assessment of efficient strains from local sites of bean production. It is an approach that is worthy of replication in other tropical agricultural areas around the world.

Outcomes of activities

One of the most important results of this study was the official recommendation of PRF 81 as a commercial Brazilian inoculant in 1998. This recommendation was based, in part, on its superior performance in tropical soils. In particular,

- Brazilian bean cultivars inoculated with PRF 81 yielded about 2 500 kg ha^{-1} of beans on nitrogen-poor soils, more than three times the national average.

This positive performance has resulted in increased inoculant use in Brazil and encouraged the identification of other genetically stable, competitive and efficient bean rhizobia from tropical areas.

Lessons learnt

This research and development effort has demonstrated that a concerted programme of strain selection in local bean cropping areas produces economically viable inoculants. If this approach is followed around the world, it might help improve poor yields and nitrogen depletion that plague tropical areas.

CASE 5

No-tillage agriculture in Southern Brazil benefits soil macrofauna and their role in soil function

A case study from
South America - Paraná, Brazil[8]

Problem to be solved

Physical disturbance is one of the principal causes of biodiversity loss in all world ecosystems. Soil macrofauna (invertebrates important for soil structure, function and fertility) are also susceptible to physical disturbances, especially those associated with tillage practices. Adapting tillage and planting regimes to minimize disturbance should provide better environments for soil macrofauna and their functions, thus benefiting sustainable agriculture.

Soil macrofauna are large invertebrates (termites, ants, earthworms, true bugs, snails, millipedes, centipedes, spiders and other arachnids, crickets, beetle grubs and other insect larvae) that spend all or a portion of their life cycle in the soil. Their activity is essential to the physical, chemical and biological integrity of soils, and important for soil fertility. Soil macrofauna include:

8   George Brown, Embrapa-Soja, Londrina, PR Brazil, in collaboration with Lenita Oliveira, and Eleno Torres of Embrapa-Soja; Norton Benito and Amarildo Pasini, Universidade Estadual de Londrina; M. Elizabeth F. Correa and Adriana M. de Aquino of Embrapa-Agrobiologia.

- decomposers that cycle organic matter and release plant nutrients;
- bioturbators that mix and move soil, affecting physical structure, aggregate formation, hydrological processes and gas exchange;
- pests that cause adverse effects on agricultural crops;
- predators that can act as bio-control agents to regulate pest and parasite outbreaks.

However, biological diversity around the world is susceptible to physical environmental disturbances, and soil fauna populations, activity and diversity can be reduced by repeated physical disturbances associated with conventional tillage (especially when pesticides and other agro-chemicals are also used). *For this reason, it is believed that minimal disturbances to agro-ecosystem soils should provide better environments for soil macrofauna and their functions than conventional tillage regimes.*

Objectives, actors and actions

Because of the importance of understanding physical disturbance in agro-ecosystems, several experiments were begun in 1979 in several locations in Southern Brazil, to compare no-tillage (NT) and conventional tillage (CT) practices in terms of long-term trends in productivity, nutrient use, carbon inputs, decomposition rates and soil conservation. In 1998, soil macrofauna comparisons were added to the sampling protocol at NT, CT and minimum tillage sites.

The work was undertaken by researchers (and students) of various institutions: George Brown, Lenita Oliveira and Eleno Torres of Embrapa-Soybean; Norton Benito and Amarildo Pasini of the Universidade Estadual de Londrina; and M. Elizabeth F. Correa and Adriana M. de Aquino of Embrapa-Agrobiologia. Several private farm owners participated.

Application of the ecosystem approach to soil biodiversity management

Sustainable ecosystems depend upon balanced biological interactions among a diversity of organisms. The same is true for soil ecosystems. However, in the latter, organismal diversity and biological activity are more strongly regulated by C availability than in other ecosystems.

Soil organic carbon (SOC) is derived primarily from the decomposition and recycling of dead plant and animal material. Its transformation is carried out principally by soil organisms. In natural ecosystems, these carbon transformations (represented by inputs and outputs) are generally balanced over time, but in agro-ecosystems, *the amount and quality of SOC can be sharply reduced by tillage disturbances.* SOC depletion is especially a problem in exposed tropical soils, where high temperatures and rainfall can drive nutrient cycles at great velocities and leaching can quickly reduce available nutrients in the system.

There is an important relationship between soil disturbance, SOC and soil biodiversity (including macrofaunal diversity).

- Specifically, soil biodiversity and SOC are higher in physically, less disturbed systems than in more disturbed systems.
- For example, pastures and planted fallows (less disturbed) show greater soil biodiversity and higher SOC than cropping systems (more disturbed).
- Among cropping systems, CT sites are more disturbed than NT sites. Accordingly, they would be expected to lose their SOC and soil biodiversity more easily.

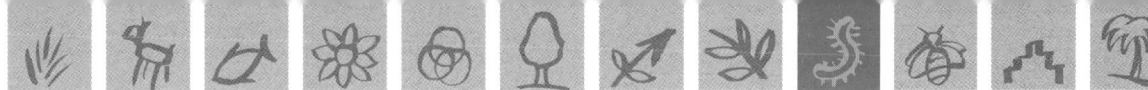

In this case study, the comparison of disturbance characteristics associated with NT and CT practices, as well as their respective abilities to conserve soil macrofauna, demonstrates **Principles 3 and 5 of the ecosystem approach. Principle 5** recognizes that *the sustained functioning and resilience of ecosystems depend on conserving relationships among diverse and interacting species.* Thus, management objectives should modify or substitute any practices that seriously limit functioning. **Principle 3** acknowledges that *management interventions, such as tillage, can have undesirable effects on soil organisms and their functions.* Careful consideration and analysis are required to avoid negative impacts.

Results showed that NT systems generally improved soil environmental conditions for plants and soil animals compared to CT systems. Improvements (see Table 1) included reduced erosion, enhanced nutrient- and water use-efficiency by crops, and improved crop yields and profitability, especially after a transition period of a few years.

NT practices also increased soil macrofaunal diversity, according to several indicators, and speeded population recovery after the cessation of CT practices. Soil organisms (Table 1) that especially benefited from NT were natural predators (important for the biological control of pests), bioturbators (important for improving soil physical structure) and decomposers (important for recycling plant residues).

Finally, the lack of soil disturbance at NT sites led to increased soil OM in the top-most soil layers, increased protection of the soil surface with plant residues and increased populations of beneficial soil invertebrates, despite the soil compaction that often accompanies NT methods.

PARAMETER MEASURED	PARAMETER MEASURED
Erosion	Greatly reduced
Compaction	Generally increased
Soil C stocks	Increased in upper layers
Productivity	Equal or higher, esp. in dry years
Profitability	Generally higher after transition period
Predators (bio-control) Spiders and other arachnids, Diplura, Beetles, Centipedes and Mermithid nematodes Ants	Generally more abundant Variable
Saprophages (decomposers) Beetle grubs, Millipedes and Pill-bugs Enchytraeids	More abundant Fewer
Bioturbators Earthworms and Termites	More abundant
Pests Beetle grubs True bugs Ants	Variable, often more abundant Variable Variable, generally fewer
Macrofauna Diversity	Greater
Taxonomic Richness	Greater

TABLE 1

Some agro-ecological and economic consequences of adopting NT practices

Nevertheless, the authors of this case study felt that more detailed research was necessary to properly link the increase in diversity and abundance of selected groups of soil macrofauna in NT with the improvement and maintenance of soil functions (e.g. water infiltration, soil aggregation, soil protection, decomposition, nutrient cycles, carbon sequestration, pest control, plant growth and yields) that are critical to the sustainability of NT systems.

This case study briefly touches upon themes relevant to **Principle 12,** that the ecosystem approach should *involve as many relevant sectors of society as possible.* In other words, it should be a multi-stakeholder approach. After 30 years of implementation in Brazil, NT practices have gained wide acceptance by farmer associations, cooperatives, researchers and extension agents, agroindustry leaders and agricultural policy makers.

Providing that economic benefits of utilizing NT methods and conserving soil macrofauna are also realized, it is likely that this stakeholder support network will rapidly disseminate and adopt a wide array of soil biological management practices amenable to NT systems.

Relevance to the Programme of Work on Agricultural Biological Diversity

In terms of *adaptive management,* it is very likely that NT practices, when combined with the use of cover crops in rotations, will be highly compatible with new and existing techniques for the conservation of soil macrofauna and the sustainable use of their ecosystem functions. The potential synergy between these management objectives deserves special attention by investigators and managers; it is highly relevant to the future success of sustainable agriculture.

As discussed above, there is a potential for forming extensive *capacity-building partnerships* in Brazil to simultaneously promote NT methods and soil macrofauna conservation. This possibility is based upon the compatibility of some of their management goals, and the existing level of coordination among current NT stakeholder groups. Nevertheless, proven techniques and economic benefits from direct soil macrofauna management remain to be demonstrated, and this will ultimately determine acceptance by farmers and other stakeholders in Brazil.

Outcomes

A notable outcome of this study was the growing confidence that:

- No-tillage systems are biologically, taxonomically and functionally more diverse than conventional tillage systems.
- The diversity of soil macrofauna plays an integral role in the successful and sustainable functioning of NT systems.

Lessons learnt

The principal lessons learnt from this case study were that:

- NT systems provide a favourable environment for re-establishing soil macrofauna following CT practices, because NT modifies soil ecosystems and the soil-litter interface, benefiting soil biological communities.
- Higher species' diversity, larger population sizes, and enhanced functional activities of selected taxa of the soil macrofauna in NT systems may be linked to better performance of NT compared to CT.
- The conservation of soil macrofauna and their biological functions may contribute to the resilience of soil

ecosystems and their capacity to withstand stressful environmental conditions, a critical factor for sustainable tropical agriculture.

CASE 6

Management practices to improve soil health and reduce the effects of detrimental soil biota associated with yield decline of sugarcane in Queensland, Australia

A case study from Queensland, Australia[9]

Problem to be solved

Yield decline of sugarcane is a widespread problem throughout the Australian sugar industry. It results from loss of productive capacity of soil under long-term monocultures of sugarcane, with lack of rotations, excessive tillage of the soil at planting and severe soil compaction from the use of heavy machinery during harvesting. Collectively, these management practices have led to soils that are low in organic C and cation exchange capacity, have a high bulk density and a low microbial biomass. This in turn is associated with a build-up of populations of detrimental soil organisms, which affect the growth and health of the sugarcane root system.

Actors and activities

Sugarcane is a major Australian crop earning AUS$ 11 billion (US$ 0.7 billion) in export income in 1999. It is produced on some 450 000 hectares of land along the coast of Queensland and northern New South Wales and in the Ord River valley in Western Australia. In 1993, a multi-disciplinary research programme, known as the Sugar Yield Decline Joint Venture, was established among concerned institutes (see authors and institutes cited above) to investigate the causes of sugarcane yield decline and develop solutions to revive a viable, productive and sustainable sugar industry.

Conventional sugarcane management

In most sugarcane-growing districts sugarcane is grown largely as a monoculture and is normally harvested 12-18 months after planting. The next or first-ratoon crop is produced from the buds remaining on the underground portions of the stalks. Normally four or five ratoon crops are produced before the stool (rootstock) is ploughed out. Traditionally, a 4-6 month fallow (either bare, as weeds or a sown legume) was applied before the cycle was repeated. However, since the early 1970s a system of plough-out/re-plant has been increasingly practised, resulting in virtually no fallow period. This increased intensity has been accompanied by more extensive use of inorganic fertilizers, insecticides and herbicides.

It has also been accompanied by green cane harvesting in many areas, in place of the more traditional burnt cane harvesting, and by the use of heavy machinery to harvest and transport the cane. Whilst these changes have increased the efficiency of cane production they have also been associated with the development of a plateau in cane production in the sugar industry over the period 1970-1990. This is thought to be due to a combination of factors including climate, the growth of cane on poorer quality soils,

9 Prepared from a paper by C.E. Pankhurst[a*], R.C. Magarey[b] G.R. Stirling[c], B.L. Blair[d], M.J. Bell[e] and A.L. Garside[f], based on the Sugar Yield Decline Joint Venture,
   [a] CSIRO Land and Water, Davies Laboratory, Aitkenvale, Queensland 4814, Australia
   [b] Bureau of Sugar Experiment Stations, Tully, Qld 4854,
   [c] Biological Crop Protection, 3601 Moggill Road, Moggill, Qld 4070,
   [d] Queensland Department of Primary Industries, Tully, Qld 4854,
   [e] Farming Systems Institute, Queensland Department of Primary Industries, Kingaroy, Qld 4610,
   [f] Bureau of Sugar Experiment Stations, Davies Laboratory, Aitkenvale, Qld 4814.

No. 9

widespread introduction of mechanical harvesting and intensification of the monoculture system.

Effects of conventional sugarcane practices on soil health and yield

In regard to the long-term impact on soil health, it was recognized as early as 1930 that cane growth on suitably prepared virgin land often out-yielded that on old cane land (i.e. under cane for several years), and that cane growth in old cane land was improved following soil pasteurization. There was a growing recognition that a combination of management factors-monoculture, inappropriate use of inorganic fertilizers, heavy harvesting machinery-has led progressively to the development of soil physical, chemical and biological properties that constrain plant growth. This catalyzed the establishment in 1993 of a multi-disciplinary research programme, the Sugar Yield Decline Joint Venture, to investigate the causes and develop solutions. It was clear that a holistic "ecosystem approach" was necessary to address the yield decline problem.

Research on the physical, chemical and biological properties of soils under monoculture cane revealed the extent to which the soils have become degraded under the current cane management system, affecting the health of the sugarcane root system. Whilst not consistent in magnitude across all sites, the soils under cane monoculture were characterized by having high bulk density, low available water, low labile organic carbon, low pH, low CEC, high exchangeable Al and Mn and low Cu and Zn.

Microbial biomass was also seen to be rapidly reduced after the introduction of sugarcane to new land. Plant growth responses to pasteurization or fumigation and fungicides were accompanied by a significant improvement in the health of the cane root system, implicating the role of detrimental soil

organisms, such as pathogenic soil fungi and plant parasitic nematodes, in yield decline. *Pachymetra* root rot *(P. chaunorhiza)* has been associated with significant yield losses in plant and first-ratoon crops and has been controlled largely through selection of resistant cultivars. *Pythium* species *(P. arrhenomanes, P. myriotylum* and *P. graminicola)* have also been shown to be pathogenic towards sugarcane. A possible detrimental effect of arbuscular mycorrhizal fungi (AMF) on yield decline has been suggested, though not yet demonstrated, in view of the relatively high use of phosphate fertilizers by the Australian sugar industry. More than 30 pest species of nematode have been associated with sugarcane roots in Australia including five plant-parasitic species: lesion nematode *(Pratylenchus zeae)*, root knot nematode *(Meloidogyne javanica)*, stubby root nematode *(Paratrichodorus minor)*, spiral nematode *(Helicotylenchus dihystera)* and stunt nematode *(Tylenchorhynchus annulatus)*. From studies *P. zeae* and *M. javanica* are considered to cause the most root damage. Yield responses of 10-50 percent were obtained across the sites when nematicides were applied to plant and ratoon cane crops.

Management practices to reduce the effects of detrimental soil biota on yield decline

Until recently, research into detrimental soil biota associated with sugarcane yield decline has provided few viable options for cane growers. The selection of cane varieties resistant to *Pachymetra* root rot has been successful but has had only a minor impact on yield decline, most likely because other fungal root pathogens have occupied that niche. Similarly, the use of fungicides to control known and unknown pathogenic fungi or nematicides and the lesion and root knot nematode would be considered uneconomic

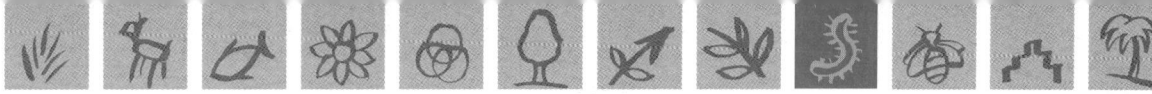

or environmentally unsustainable by both the industry and the wider community. Such treatments would also reduce populations of beneficial fungi and nematodes in the soil. More sustainable options centred on changing the current cane management system are therefore required to address the problem. Such options include the incorporation of rotation breaks and the more general use of organic amendments in the cropping system. Other options focus on reducing tillage and compaction of the soils during the planting and harvesting operations in order to reduce damage to soil structure and to improve water infiltration. Each of these options has the potential to contribute substantially to improving soil physical and chemical properties and hence influencing the balance between beneficial and detrimental organisms in the soil.

In this case study, recognition was made of **Principles 3 and 5 of the ecosystem approach: Principle 3** acknowledges that *management interventions, such as tillage, can have undesirable effects on soil organisms and their functions;* **Principle 5** recognizes that *the sustained functioning and resilience of ecosystems depend on maintaining a dynamic relationship and restoring interactions among diverse and interacting species and their environment*. Thus negative impacts of different practices on soil functioning were identified and analyzed with a view to adapting and testing improved management options. Moreover, in addressing the problems of productivity and sustainability of the sugarcane industry with stakeholders, it directly addresses **Principle 4** of the ecosystem approach: *Recognizing potential gains from management, there is usually a need to understand and manage the ecosystem in an economic context*. Understanding that economics tends to override sustainable agricultural practices, the aim was to identify viable alternatives to the conventional monoculture and high-input system through diversification and the management of natural agro-ecological interactions.

Rotation breaks

Rotation breaks as a means of breaking disease cycles and for improving soil fertility have generally not been considered a viable option by the Australian sugarcane growers for economic reasons. These include pressure from the sugar mills for a constant and high volume of cane for processing, perceived economic losses associated with taking land out of cane production for a short period, lack of evidence of perceived benefits from a break crop, and the lack of suitable rotation crops. Efforts were therefore made to explore more fully the benefits of rotation breaks to cane production and soil health. Trials were established at five sites in Queensland on land under cane monoculture for at least 20 years, incorporating three different breaks, varying from 9 to 42 months, under a sown legume/grass pasture (e.g. a mixture of *Brachiaria decumbens* and *Arachis pintoi*), alternative crops (soybean, peanut, maize) and a bare fallow (kept free of weeds with glyphosphate) and control plots. Each of the breaks gave substantial increases in cane yield (average of 33 percent across all five sites) that were comparable to that achieved following methyl bromide fumigation of monocultured cane soil.

It was seen that significant yield responses could be achieved with both a pasture and a bare fallow break despite the fact that these breaks had contrasting effects on soil properties. The observed rotation response following the pasture and crop breaks was probably due to a combination of soil fertility factors in addition to a reduction in detrimental soil

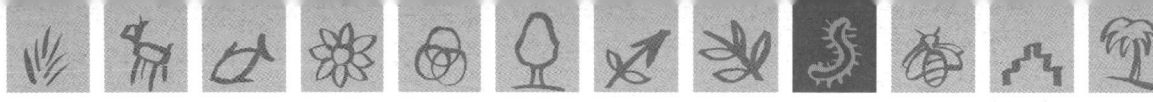

organisms. A build-up of suppressive micro-organisms in the soil under the pasture break and the increase in populations of free-living bacterial- and fungal-feeding nematodes under the pasture and crop breaks are indicative of a shift towards a more balanced soil biology in these systems. These nematodes are beneficial to plant growth through their role in nutrient cycling and in regulating organic matter decomposition; they are commonly regarded as an indicator of the biological status of the soil. There appeared to be considerable scope for improvement of the biological status of monocultured cane soils in addition to the removal of known specific root pathogens.

Organic amendments

Green cane harvesting is now practised widely throughout the Australian sugar industry. This generally results in the return of 10-15 tonnes (dry weight)/ha of cane trash to the soil surface after each harvest. More than 80 percent of the C in this trash is lost through respiration during the following year, although considerably more of the trash N is retained. Evidence suggests that while there are measurable increases in microbial biomass and organic C in the top 0-5 cm of soil under green cane trash blanketing there is no evidence that this counteracts yield decline. This may be partly due to the intensive soil cultivation prior to establishing a new crop, which accelerates erosion of microbial biomass and oxidation of organic C.

A number of sugar mill by-products such as filter mud and boiler ash are commonly applied to the soil prior to establishment of a new sugarcane crop. These products are valued for their capacity to improve soil nutrient status of the soil (N, P, Ca, Mg, Si, K, Cu and Zn) but it is not known if they have any impact on detrimental soil biota in

sugarcane in Australia. Though the many reports of successful use of compost and other organic materials to suppress fungal and nematode pathogens in other crops suggests that appropriate organic amendments could have beneficial effects on detrimental soil organisms associated with sugarcane yield decline.

Tillage

Under long-term sugarcane monoculture, characteristic poor soil structure and widespread compaction are due to a combination of intensive cultivation, use of heavy machinery during harvesting, often under wet conditions, and a mis-match of crop row spacing and machinery wheel spacing. Soil compaction has been shown to reduce water infiltration and soil hydraulic conductivity in sugarcane soils, and together with frequent damage to the sugarcane stool during harvesting, contributes to yield decline. To counter these problems, strategic tillage techniques based on minimum cultivation and the use of herbicides to remove the old sugarcane stool in the row at re-planting, coupled with maintenance of compacted inter-rows during the crop cycle, have been developed. These techniques are shown to have no adverse effect on cane yields but offer cane growers significant savings in fuel and machinery costs.

There is a traditional belief that tillage of the soil between sugarcane cropping cycles has beneficial effects in terms of controlling root diseases and pests. This may be true with some root feeding pests such as the canegrub (species of *Antitrogus, Dermolepida, Lepidiota* and *Rhopaea)*. However, the deployment of biopesticide products (e.g. those containing the fungus *Metarhizium anisopliae* - a natural biological control agent of the canegrub) may be more effective in a minimum tillage situation. The premise is that natural biological control systems will have a greater chance of developing if tillage is kept to a minimum, though little is known about the effects of minimum tillage practices on soil organisms associated with sugarcane yield decline.

Strategy for promoting uptake of improved management practices

For raising awareness of impacts of management practices among cane growers and bringing about changes in the sugarcane industry, soil health report cards are being used. These monitor changes in soil properties using a minimum set of seven indicators of soil fertility and quality and illustrating effects of improvement from legume-cane rotations, organic matter management and minimum tillage: reduced nematodes and diseases, improved soil structure and increase in beneficial disease-suppressive organisms. This is backed up by an information and communication strategy and demonstrations and field trials of best management practices (see www.landcareresearch.co.nz).

This process addresses **Principle 12** of the ecosystem approach *recognizing the need to involve all relevant sectors of society and scientific disciplines*. Through the Sugar Yield Decline Joint Venture, the cane growers, representatives of industry, and various land and water, crop protection, sugar and farming systems research bodies, are directly involved in the soil health reporting and analysis. Their views and suggestions are taken into account in the development of the most appropriate soil and crop management system. Research is continuing on certain aspects with a view to enhancing adoption, such as the duration of the rotation break required to achieve a significant benefit and the longevity of the benefit from the break: important economic considerations for cane-growers.

Summary

In order to circumvent yield decline and improve productivity and long-term sustainability, major changes to the sugarcane cropping system need to be promoted. Efforts are ongoing to promote adoption of a legume-based rotation break at the end of the cropping cycle to reduce populations of detrimental biota, particularly nematodes, and to help restore soil fertility (e.g. N), supplemented by the incorporation of organic amendments prior to planting, in view of their beneficial impact on soil biota. Adoption of minimum tillage and controlled traffic practices to increase the capture of potential benefits from green cane harvesting and to reduce the amount of soil subjected to compaction are also encouraged.

SUMMARY GUIDE TO CASE STUDIES ON THE MANAGEMENT OF SOIL BIOLOGICAL DIVERSITY

CASE STUDY: 1 FARMER-TO FARMER METHODS CAN PROMOTE SOIL AND AGRICULTURAL BIODIVERSITY

Vicente Guerrero Group with 20+ years of success

CONTINENT, COUNTRY, AND STATE OR REGION	PROBLEM TO BE SOLVED	OBJECTIVES	ACTORS ACTIONS	RELEVANCE TO 12 CBD ECOSYSTEM PRINCIPLES	RELEVANCE TO 4 AREAS OF FAO WORK PROGRAMME	OUTCOMES	LESSONS LEARNT
North America **Mexico** **Central Highlands**	Poor soil quality, quantity and biological diversity on degraded lands	To promote integrated crop, soil and soil biological management through farmer-to-farmer methods	Farmer-to-farmer promoters from local community Rural participatory methods for basic grain production and landscape management	P1- sharing of benefits equitably P2-decentralized management P5- conserve ecosystems and their services P9- adaptive management	Adaptive management Capacity building	More than 2000 farmers in Mexico and Latin America have been trained in agro-ecological principles and practices during more than 20 years	Long-term success in crop and soil management depends more upon shared values and local conviction, than on technological trends, which can change rapidly

CASE STUDY: 2 MANAGING TERMITES AND ORGANIC MULCH CAN INCREASE SOIL PRODUCTIVITY

Severely crusted soils are restored

CONTINENT, COUNTRY, AND STATE OR REGION	PROBLEM TO BE SOLVED	OBJECTIVES	ACTORS ACTIONS	RELEVANCE TO 12 CBD ECOSYSTEM PRINCIPLES	RELEVANCE TO 4 AREAS OF FAO WORK PROGRAMME	OUTCOMES	LESSONS LEARNT
Africa **Burkina Faso** **Sahel Region**	Restore soils in order to extend arable lands and increase productivity	To manage termites and local organic matter to rehabilitate crusted soils	Academic researchers and their institutions Mulch applied to soils, thus stimulating termites to improve soil structure and soil processes	P6- functional constraints P9- adaptive management P10- flexible conservation and use objectives	Technical assessments Adaptive management	Termites can be highly beneficial agents whose soil mixing and decomposing activities can be managed by organic matter additions in order to enhance soil productivity.	Soil structure degradation results from eradicating native soil organisms (termites). Applying surface OM feeds termites and promotes their regenerative activities.

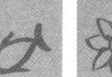

CASE STUDY: 3 MANAGING EARTHWORMS AND ORGANIC MATTER CAN IMPROVE CROP AND SOIL PRODUCTIVITY

Renewal of soil fertility, even at sites of intensive agriculture

CONTINENT, COUNTRY, AND STATE OR REGION	PROBLEM TO BE SOLVED	OBJECTIVES	ACTORS ACTIONS	RELEVANCE TO 12 CBD ECOSYSTEM PRINCIPLES	RELEVANCE TO 4 AREAS OF FAO WORK PROGRAMME	OUTCOMES	LESSONS LEARNT
Asia India Tamil Nadu	Rehabilitate plantation lands degraded by decades of intensive tea cultivation	To restore soil fertility and improve stagnant tea yields using soil fauna and local organic matter	Agro-industry representatives, farm managers and academic researchers and their institutions Tea prunings and other organic materials were trenched with earthworms, to improve soil OM and structure.	P4- economic context P8- consider lag effects & long-term objectives P9- adaptive management P11- consider relevant sources of knowledge	Technical assessments Adaptive management Mainstreaming	A practical, economical, and conservation-minded solution to soil degradation Patented methods that are being mainstreamed world-wide as Bio-organic fertilization (FBO)	FBO must be tailored to each site and needs the regular attention of trained personnel, FBO best suited for countries with inexpensive and readily available labour force

CASE STUDY: 4 SYMBIOTIC NITROGEN FIXATION WITH THE COMMON BEAN CROP

CONTINENT, COUNTRY, AND STATE OR REGION	PROBLEM TO BE SOLVED	OBJECTIVES	ACTORS ACTIONS	RELEVANCE TO 12 CBD ECOSYSTEM PRINCIPLES	RELEVANCE TO 4 AREAS OF FAO WORK PROGRAMME	OUTCOMES	LESSONS LEARNT
South America Brazil Paraná	Improve low bean crop yields on nitrogen-poor, tropical soils	To improve common bean response to nitrogen-fixing soil bacteria	Academic researchers and their institutions Select efficient Rhizobia strains from local bean production sites	P3- impact of interventions P4- economic context P6- functional constraints	Technical assessments	Common beans inoculated with competitively superior, native Rhizobia produce high yields in nitrogen-poor, tropical soils	Superior strains of Rhizobia can be selected from the diversity of native soil bacteria with no need for genetic modifications

No. 9

CASE STUDY: 5 NO-TILLAGE AGRICULTURE BENEFITS SOIL MACROFAUNA

CONTINENT, COUNTRY, AND STATE OR REGION	PROBLEM TO BE SOLVED	OBJECTIVES	ACTORS ACTIONS	RELEVANCE TO 12 CBD ECOSYSTEM PRINCIPLES	RELEVANCE TO 4 AREAS OF FAO WORK PROGRAMME	OUTCOMES	LESSONS LEARNT
South America **Brazil** **Paraná**	Restore and maintain soil fertility on severely eroded agricultural lands	To provide the best environment for macrofauna and their soil fertility functions	Private farm owners with academic researchers and their institutions Compare no-tillage (NT) vs. conventional tillage (CT) practices for conserving soil macrofauna	P3- impact of interventions P5- conserve ecosystems and their services P12- multi-stakeholder involvement	Adaptive management Capacity building	NT systems provided better conditions for macrofauna than CT systems NT systems had higher soil macrofauna diversity than CT systems Macrofauna diversity was important to soil functioning in NT systems	NT can help re-establish soil fauna after CT disturbances Highly varied soil biological activity suggests that NT systems are ecologically resilient and stress-resistant

CASE STUDY: 6 MANAGEMENT PRACTICES TO IMPROVE SOIL HEALTH AND REDUCE EFFECTS OF DETRIMENTAL SOIL BIOTA ASSOCIATED WITH YIELD DECLINE OF SUGARCANE IN AUSTRALIA

CONTINENT, COUNTRY, AND STATE OR REGION	PROBLEM TO BE SOLVED	OBJECTIVES	ACTORS ACTIONS	RELEVANCE TO 12 CBD ECOSYSTEM PRINCIPLES	RELEVANCE TO 4 AREAS OF FAO WORK PROGRAMME	OUTCOMES	LESSONS LEARNT
Australia **Queensland**	Widespread sugarcane yield decline under monocultures with excessive tillage and soil compaction by heavy machinery	Adapt practices to improve crop growth and soil health increase soil OM and CEC increase activity of beneficial soil organisms for fertility and soil structure and reduce detrimental soil organisms	The Sugar Yield Decline Joint Venture involves the cane-growers, representatives of industry and various land and water, crop protection farming systems and sugar research bodies.	P4- understand and manage the ecosystem in an economic context P5- conservation of ecosystems and their services P12- involves all relevant sectors of society and scientific disciplines	Adaptive management of soil-crop system and practices Technical impact assessment of management options on soil ecosystem functions and ways to optimize benefits/reduce harmful effects	Industry promoting uptake and monitoring of better practices: legume-based rotation break; OM; minimum tillage to enhance positive/reduce negative effects of soil biota restore soil fertility+structure	Unsustainable practices in agro-industry can be trans-formed into sustainable and productive systems

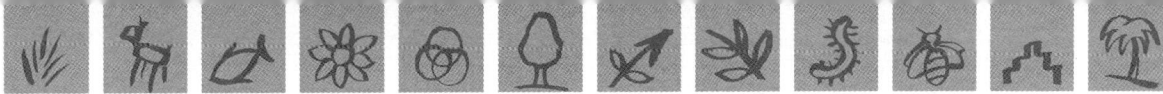

No. 9

RESPONSIBLE
TECHNICAL DIVISION

Plant Production and Protection Division
Seed and Plant Genetic Resources Service

Linda Collette

CASH CROP FARMING IN THE HIMALAYAS:

THE IMPORTANCE OF POLLINATOR MANAGEMENT AND MANAGED POLLINATION

10

CASH CROP FARMING IN THE HIMALAYAS:
THE IMPORTANCE OF POLLINATOR MANAGEMENT AND MANAGED POLLINATION

AUTHORS

Uma Partap

International Centre for
Integrated Mountain Development
Kathmandu, Nepal
uma@icimod.org.np

ACKNOWLEDGEMENTS

I am thankful to ICIMOD for providing necessary facilities, Federal Chancellery of Austria through Austroprojekt for financial assistance to carry out these studies and to farmers and local institutions in sharing their insights to the problem during the surveys. I am also grateful to Dr. Tej Partap, Vice Chancellor of Himachal Pradesh Agricultural University and my ICIMOD colleagues Dr. Farooq Ahmad, Mr. Min B. Gurung and Dr. S.R. Joshi for critically reviewing this paper and providing their valuable comments, which were of great help in improving this paper.

CONTENTS

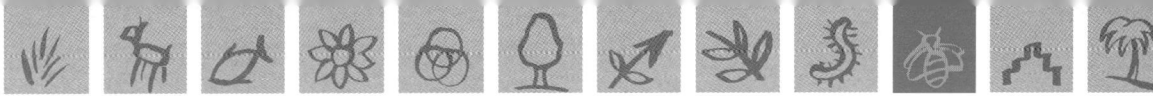

ABSTRACT

The focus of agriculture in the Himalayan region is slowly shifting from traditional cereal crops to high-value cash crops farming such as fruits and vegetables. This transformation from subsistence systems to commercial agriculture poses new challenges for improving and maintaining productivity and quality. Among these challenges are crop failures due to inadequate pollination. This is caused by several factors, the most important of which include the lack of adequate number of pollinators as a result of decline in pollinator populations and diversity due to several factors such as decline in wilderness and loss of habitat, land use changes, monoculture-dominated agriculture and excessive and indiscriminate use of agricultural chemicals and pesticides. Consequently, the need for ensuring pollination particularly through conserving pollinators and incorporating managed crop pollination has increased and will increase further. This calls for a more intensive focus on the issue from the perspective of policy, research, development and extension. Policy reorientation, improving institutional capabilities and human resources development are the key areas needing attention.

Based on our studies in apple pollination issues and farmers' concerns in Bhutan, China, India, Nepal and Pakistan, this paper presents a general picture of pollination issues faced by the farmers in the Himalayan region countries. The paper explains the importance of pollination in improving food security and livelihoods through enhancing agricultural productivity. It tries to analyze such issues as the decline in pollinator populations and its impact on agricultural productivity and implications on pollination management, and challenges to integrate pollination as a necessary input in agricultural policies and plans in the light of available information on pollination. This paper also emphasizes the need to conserve pollinator diversity to ensure pollination and at the same time it tries to present an alternative perception to beekeeping and that is "to promote" beekeeping primarily for crop pollination with honey and other bee products as by products". This new approach combines the two benefits well but institutional reorientation in the context of policies, research and extension might be necessary.

CASH CROPS FARMING IN THE HIMALAYAN REGION

Agriculture is the basis of the livelihood of over 80 percent of the rural population in the countries of the Hindu Kush-Himalayan (HKH) region. However, more than 90 percent of the farmers in the hill and mountain areas are marginal or small land-holding families, cultivating less than one hectare of land each (Banskota, 1992; Partap, 1995; Koirala and Thapa, 1997; Partap, 1999). Most agricultural land in the mountain areas is not only marginal in terms of potential productivity, but its quality also appears to be deteriorating as indicated by declining soil fertility and crop productivity. As a result, many mountain families face food shortages of varying degrees, that contribute to the chain reaction process of poverty–resource degradation–scarcity–poverty (Jodha and Shrestha, 1993). Therefore, it is necessary to explore all possible ways of increasing the sustainable productivity and carrying capacity of the farming systems in the mountains in order to improve the livelihoods of marginal mountain households (Partap, 1998, 1999).

No. 10

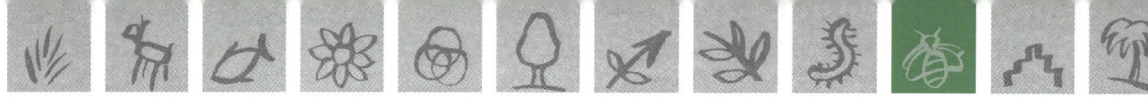

This, however, cannot be done by emphasizing the cultivation of cereal crops alone. If the poor mountain farmers are going to compete favourably in the modern world, they must be given options and alternatives that are not already captured by the competition. Development efforts tend to focus on exploring farming approaches to increase the productivity and carrying capacity of farms (Partap and Partap, 1997; Partap, 1999). Cash crops farming – fruit and vegetable crops suitable to specific agro-climatic conditions – is one comparative advantage that can be exploited by these farmers. For example, in uplands of the Himalayan region, off-season vegetables and fruits provide the comparative advantage to the farmers. As a result the focus of mountain agriculture is shifting from traditional cereal crops farming to high value cash crops and the cultivation of such crops as apples, almonds, pear, peaches, plums and cherries and off-season vegetables, both for local and export markets is increasing (Table 1).

TABLE 1

Cash crop farming in the Himalayan region

CASH CROPS	PROVINCE, COUNTRY	AREA (000 HA)	ANNUAL PRODUCTION (000 TONNES)
Apple	Indian Himalayas	227.61	1320.49
	Nepal	4.97	30.46
	Bhutan	2.03	13.00
	Chinese Himalayas[4]	82.86	208.22
	Pakistan[5]	49.46	637.97
Citrus	Bhutan[3]	8.00	77.00
	Nepal[2]	1.20	6.00
	Northwest Indian Himalayas[1]	39.80	100.00
	Chinese Himalayas[4]	0.20	0.80
	Pakistan		
Other fruit crops	Bhutan	0.13	-
	Chinese Himalayas (Sichuan, Yunnan)[4]	61.60	354.20
	Indian Himalayas[6]	530.00	1,595.00
	Pakistan[7]	39.10	386.20
Vegetable Crops	Bhutan[3]	6.00	22.30
	Chinese Himalayas[6]	14.50	26.30
	Indian Himalayas[6]	318.10	1,354.40
	Nepal[6]	140.00	741.60
	Pakistan[6]	282.90	1,418.80
Oilseed crops	Chinese Himalayas (Sichuan, Yunnan)[4]	1172.60	1756.30
Other crops (chilli, ginger, pulses, oilseeds, tea, cardamom, cotton, potato, tomato, etc.)	Bhutan	105.60	125.40
	Chinese Himalayas (Sichuan, Yunnan)[4]	2276.30	16,688.20
	Himachal Pradesh, India	-	-
	Uttaranchal, India[7]	19.00	392.00
	Balochistan, Pakistan[7]	5.00	765.00
	NWFP, Pakistan[7]	9.20	90.50

Sources:

[1] National Horticulture Board, New Delhi, India 1998, Department of Horticulture, Himachal Pradesh 1998, [2] Agricultural Statistics of Nepal, (1998/99), Department of Agriculture, HMG, Nepal [3] Policy and Planning Division, Ministry of Agriculture, Royal Government of Bhutan, 1999

[4] Agricultural Statistics of China, 1997; and Agricultural Census of Tibet, 1997 [5] Agricultural Statistics of Pakistan. Ministry of Food, Agriculture and Livestock, Economic Wing, Islamabad Government of Pakistan. 1998-99; and Khan, 1998. [6] Partap and Partap, 1997 [7] Tulachan, 2001.

THE ROLE OF POLLINATION IN IMPROVING FOOD SECURITY AND LIVELIHOODS

For a farmer, the most desired goal in agriculture is to get the maximum possible crop yields and better quality fruit and seeds under given inputs and ecological settings. It is particularly important to get a premium price for the produce when farmers are engaged in cash crop farming. There are two well known methods for improving crop productivity. The first method is making use of agronomic inputs, including plant husbandry techniques such as the use of good quality seeds and planting material, and practices to improve yields, for example, providing good irrigation, organic manure and inorganic fertilizers and pesticides. The second method includes the use of biotechnological techniques, such as manipulating rate of photosynthesis and biological nitrogen fixation, etc. These conventional techniques ensure healthy growth of crop plants, but work up to a limit. At some stage crop productivity becomes stagnant or declines with additional inputs for the known agronomic potentials of crop will have been harnessed (Partap and Partap, 1997).

The third and relatively less known (particularly in the countries of the Himalayan region) method of enhancing crop productivity is through managing pollination of crops using friendly insects, which in the process of searching for food perform this useful service to farmers (Partap and Partap, 1997). Pollination is an ecological process based on the principle of mutual interactions or inter-relationships (known as proto-cooperation) between the pollinated (plant) and the pollinator. Pollinators visit the flowers of the plants to obtain their food (i.e. nectar and pollen) and in return pollinate them. In many cases it is the result of the intricate relationship between plants and its pollinators and the reduction or loss of either affects the survival of both. In recent years the Convention on Biological Diversity (CBD) has recognized pollination as a key driver in the maintenance of biodiversity and ecosystem function.

The pollination process involves the transfer of pollen from the male part of the flower called 'anthers' to the female part called 'stigma' of the same flower (self-pollination) or another flower of the same or another plant of the same species (cross-pollination). Pollination is vital for completing the life cycle of plants and ensuring production of fruit and seed whether agricultural crops or natural vegetation/flora. This ecological process is an essential prerequisite for fertilization and fruit/seed set. If there is no pollination, there will be no fertilization, no fruits or seeds will be formed and farmers will harvest no crop. Pollination is therefore the most crucial process in the life cycle of the plants and is essential for crop production and biodiversity conservation and helps enhance farm income and rural livelihoods. Figure 1 shows the relationship of pollination to improved livelihoods through enhancing agricultural productivity and biodiversity conservation.

Many cash crops are actually self-sterile and require cross-pollination to produce seeds and fruit (McGregor, 1976; Free, 1993). But it is not only self-sterile varieties that benefit from cross-pollination, but self-fertile varieties also produce more and better quality seeds and fruits if they are cross-pollinated (Free, 1993). Logically, the increase in the cultivation of cross-pollinated cash crops will also increase the need for managed pollination. Equally

No. 10

interesting is the adoption of apiculture as a new enterprise by many people. Promoting use of beekeeping for pollination of cash crops will be of benefit to both the beekeeper who will receive money for the pollination services of his honeybees and harvest honey and to the farmer whose income will be increased through boosting crop productivity as a result of pollination services of bees. This will help ensure food security and enhance the livelihoods of both the farmers and the beekeepers (Figure 1). This system of hiring and renting honeybee colonies for apple pollination is being practiced in Himachal Pradesh in northwest Indian Himalayas. In Maoxian county in the Hengduan Mountains a somewhat similar but rather unsustainable system of apple pollination is prevalent. Here, farmers hire 'human pollinators' for pollinating apple and pear trees by hand.

FIGURE 1

Contribution of pollination to agricultural productivity and improving rural livelihoods

INADEQUATE POLLINATION AS A FACTOR AFFECTING CROP PRODUCTIVITY

The ongoing transformation from subsistence to cash crop farming poses new challenges for maintaining crop productivity and quality. There are signs that across the HKH region the overall productivity of many mountain crops is going down. Possibly the worst affected crops are the cash crops like fruit, particularly apples, and off-season vegetables that are the hope of the region in terms of providing farmers with cash income and underpinning development efforts. This reduction in productivity is taking place despite extensive efforts at extension and information to support improvements in a range of management practices, and strong support for the introduction of successful commercial varieties. The studies revealed that among the several factors affecting

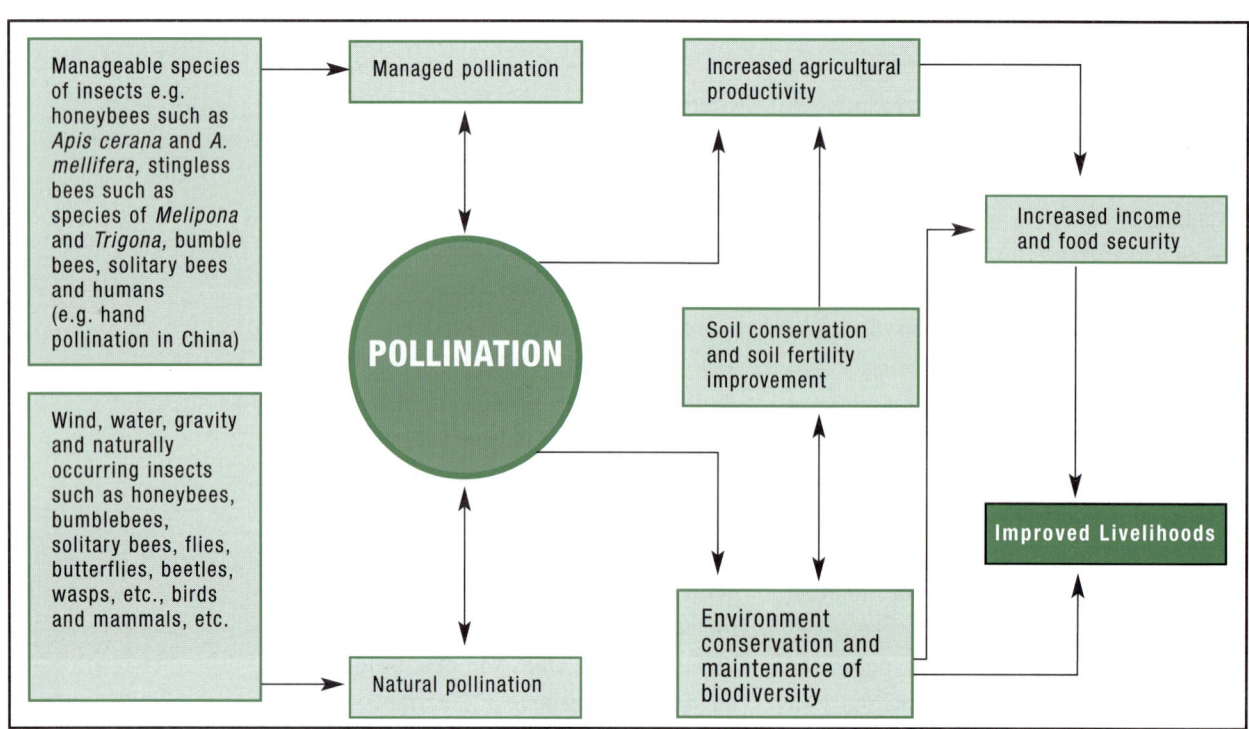

mountain crop productivity pollination plays an important role. Evidence of this emerging pollination problem has been documented in a series of field studies carried out by ICIMOD across the Himalayan region (Partap, 1998; Partap and Partap, 2000, 2001; Partap *et al.,* 2000). These studies investigated the state of inadequate pollination, its cause factors and its impact on crop productivity.

POLLINATOR DIVERSITY AND ITS ROLE IN ENHANCING CROP PRODUCTIVITY

Pollinators provide an essential ecosystem service that contributes to the maintenance of biodiversity and ensures the survival of plant species including crop plants. Two types of pollinators occur in nature. These include abiotic pollinators such as wind, water and gravity, and biotic pollinators such as insects, birds and various mammals. It has been estimated that over three quarters of the world's crops and over 80 percent of all flowering plants depend on animal pollinators, especially bees (Kenmore and Krell, 1998). Globally the annual contribution of pollinators to the agricultural crops has been estimated at about US$ 54 billion (Kenmore and Krell, 1998).

Insects are the most commonly occurring pollinators of many agricultural and horticultural crops. Different kinds of insect pollinators such as bees, flies, beetles, butterflies, moths and wasps are important pollinators of many crops. Among insects, bees are more effective pollinators than other insects because, unlike other insects, they are social and collect nectar and pollen

not only to satisfy their own needs but to feed their young; their body hairs help transfer pollen from one flower to another; they show flower constancy and move from one flower to another of the same species; and many species can be reared and managed for pollination.

Over 25 000 species of bees are found in the world. These include honeybees, bumble bees, stingless bees and solitary bees. Bees are the most effective pollinators of crops and natural flora and are reported to pollinate over 70 percent of the world's cultivated crops. It has also been reported that about 15 percent of the hundred principal crops are pollinated by domestic bees (i.e. manageable species e.g. hive-kept species of honeybees, bumble bees, alfalfa bees, etc.), while at least 80 percent are pollinated by the wild bees (Kenmore and Krell, 1998). These non-honeybee pollinators are estimated to provide the pollination services worth US$ 4.1 billion per year to the US agriculture (Prescott-Allen and Prescott-Allen, 1990).

THE ISSUE OF DECLINING POLLINATOR POPULATIONS

In recent years there is a world-wide decline in pollinator populations and diversity. The factors causing this decline could be the decline in the habitat, with the accompanying decrease in their food (nectar and pollen) supplies as a result of decline in pristine areas, land use changes, increase in monoculture-dominated agriculture, and negative impacts of modern agricultural interventions, e.g. use of chemical fertilizers and pesticides (Verma and Partap, 1993; Partap and Partap, 1997; Partap and Partap, 2002). Earlier, farmers

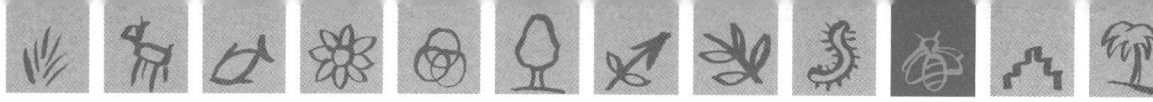

used to grow a variety of crops, which bloomed during different months of the year and provided food and shelter for a number of natural insect pollinators and hence the pollination problem never existed. Monocropping also requires pesticide use to control various pests and diseases. Thus, it not only reduced the diversity of food sources of pollinator but also led to the killing of many pollinators due to pesticides. The insecticides have contributed to the extermination of both the diversity and abundance of pollinating insects. Changes in climate might also be affecting insect numbers (Partap and Partap, 2002).

IMPACT OF DECLINE IN POLLINATOR POPULATION AND DIVERSITY

The decline in pollinator population and diversity presents a serious threat to agricultural production and conservation and maintenance of biodiversity in many parts of the world. One indicator of the decline in natural insect pollinators is decreasing crop yields and quality despite necessary agronomic inputs. Examples can be found in Himachal Pradesh in northwest India, northern Pakistan and parts of China where despite all agronomic inputs, production and quality of fruit crops, such as apples, almonds, cherries and pears, is declining. Extreme negative impact of declining pollinator populations can be seen in other areas, for example in northern Pakistan where both farmers and institutions have failed to understand the importance of managed pollination. Disappointed with the very low yields and quality of apples as a result of poor pollination several farmers in Azad Jammu and Kashmir of Pakistan have chopped off their apple trees (Partap, 2001).

One implication of the decline in the pollinator populations as well as diversity is that it has created the need for managed pollination in order to maintain crop yields and quality. In fact, farmers engaged in cash crops' farming in those areas where pollinator populations have declined are forced to manage pollination of their crops through different ways. For example, farmers in Himachal Pradesh in northwest India are using honeybees for pollination of their apples, while those in Maoxian county in Hengduan Mountains of China are pollinating their crops, e.g. apples and pears, through hand pollination using human beings as pollinators (beekeepers do not rent their honeybee colonies for pollination of these crops because farmers make excessive use of pesticides even during flowering season). Hand pollination is an interesting method of pollinating crops and provides employment and income generating opportunities to many people during apple flowering season. But at the same time it is an expensive, time-consuming and highly unsustainable proposition of crop pollination owing to the increased labour scarcity and costs. Moreover, a large part of farmers' income is used in managing pollination of their crops.

THE IMPORTANCE OF POLLINATOR MANAGEMENT FOR CASH CROP POLLINATION IN THE HIMALAYAS

As explained earlier in this paper the populations of these pollinators are declining in several intensively cultivated areas. Thus, the need to conserve, promote and diversify pollinator resources is pressing in several countries of the developing world. This calls for initiating research and extension activities in this direction and developing strategies to promote conservation and sustainable use of pollinators. This will require much wider understanding of the multiple services provided by the pollinator diversity and the factors that influence them, including farmers, in order to secure sustained pollinator services in agricultural ecosystems. This calls for initiating efforts at awareness, research and extension level. Certain measures suggested for increasing the number of insect pollinators include habitat conservation, discouraging over-use of pesticides, promoting integrated pest management (IPM), awareness raising, formulating policies to include managed crop pollination as an input in agricultural development packages and strengthening R&D systems.

Many species of bumblebees (*Bombus* spp.) and solitary bees like *Amegilla, Andrena, Anthophora, Ceratina, Halictus, Lasioglossum (Evylaeus), Megachile, Nomia, Osmia, Pithis,* and *Xylocopa* can be reared on a large scale and managed for crop pollination. In fact in many developed countries various insect pollinators, including some species of bumblebees and solitary bees, are being reared and managed commercially for pollination of various crops, particularly those that are less or uneffectively pollinated by

honeybees. Bumblebees, for example, are used for the pollination of potatoes, tomatoes, strawberries and other crops grown in glasshouses, alkali bees and leaf cutter bees for the pollination of alfalfa, horn-faced bees for apples, almonds and other fruit trees, and other species of solitary bees for pollination of cotton, mustards, lucerne and berseem. In Japan the solitary bee *Osmia cornifrons* Rad. is being reared and managed on a large scale to pollinate about one-third of all apple crops (Batra, 1995; Sekita, 2001).

There is good potential for the managed use of non-*Apis* pollinators in the developing countries. There are thousands of hectares of land under crops that need cross-pollination. In cold and arid areas, for example Balochistan (Pakistan), Mustang (Nepal) and Lahul (Himachal Pradesh), where stationary beekeeping cannot be practised because of the prevailing cold and dry climatic conditions and lack of forage during the larger part of the year, conserving and managing non-honeybees for pollination can be a good option. Their conservation can be ensured simply by avoiding use of pesticides during the period when crops and other plants are blooming. This could be of great help in saving these pollinators from the hazardous effects of pesticides because the period of adult life of these insects coincides with the flowering of crops.

Even though both the need and the potential exist, the practice of rearing and managing natural pollinators for crop pollination is practically absent in the developing countries. The reason is that most institutions do not have the mandate and necessary expertise in this field. Thus, development and use of these insects in this part of world will take a long time. Major research and extension efforts will be needed before such insects can be reared and managed for pollination of crops in the

No. 10

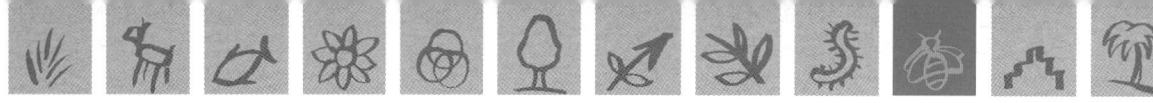

region. However, efforts towards the conservation of these non-*Apis* pollinators can be initiated. The first step in this direction could be to save them from the harmful impacts of pesticides. For this, there is need to raise awareness about the harmful effects of agricultural chemicals and pesticides on these invaluable pollinators. There is also need to train farmers and extension workers to make safe use of carefully selected, less toxic pesticides outside the blooming period of crops.

MANAGED POLLINATION AS A SOLUTION TO ADDRESS THE IMMEDIATE PROBLEM OF CASH CROP POLLINATION

Promoting conservation and management of naturally occurring insect pollinators is very important for sustaining agricultural productivity in the long run. Efforts – both at research and extension level-must be continued to identify, assess and develop techniques/methodology to rear and manage them for enhancing crop pollination. But as already explained the population of these bees are declining and there is a lack of scientific manpower and institutional infrastructure to rear and promote them in the countries of the Himalayan region. Moreover, the problem of pollination has already started in several areas. Therefore, promoting managed pollination is essential to address this immediate problem of inadequate pollination in several cash crops, for example, apples and pears. Our findings revealed two different cases of managed pollination of apple crops in the Himalayan region; one in Himachal Pradesh a small province in north-

western Indian Himalayas where farmers are using manageable species of honeybees (*Apis cerana* and *Apis mellifera*) for pollination of their apple crop, and another in Maoxian valley located in the northwest of Sichuan Province of China where farmers employ "human bees" to pollinate their apples by hand (hand pollination). The details of these case studies are given in the following text.

HONEYBEES AS THE MOST EFFICIENT AND MANAGEABLE POLLINATORS

As explained, many varieties of these cash crops are partially or completely self-incompatible and cannot produce fruit or seed without cross-pollination of their flowers. Moreover, it is not only self-incompatible varieties that benefit from cross-pollination, but self-fertile varieties also produce better quality fruit and seeds if they are cross pollinated (Free, 1993). While other agronomic inputs, such as the use of manure, fertilizers, pesticides and irrigation are important, without cross-pollination desired crop yield and quality of harvest cannot be achieved.

Honeybees are the most widely known of all the bees because they provide honey, beeswax and other products and beekeeping is a prevailing tradition among mountain farming communities. They are the most efficient pollinators of cultivated crops because their body parts are especially modified to pick up pollen grains, they have body hair, have potential for long working hours, show flower constancy, and adaptability to different climates (Free, 1964, 1966; McGregor, 1976). Research has shown that pollination by honeybees increases fruit set, enhances fruit quality and reduces fruit drop in apple (Dulta

and Verma, 1987). Among different species of honeybees, the hive-kept species (*Apis cerana* and *Apis mellifera*) are of special value because they can be managed for pollination and moved to fields/orchards where and when necessary for pollination. Pollination using honeybees is the most cost-effective method for pollinating apple and other fruit crops. Use of beekeeping is, therefore, the most promising method of cash crop pollination in the Himalayan region.

In fact, the main significance of honeybees and beekeeping is pollination, whereas hive products are of secondary value. It has been estimated that the benefit of using honeybees for enhancing crop yields through cross-pollination is much higher than their role as produces of honey and beeswax. Various estimates have been made to prove the economic value of honeybees in agriculture in developed countries. Recent estimates by Morse and Calderone (2000) show that the value of honeybee pollination to crop production in the US is US$ 14.6 billion. Similar estimates have been made for other countries. For example the value of honeybee pollination has been estimated at CAN$ 1.2 billion in Canadian agriculture (Winston and Scott, 1984), US$ 3 billion in EEC (Williams, 1992), and US$ 2.3 billion in New Zealand (Matheson and Schrader, 1987). Cadoret (1992) estimated that the direct contribution of honeybee pollination to increase farm production in 20 Mediterranean countries was US$ 5.2 billion per year – 3.2 billion in developing countries and two billion in other countries. Similarly, Chen (1993) estimated the value of honeybees to four major crops in China, including cotton, rapeseed, sunflower and tea, at US$ 0.7 billion.

Experimental research on the impact of honeybee pollination on crop productivity in the Himalayan Region

Honeybees are reported to play a vital role in enhancing the productivity levels of different crops such as fruit and nuts, vegetables, pulses, oilseeds and forage crops. A number of studies have been done to show the impact of honeybee pollination on different cash crops. However, the role of honeybees is not very well understood in the countries of the Himalayan region. Most of the research work has been done in developed countries of the world where honeybees are being used for the pollination of various crops. However, the limited research carried out in the countries of the Himalayan region has proved that bee pollination increases the yield and quality of various crops (Table 2). These experiments showed that bee pollination increased yield and fruit quality in apple (Dulta and Verma, 1987; Gupta *et al.*,

CROP	INCREASE IN FRUIT SET (%)	INCREASE IN FRUIT WEIGHT (%)	INCREASE IN FRUIT SIZE (LENGTH, DIAMETRE) (%)	REFERENCES
Apple	10	33	15, 10	Verma and Dulta, 1998
Peach	22	44	29, 23	Partap et al., 2000
Plum	13	39	11, 14	Partap et al., 2000
Citrus	24	35	9, 35 Also, premature fruit drop decreased by 46 percent, increased juice by 68 percent and sugar contents in juice by 39 percent	Partap, 2000
Strawberry	112	48	Misshapen fruits decreased by 50 percent	Partap, 2000

TABLE 2

Impact of honeybee *(Apis cerana)* pollination on fruit productivity

No. 10

1996), peach, plum, citrus, kiwi (Gupta *et al.*, 2000) and strawberry (Partap, 2000; Partap *et al.*, 2000). Bee pollination did not only increase the fruit set but also reduced fruit drop in apple, peach, plum and citrus (Dulta and Verma, 1987; Partap, 2000; Partap *et al.*, 2000). Reports have also indicated an increase in fruit juice and sugar content in citrus fruits (Partap, 2000). In strawberry, bee pollination reportedly reduces the percentage of misshapen fruits (Partap, 2000).

Studies have shown that honeybee pollination enhanced seed production and quality of seed in various vegetable crops such as cabbage, cauliflower, radish, broad leaf mustard and lettuce (Partap and Verma, 1992; 1994; Verma and Partap, 1993; 1994). These results confirm the usefulness of bee pollination and its role in increasing crop productivity and improving the quality of fruits and seeds (Table 3).

Scientific evidence confirms that bee pollination also improves the yield and quality of other vegetable crops such as asparagus, carrots, onion, turnips and several other crops (Deodikar and Suryanarayana, 1977). Recent experiments carried out in different parts of the northeast Himalayan region show that honeybee pollination does not only increase fruit set in rapeseed and sunflower but also increases the oil contents in these oilseed crops (Singh *et al.*, 2000).

The quality of pollination is determined by the number of colonies per unit area, strength of bee colonies, placement of colonies in the field, time of placement of bee colonies, and the weather conditions. Experiences from pilot experiments have shown that the best results are achieved by placing strong bee colonies, having large amount of unsealed brood, free of diseases, at the time of 5-10 percent flowering in the crop (Free, 1993; Verma and Partap, 1993).

The significance of honeybee diversity for pollination

The Himalayan region is one of the richest in honeybee species' diversity in the world. There are five species of honeybees: three wild species that cannot be kept in hives — the giant honeybee *(Apis dorsata)*, the little bee *(Apis florea)*, and the rock bee *(Apis laboriosa)* — and two hive-bee species, the Asian hive bee *(Apis cerana)*, and the introduced bee *(Apis mellifera)*. All honeybees are good crop pollinators, but because the wild species cannot be kept in man-made hives they cannot be transported to the sites where bees are needed for crop pollination. The honeybee species' diversity in the Himalayan region holds much potential for wider use in managing

TABLE 3

Impact of honeybee *(Apis cerana)* pollination on vegetable seed production

CROP	INCREASE IN POD SETTING (%)	INCREASE IN SEED SETTING (%)	INCREASE IN SEED WEIGHT (%)
Cabbage	28	35	40
Cauliflower	24	34	37
Radish	23	24	34
Broad leaf mustard	11	14	17
Lettuce	12	21	9

Source: Partap and Verma, 1992; 1994, Verma and Partap, 1993; 1994

crop pollination in ways suited to the conditions in specific areas. In particular, the native hive honeybee *Apis cerana* offers clear advantages as a pollinator in remote and higher altitude area (Partap and Partap, 1997, 2001, 2002). Partap and Partap (2002) suggested an area-based approach to use the existing honeybees' diversity for pollination.

Managed pollination through using honeybees for apple pollination in the Himalayan Region: A case study from Himachal Pradesh, India

In the developed countries like the US, Canada, Europe and Japan honeybees are used as one of the inputs in agriculture. These countries are for long using honeybees for pollination of crops such as apples, almonds, pears, plums, cucumbers, melons, watermelons, and a number of berries. But the Himalayan region lags far behind in making use of honeybees for crop pollination. Even though plenty of scientific evidence is now available to prove that honeybees increase the productivity of various cash crops, still the practice of using honeybees for crop pollination does not exists in the Himalayan region. While in the US first colonies of honeybee, *Apis mellifera* were rented for pollination of pears in Virginia in 1895 (Waite, 1895) and for apple pollination in 1909 in New Jersey (Morse and Calderone, 2000), in the Himalayan region (in Himachal Pradesh) first colonies of honeybees were rented for apple pollination only recently in 1996.

A recent survey carried out by the author in apple farming areas of Bhutan, China, India, Nepal and Pakistan revealed that it is only in Himachal Pradesh in northwestern Indian Himalayas where honeybees are being used for apple pollination (Partap, 1998). Here, some farmers keep their own honeybee colonies while others rent them from the Department of Horticulture or from the private beekeepers. The fees for renting bee colonies either *Apis cerana* or *A. mellifera* is Indian rupees 800 (US$ 16) per colony for two weeks. This includes Rs 500 (US$ 10) as refundable security deposit and Rs. 300 (US$ 6) per colony per two weeks of rent. *Apis mellifera* is the main bee species made available to farmers from government institution and private beekeepers for pollination purpose.

At present, Himachal Pradesh is the only place in the whole of the HKH region where a well-organized system has been established for hiring and renting honeybee colonies. A number of pollination entrepreneurs (beekeepers who rent honeybee colonies for crop pollination) have now started up in the state to complement the official services. The findings also revealed that in addition to increasing the number of insect pollinators by renting colonies of honeybees, some farmers are trying to save the populations of existing pollinators by making judicious use of carefully selected, less toxic pesticides and spraying outside the flowering period of apple. Even though beekeeping is a common tradition throughout the Himalayan region, yet renting honeybees for crop pollination is not much known in other countries. In Pakistan, Bhutan and Nepal farmers are not aware of the importance of pollination and the existing pollination crises in their orchards. Thus, any kinds of management efforts are also absent.

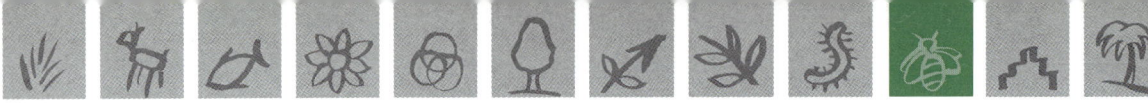

HAND-POLLINATION (USING HUMANS AS POLLINATORS): A CASE STUDY FROM MAOXIAN VALLEY, CHINA

Among other countries of the Himalayan region, it is in China where farmers have understood the value of managed pollination. There is a serious problem of apple and pear pollination in Maoxian county of China because pollinators have been killed by the overuse of pesticides and beekeepers do not rent their honeybees for pollination for the same reason. But unlike Himachal Pradesh, farmers pollinate their apples by hand. Hand-pollination is a common practice of managing pollination in apple crop in this valley where every family member - men, women and children are engaged in pollination of apple flowers making it a community effort (Partap and Partap, 2000). Here people do the work that can otherwise be done more efficiently by honeybees and other insect pollinators. They pollinate large areas of apples and pears by hand to make sure that each flower is properly pollinated (Partap and Partap, 2000). Therefore, hand-pollination has been promoted by the local government and is now a common practice of managing pollination in apple crop in this valley.

Various cooperation mechanisms among farmers have also evolved for sharing labour and skills. Farmers having larger orchards generally employ labourers for this purpose. Even though beekeeping is common in the area, the practice of renting honeybee colonies for pollination is surprisingly absent. Two reasons were assigned for it; one, it was not promoted in the first place; and second, beekeepers are hesitant to rent their bee colonies because of excessive use of pesticide sprays on apples.

Hand pollination is a laborious and time-consuming method of crop pollination. Even though it is the most reliable method of ensuring apple pollination today, it will not be sustainable as a long-term solution, largely because of increasing labour scarcity and costs. Therefore, in areas where agriculture is diversifying to new cash crops there is a need to raise awareness among people and local research and extension systems about not only the significance of managing pollination but also for using bee pollinators as an alternative to the prevalent practice of pollination by hand. The risk of pesticides can be minimized through judicious use, as well as by adopting practices like integrated pest management practices (Partap *et al.*, 2001; Partap and Partap, 2002).

CHALLENGES IN MANAGED CROP POLLINATION

As reported in the earlier sections of this paper, insect pollinators including manageable species of honeybees, stingless bees, bumblebees and solitary bees can play an important role in pollination and in areas like N. America, Europe, and Japan they are used extensively to ensure pollination of fruit and vegetable crops. However, although both the need and the potential exist in the developing countries, the practice of managed pollination is practically absent. Forget about rearing and using species of bumblebees and solitary bees; here there is even no practice of using hivebees such as *Apis cerana* and *Apis*

mellifera and stingless bees even though beekeeping is a tradition throughout the developing countries. Development and use of insects other than hivebees in this part of world will take a long time and need major research and extension efforts before such insects can be reared and managed for pollination of crops in the region.

This section discusses the issues and challenges in ensuring crop pollination through using manageable species as well as promoting conservation and sustainable use of natural pollinators as a sustainable solution to enhance agricultural productivity. Figure 2 presents the challenges to integrate pollination with farming systems and enhancing rural livelihoods through promoting managed pollination and conserving pollinator populations. The main constraints to promoting managed pollination by using honeybees and other pollinators are lack of awareness and understanding among farmers, extension workers, planners and policy-makers about the importance of pollinators and pollination, lack of integrating pollination in agricultural development packages, scarcity of managed colonies of honeybees, and lack of knowledge about conservation, rearing and use of pollinators and their pollination behaviour.

Awareness raising

Lack of awareness at all levels - be it farmers, extension workers, and professionals at policy and planning level-is one of the main problems in promoting managed pollination. With a few exceptions of farmers in those areas where there is a pollination problem, people are not aware of the value of honeybees (including other pollinators) for agricultural production. This is both because beekeeping has always been promoted exclusively as an enterprise for honey production and because cash crops' farming is a new activity in many developing countries, and there is no indigenous knowledge on the need for managed crop pollination for enhancing cash crop production. Raising awareness at all levels about the importance of managed crop pollination through beekeeping and other pollinators is the first step as part of development efforts.

Including pollination as a technological input to agricultural development packages

Pollination has been overlooked in agricultural development strategies and is not included as a technological input in agricultural development packages. High value agriculture is being promoted in several areas and extension institutions offer packages of practices for each type of crop, but the importance of managing pollination to achieve higher yields has been overlooked. Thus farmers have no way of knowing how essential it can be. This weakness in the agricultural extension system needs to be addressed.

Since pollination is essential for the production of fruits and seeds, it should be included in agricultural development packages by promoting beekeeping for crop pollination as a 'double benefit approach'. Thus the most important step in promoting the wider use of honeybees for crop pollination is to include beekeeping as part of agricultural development efforts. Including managed crop pollination in agricultural development packages will also help develop strategies to conserve, promote and use other pollinators.

No. 10

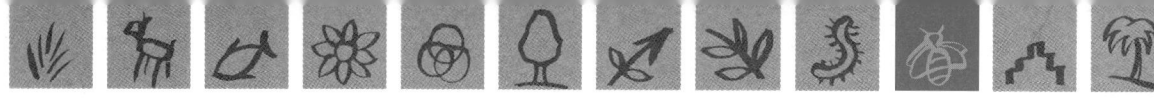

F I G U R E   2

Awareness raising, reorientation of agricultural development policies to include pollination as an input, institutional strengthening R&D Institutions, human resources development and capacity building are necessary to integrate pollination in farming systems and enhance agricultural productivity and livelihoods of rural people

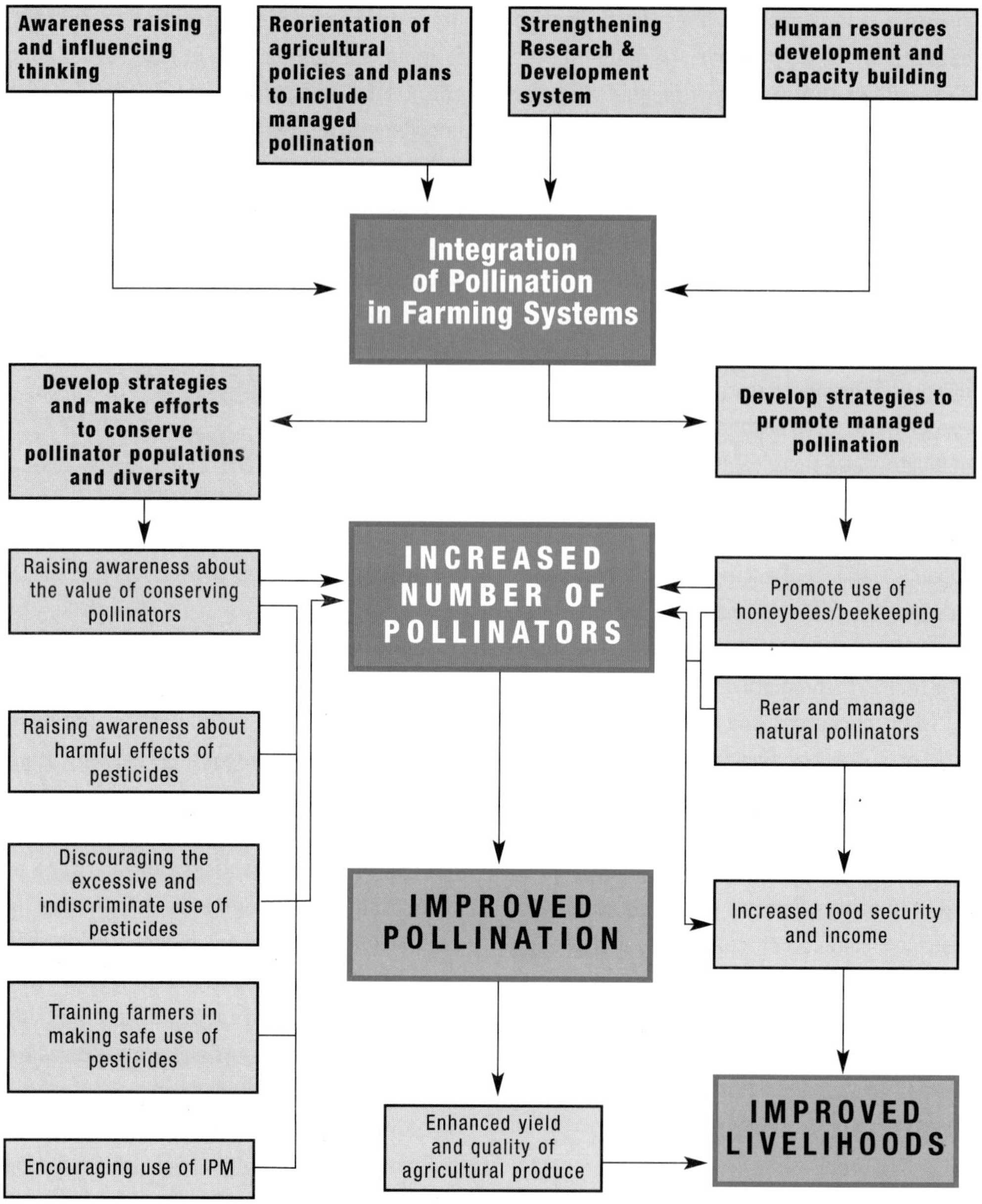

Awareness raising and influencing thinking

Reorientation of agricultural policies and plans to include managed pollination

Strengthening Research & Development system

Human resources development and capacity building

Integration of Pollination in Farming Systems

Develop strategies and make efforts to conserve pollinator populations and diversity

Develop strategies to promote managed pollination

Raising awareness about the value of conserving pollinators

INCREASED NUMBER OF POLLINATORS

Promote use of honeybees/beekeeping

Raising awareness about harmful effects of pesticides

Rear and manage natural pollinators

Discouraging the excessive and indiscriminate use of pesticides

IMPROVED POLLINATION

Increased food security and income

Training farmers in making safe use of pesticides

Encouraging use of IPM

Enhanced yield and quality of agricultural produce

IMPROVED LIVELIHOODS

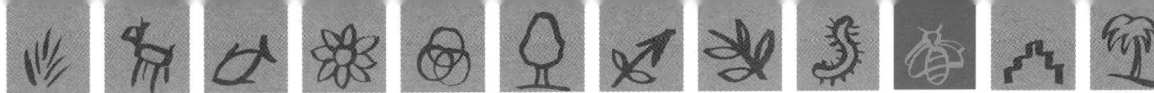

Influencing thinking about bees and beekeeping

Traditional thinking is that beekeeping is for honey production, its role in crop pollination is rarely considered. Today, most government agencies are only engaged in promoting beekeeping for honey production. The move towards introduction of *Apis mellifera* to increase honey production is an example of this. Thus there is need to change the general 'mindset' about honeybees and beekeeping, and to raise awareness about the importance of managed crop pollination.

Strengthening research and development institutions

Managed crop pollination is a relatively new area. There are few institutions with explicit mandates or expertise for research and extension in this area. Most institutions are working only with beekeeping and promoting it as a cottage industry to increase family income through the sale of honey. Promoting the value of honeybees as reliable pollinators of agricultural crops will require special efforts to strengthen research and extension systems. This is necessary in order to underline applied research in key areas of managed crop pollination. Issues such as decline of pollinator populations and the need to conserve them need to be addressed by the institutions.

Human resources development and capacity building

Lack of knowledge among farmers about the pollination behaviour of honeybees is another constraint hindering the use of honeybees for crop pollination. Even those farmers, who do know that they can use honeybees to increase apple pollination and yield, don't always know how to use the bees. Though linked with the institutional strengthening, it requires more focus to build the capacities of individual farmers, development workers and farmer-led organizations that are the agents of change. There is need to train farmers and beekeepers in managing honeybees for crop pollination. There is also need to develop human resources and build their capacities to initiate activities in the area of conserving, rearing and using pollinators to improve pollination and thus agricultural productivity.

CROP POLLINATION INVESTMENT PROSPECTS

The inputs of pollinators in agriculture husbandry and biodiversity conservation have not been recognized by policy makers, planners, development workers and farmers. There is no conceptual clarity and recognition of the value of pollinators. There is also need for a change in thinking about the value of honeybees as crop pollinators at all levels: policy, planning, research, beekeeping and farming. The initial thrust of the pollination programme should be to raise awareness about the significance of managing pollination through honeybees and generate knowledge and information to facilitate the formulation of strategies to

No. 10

ensure the wider use of beekeeping for pollination. Honeybees should be seen as crop pollinators first, and as honey producers second. Changes in research and development investment policies may be needed to encourage this. It is also necessary to evolve strategies to promote investment in research and development that will enhance the use of honeybees and other pollinators for pollination. This means developing area-based approaches, making full use of the existing diversity of pollinators including honeybees.

GENDER CONCERNS IN POLLINATOR MANAGEMENT AND MANAGED POLLINATION

Women play an important role in agriculture and food production in several developing countries. They are the dominant labour force in agriculture and make a crucial contribution through engaging themselves in all agricultural activities from preparation of the soil to post-harvest operations. Development of rural women and encouraging their full participation as equal partners in the social and economic mainstream is one of the greatest challenges being faced by several developing countries today.

Pollinator management and managed pollination have direct impact on improving women's lives in terms of increasing the economic/food security and reducing their drudgery. Information on the role and significance of women and how their livelihood is affected by failure of pollination and better management of crop pollination is presented here to make a case why future strategies relating to managing

pollination should give due attention to gender roles and capacity building.

Let us first analyze how women are affected by managed pollination. As reported, the most visible impact of crop pollination failure is seen in cash crops. Cash crops have played a major role in improving food security and livelihoods of farmers in hills and mountains. Increased productivity through managed pollination has direct implications for women's lives in terms of increasing their economic security. Better pollination leads to increased agricultural production resulting in increased family income leading to enhanced food security and livelihoods. This also ensures better health, nutrition and education for women. On the other hand declining crop yields through lack of adequate pollination increases drudgery as the women have to work extra hard to achieve food security. Therefore, involvement of women in managed pollination should be encouraged.

Our studies also show that women are key to successful management of pollinators (Partap, 1998; Partap and Partap, 2000; Partap and Partap, 2002; Partap et al., 1991). In Himachal Pradesh, where honeybees are being used for pollination of fruit crops, women farmers manage colonies for use in their own orchards as well as for renting (Partap, 1998; Partap and Partap, 2002). There are numerous local women farmers' associations in Himachal Pradesh, known as *Mahila Mandals,* which are actively engaged in beekeeping for renting bee colonies for pollination. These hill women farmers are being encouraged to raise honeybees and rent them for apple pollination. As a result a number of women beekeepers' groups are coming up. This has not only increased the income of these 'women-headed pollination entrepreneurs'

through renting bee colonies for pollination but also through sale of honey. In Hengduan mountains of China women are the backbone of the hand pollination process of fruit trees. Thus, it is necessary to evolve strategies for improving the skills of women in this field.

It is necessary to encourage women's involvement in management of pollinators and pollination in other countries also by creating a conducive environment through extension and demonstration activities, empowering them through training, research, and involving them in projects at national and international levels. While designing training programmes and formulating policies on pollination and conservation and sustainable use of pollinators special consideration should be given to the training of women and building their capacities. For example, women can be encouraged to take up beekeeping for pollination as an income generating activity. Programmes to provide training and support to such pollination entrepreneurs headed by the women can be launched. Such programmes may include capacity building, training and transfer of knowledge and appropriate technology. This will also help in bringing them into the mainstream of development.

CONCLUSION

Like soil, water and nutrients, pollination is also a limiting factor in crop productivity. The declining agricultural productivity can be attributed to a number of factors, but pollination plays a crucial role. We may make use of plant husbandry techniques, such as the use of better quality seed and planting material and provide all agronomic inputs including, good irrigation, use of organic and inorganic fertilizers and biocides, but if there is no pollination, no fruit or seed will be formed.

The pollination problem is relatively new and needs due attention at this early stage. Since pollinator scarcity is the main factor responsible for inadequate pollination, solutions to this lie in increasing the number of pollinators. This can be done by conserving populations of natural insect pollinators by promoting integrated pest management and making judicious use of chemical fertilizers and pesticides, however, the most practical and preferred solution to increase the number of pollinators would be by promoting manageable species of honeybees for pollination. There is need to formulate policies that include pollination as an integrated input to agricultural production technologies. Other challenges include strengthening research and extension institutions and human resources development.

REFERENCES

Batra, S.W.T., 1985. Bees and pollination in our changing environment.' *Apidology,* 26: 361–370.

Batra, S.W.T., 1997. Solitary bees for orchard pollination. *Pennsylvania Fruit News,* April 1997.

Cadoret, J.P., 1992. Doorway doses help defeat honeybees' vampire enemy. Ceres: The *FAO Review* 24: 8-9.

Crane, E., 1992. The past and present status of beekeeping with stingless bees. *Bee World* 73: 29-42.

Deodikar, G.B.; Suryanarayana, M.C., 1977. Pollination in the services of increasing farm production in India. In P.K.K. Nair (ed) *Advances in Pollen Spore Research,* pp 60-82. New Delhi: Today and Tomorrow Printers and Publishers.

Dulta, P.C.; Verma, L.R., 1987. Role of insect pollinators on yield and quality of apple fruit. *Indian Journal of Horticulture* 44:274-279.

Free, J.B., 1993. *Insect Pollination of Crops.* (2nd edn). London: Academic Press.

Gupta, J.K; Goyal, N.P.; Sharma, J.P.; Gautam, D.R., 1993. The effect of placement of varying numbers of *Apis mellifera* colonies on the fruit set in apple orchards having different proportions of pollinisers. In Veeresh, G.K.; Uma Shankar, R.; Ganeshaiah, K.N. (eds), *Proceedings of the International Symposium on Pollination in the Tropics, India:* International Union for Studies on Social Insects.

Gupta, J.K., Rana, B.S. and Sharma, H.K., 2000. Pollination of kiwifruit in Himachal Pradesh. In (M. Matsuka, L.R. Verma, S. Wongsiri, K.K. Shrestha and Uma Partap (eds.) *Asian Bees and Beekeeping: Progress of Research and Development.* Proceedings of the Fourth International Conference 23-28 March 1998, Kathmandu. Oxford and IBH Publishing Co. Pvt. Ltd. New Delhi, 274pp.

Jodha, N.S.; Shrestha, S., 1993. Sustainable and more productive mountain agriculture: problems and prospects. In *Mountain Environment and Development –Part 3 (Thematic Papers),* pp 1-65. Kathmandu: ICIMOD.

Kenmore, P. and Krell, R., 1998. Global perspectives on pollination in agriculture and agroecosystem management. *International Workshop on the Conservation and Sustainable Use of Pollinators in Agriculture, with Emphasis on Bees. October 7-9, 1998, Sao Paulo, Brazil.*

McGregor, S.E., 1976. *Insect Pollination of Cultivated Crop Plants.* Washington, DC: United States Department of Agriculture (USDA).

Morse, R.A.; Calderone, N.W., 2000. *The Value of Honey Bees as Pollinators of US Crops in 2000.* Electronic Document Downloaded from the Internet Page http://bee.airoot.com/beeculture/pollination2000.

Partap, T., 1998. Agricultural sustainability challenges in upland areas of semi arid and humid Asia.' In: *Proceedings of the Study Meeting on Sloping Land Agriculture and Natural Resources Management,* pp 39–84. Tokyo: Asian Productivity Organisation.

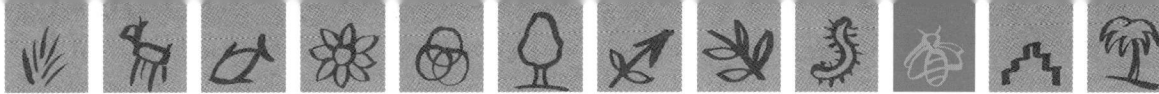

Partap, T., 1999. Sustainable land management in marginal mountain areas of the Himalayan region. *Mountain Research and Development* 19: 251-260.

Partap, U., 2001. *Warning Signals from the Apple Valleys.* Video Film, 31 minutes and 7 seconds, VHS Format. Kathmandu: ICIMOD

Partap, U., 2000a. Foraging behaviour of *Apis cerana* on citrus (*Citrus sinensis* var. Red Junar) and its impact on fruit production. In (M. Matsuka, L.R. Verma, S. Wongsiri, K.K. Shrestha and Uma Partap (eds.) *Asian Bees and Beekeeping: Progress of Research and Development.* Proceedings of the Fourth AAA International Conference 23-28 March 1998, Kathmandu. Oxford and IBH Publishing Co. Pvt. Ltd. New Delhi, 274pp.

Partap, U., 2000b. Pollination of strawberry by the Asian hive bee, *Apis cerana* F. In (M. Matsuka, L.R. Verma, S. Wongsiri, K.K. Shrestha and Uma Partap (eds.) *Asian Bees and Beekeeping in Asia: Progress of Research and Development.* Proceedings of the Fourth AAA International Conference 23-28 March 1998, Kathmandu. Oxford and IBH Publishing Co. Pvt. Ltd. New Delhi, 274pp.

Partap, U., 1998. Successful pollination of apples in Himachal Pradesh. *Beekeeping and Development* 48: 6-7.

Partap, U. and Partap, T., 1997. *Managed Crop Pollination: The Missing Dimension of Mountain Agricultural Productivity.* Mountain Farming Systems' Discussion Paper Series No. MFS 97/1. Kathmandu: ICIMOD.

Partap, U. and Partap, T., 2000. Pollination of apples in China. *Beekeeping and Development* 54: 6-7.

Partap, U. and Partap, T., 2002. *Warning Signals from Apple Valleys of the HKH Region: Productivity Concerns and Pollination Problems.* Kathmandu: ICIMOD, pp 106.

Partap, U. and Partap, T. and Yonghua, H., 2001. Pollination failure in apple crop and farmers' management strategies. *Acta Horticulturae 561:* 225-230

Partap, U.; Verma, L.R., 1992. Floral biology and foraging behaviour of Apis cerana on lettuce crop and its impact on seed production. *Progressive Horticulture* 24: 42-47.

Partap, U.; Verma, L.R., 1994. Pollination of radish by *Apis cerana. Journal of Apicultural Research* 33: 237-241.

Partap, U., Shukla, A.N. and Verma, L.R., 2000. Impact of *Apis cerana* pollination on fruit quality and yield in peach and plum in the Kathmandu valley of Nepal. In (M. Matsuka, L.R. Verma, S. Wongsiri, K.K. Shrestha and Uma Partap (eds.) *Asian Bees and Beekeeping: Progress of Research and Development.* Proceedings of the Fourth AAA International Conference 23-28 March 1998, Kathmandu. Oxford and IBH Publishing Co. Pvt. Ltd. New Delhi, 274pp.

Prescott-Allen, R. and Prescott-Allen, C., 1990. How many crops plants feed the world? *Conservation Biology* 4: 365-374.

Sekita, N., 2001. Managing *Osmia cornifrons* to pollinate apples in Aomori Prefecture, Japan. *Acta Horticulturae* 561: 303-308.

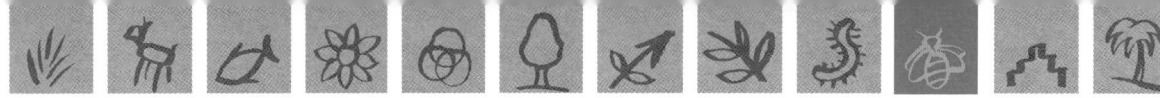

Singh, M.P.; Singh, K.I. and Devi, C.S., 2000. Role of *Apis cerana* pollination on yield and quality of rapeseed and sunflower crops. In (M. Matsuka, L.R. Verma, S. Wongsiri, K.K. Shrestha and Uma Partap (eds.) *Asian Bees and Beekeeping in Asia: Progress of Research and Development.* Proceedings of the Fourth AAA International Conference 23-28 March 1998, Kathmandu. Oxford and IBH Publishing Co. Pvt. Ltd. New Delhi, 274pp.

Verma, L.R.; Partap, U., 1993. *The Asian Hive Bee, Apis cerana, as a Pollinator in Vegetable Seed Production.* Kathmandu: ICIMOD, 52pp.

Verma, L.R. and Partap, U., 1994. Foraging behaviour of *Apis cerana* on cabbage and cauliflower and its impact on seed production. *Journal of Apicultural Research* 33: 231-236.

Waite, M.B., 1895. *The Pollination of Pear Flowers.* U.S. Department of Agriculture, Vegetable Pathology Bulletin 5, 86 pp.

Williams, I.H., 1992. Apiculture for agriculture. *British Beekeepers' Association, Spring Convention.* Stone Leigh, pp 14-18, UK.

Winston, M.L. and Scott, C.D., 1984. The value of bee pollination to Canadian agriculture. *Canadian Beekeeping* 11: 134.

RESPONSIBLE
TECHNICAL DIVISION

Plant Producion and Protection Division
Plant Protection Service

William Settle

ECOSYSTEM MANAGEMENT IN AGRICULTURE:

PRINCIPLES AND APPLICATION OF THE ECOSYSTEM APPROACH

11

ECOSYSTEM MANAGEMENT IN AGRICULTURE:
PRINCIPLES AND APPLICATION OF THE ECOSYSTEM APPROACH

AUTHORS

William Settle

FAO Plant Protection Service.
william.settle@fao.org

ACKNOWLEDGEMENTS

Case studies for this document
were commissioned by
Nadia Scialabba,
FAO - Environment and Natural
Resources Service.
Authors for the case studies were:
Uganda - Charles Walaga
(World Conservation Union IUCN)
and Mary Jo Kikinda
(Africa 2000);
Thailand - Vitoon Ruenglertpanyakul
(Green Net);
Iran - Taghi Farvar
(Centre for Sustainable
Development - CENESTA);
India - Bernward Leclerq
(Auroville).
Special thanks to
Caroline Hattam for editing the
drafts of the case studies, and
translation from Spanish for the
Honduran case study.
Both Caroline Hattam and
William Settle were supported by
the FAO/Netherlands Partnership
Programme for Biodiversity.

CONTENTS

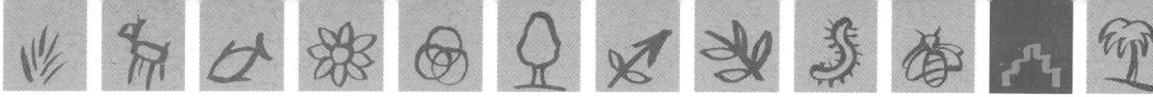

INTRODUCTION

"...in the sort of issue-driven science relating to environmental debates, facts are uncertain, values in dispute, stakes high, and decisions urgent."

(Funtowicz et al., 1999)

World-wide degradation of ecological systems is challenging researchers and policy makers to move beyond traditional disciplinary and institutional boundaries in order to seek practical and productive cross-disciplinary solutions to increasingly serious economic, environmental and social problems. The challenge demands the willingness to share ideas and experiences among researchers, resource managers, industry and environmental advocacy groups, policy makers, and most importantly, with the vast numbers of people who, through their everyday actions, have direct and often cumulative impacts on the systems in question.

An ecosystem approach

The Convention of Biological Diversity (CBD) describes 'ecosystem' as: "a dynamic complex of plant, animal and micro-organism communities and their non-living environment acting as a functional unit". Decision V/6 (http://www.biodiv.org/decisions/) suggests an Ecosystem Approach as defined by 12 Principles and five Operational Guidelines. The CBD considers that a general application of the Ecosystem Approach will help achieve a balance of three objectives: conservation; sustainable use; and the fair and equitable sharing of the benefits arising out of the utilization of genetic resources.

While this definition is meant to give the widest possible scope for discussions related to biological diversity, specific reference to agro-biodiversity can be found in Decision V/5. However, in comparison to the rest of the CBD documents on biodiversity, this reference to agro-biodiversity is brief and to the point stating that we lack sufficient methods and understanding of the role of biodiversity in agro-ecosystems:

"...Understanding of the underlying causes of the loss of agricultural biodiversity is limited, as is understanding of the consequences of such loss for the functioning of agricultural ecosystems. Moreover, the assessments of the various components are conducted separately; there is no integrated assessment of agricultural biodiversity as a whole. There is also lack of widely accepted indicators of agricultural biodiversity. The further development and application of such indicators, as well as assessment methodologies, are necessary to allow an analysis of the status and trends of agricultural biodiversity and its various components and to facilitate the identification of biodiversity-friendly agricultural practices...".

Although many of the details and mechanisms remain unclear, the overall trends are hardly in dispute: loss of biological diversity around the world, from a multitude of causes, is *correlated* with decreasing productivity, increasing fragility in systems and increasing exposure of farming families to uncertainty, poverty and hunger. Reversing these trends will require a huge effort to understand the ecological, economic and social problems, while at the same time educating people from all walks of life producer, consumer, scientist, policy maker and farmer. While our lack of understanding is profound, the urgency of the problems forces us to conclude that we cannot wait for the development of a

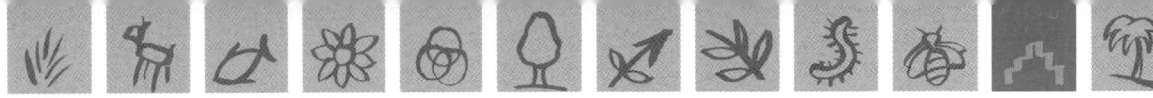

complete set of assessments and indicators before we start to take action. Policy, research and operational programmes in the field must work together; all seeking to communicate, to evaluate and to improve through 'learning-by-doing' this is a form of science in the broadest meaning of the word, and in essence what is meant by *Adaptive Management,* one of the five Operational Guidance points to the Ecosystem Approach.

Balancing theory and data

Science, as defined by the noted 20th Century philosopher, Thomas Kuhn, involves the development of a *disciplinary matrix* comprising: a *symbolic language; conceptual models* (i.e., theories); *case studies* (i.e., data), and (this being a major point of departure from previous definitions) *values* held by the scientist (Clark and Minta, 1994). Theories without data lack substance, while data without theory lack direction and meaning and are subject to misunderstanding and misapplication. Theories and data without explicitly understood values lack human context and are subject to bias and hidden assumptions of researchers and agencies. For a practical science that addresses the needs of society we need to interweave theories, data and values in a process that is responsive to the needs of people, transparent to all, and in which all people can be engaged.

The ecosystem approach set forth in Decision V/6 of the CBD is a highly-distilled set of principles and guidelines based on ecological, economic and social theories of ecosystem management. In slightly more than bullet-point form the document offers a kind of "toolbox" of concepts from which administrators, scientists and policy makers might begin to think and to discuss common issues. However, to successfully make use of

this toolbox will require people first to explore and to understand the meaning, validity and utility of these key concepts. The best way to do this is to challenge them in light of data from the field case studies inform theory, and theory illuminates case studies.

FAO's efforts to mainstream biodiversity

While the ecosystem approach was designed to address the broadest definition of ecosystems, and particularly with "natural" systems in mind, FAO has for several years been employing and making operational an ecosystem approach to a slightly more narrowly defined set of managed ecosystems, including managed forests, fisheries and all manner of cropping systems. FAO's biodiversity programme applies ecosystem approaches in community level education and experiential learning by rural people. It applies the same approaches to educate national agricultural policy makers who wish to understand how to fulfill commitments made to environmental treaties such as the CBD and the Convention to Combat Desertification and the Stockholm Convention on Persistent Organic Pollutants, while still meeting production demands. The dual processes of ecological education by and of farmers as well as policy makers is FAO's tested strategy for mainstreaming agro-biodiversity. The Objective of the Agro-biodiversity Theme is to advance agro-biodiversity considerations into the mainstream of the national programmes and policies of FAO's member countries. This recalls both the FAO-CBD-Netherlands 1998 Technical Workshop on Sustaining Agricultural Biodiversity and Ecosystem Functions, and the Programme of Work on Agricultural Biodiversity, which the

CBD in its 2000 COP Decision V/5 asked FAO to coordinate.

In the context of FAO's six-year Medium Term Plan this objective is being approached through partnerships that produce three categories of outcomes. These outcomes are:

- **Assessments** of Agro-biodiversity and which services it currently or potentially delivers to agriculture and the larger public good;
- **Case Studies** of Adaptive Management and good practices illustrating how ecosystem services can be generated; and
- **Local Capacity Building,** which empowers local communities to make and implement better decisions as they manage their agro-ecosystems to deliver services.

The theme brings together work from 13 operating Divisions of FAO in the Agriculture, Economic/Social, Fisheries, Forestry, Information, and Sustainable Development Departments and the Legal Office.

The work of the Agro-biodiversity Theme is organized into three Sub-theme Clusters; all strive in parallel to stimulate the outcomes and reach the theme's objective. Each cluster contains four sub-themes that share common approaches and goals.

- **Genetic Resources:** animal genetic resources, plant genetic resources, Genetic Use Restriction Technologies (GURTS), and access to, exchange and sustainable use of agro-biodiversity;
- **Ecosystem Approaches:** Integrated Pest Management (IPM), soil biodiversity, pollinators, and global invasive species programme; and
- **Monitoring and Appraisal:** forest ecosystem services, indicators, awareness, and remote sensing.

A current picture of progress is given in the Indicative Summary Matrix, which presents the work of the theme against the three intended outcome areas and the objective in terms of outputs delivered, activities ongoing, and activities planned.

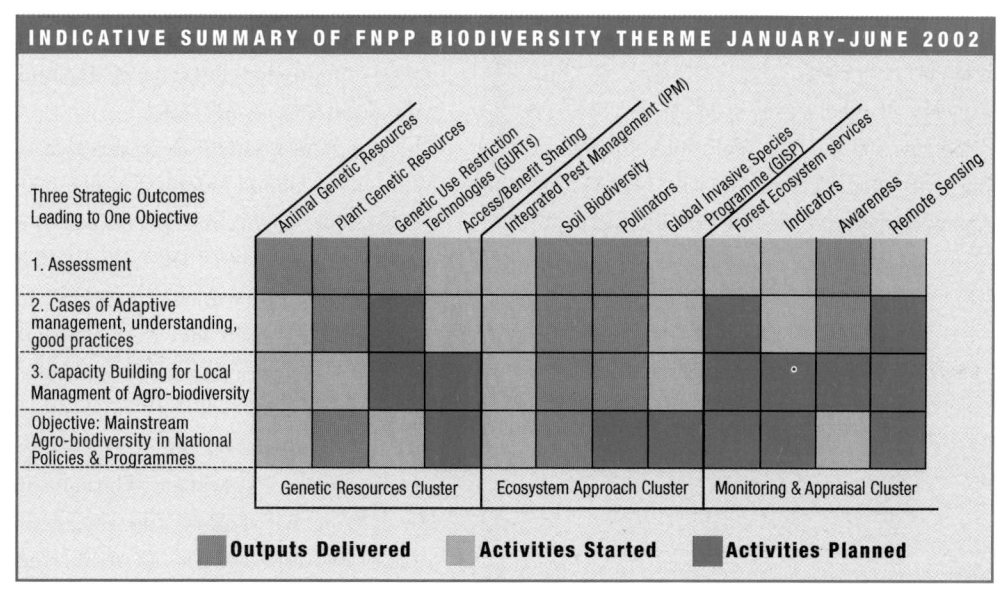

Partnerships

At the core of the Agro-Biodiversity Programme of the FAO/Netherlands Partenership Programme are about 45 officers from thirteen divisions across FAO. This is the Inter-Departmental Working Group on Biodiversity, which has been mainstreamed into FAO as a Priority Area for Inter-disciplinary Action. The members share inter-disciplinary concepts and language about genetic resources, ecosystem approaches, awareness, policy reform, case studies and, in particular, local community empowerment for sustainable use of agro-biodiversity. These officers are linked to a number of external partnerships with NGOs, field projects and innovative government programmes. The five case studies presented below are just a recent small sample taken from these partnerships.

FIVE CASE STUDIES

We present here a brief summary of five case studies from very different ecological regions of the world. Interspersed throughout the summaries we provide comments that attempt to link aspects of the case studies to the 12 principles and the five Operational Guidelines for the application of the Ecosystem Approach as set out by the CBD Decision V/6.

CS-1 Uganda: poverty eradication in the Iganga district

The "Poverty Eradication through Environmentally Sustainable Technologies" (PEEST). The Iganga district, is one of the most densely populated districts of Uganda with approximately 200 people per Km^2. The project has been supported since inception in 1997 by Cordaid (formally Bilance) of the Netherlands. The lowland project area comprises a network of permanent swamps and wetlands, which drain into lake Victoria to the south and Lake Kioga to the north. Farmers combine perennial tree crops with rain-fed annual crops within a mosaic of tropical forest remnants and encroaching savannah shrubs and grasslands. Households keep small numbers of farm animals, mostly under free-range conditions with some supplementary feeding. Animal traction is rare.

Project Location and Coverage

The district, with a population estimated at over one million people is growing at a rate of 3.5 percent per year, which is well above the national average of 2.5 percent per year. It is one of the most densely populated districts of Uganda with about 200 people /Km^2.

The population situation is aggravated by the polygamous practices of many of the families. Eighty-five percent of the households depend on subsistence farming. Mean annual rainfall is approximately 1250 mm occurring on 100–130 days per annum. It is mainly conventional with two peaks associated with the equatorial trough in April–May and September–November. The soils of the district are shallow and represent almost the final stages of weathering and prolonged leaching and are therefore of low to medium productivity. The district is characterized by gently undulating flat topped hills with gradients ranging from two degrees on the lower slopes, five degrees in the middle to 12 degrees at the top. The undulating hills are separated by wide valleys which, are either occupied by wetlands of impeded drainage or drained by sluggish streams.

Farm holdings average two hectares and support an average family size of eight people. A wide variety of crops are grown under traditional farming systems to provide food and income. The most important crops are sweet potatoes, cassava, maize, bananas, rice, yams, arrowroot, millet, sorghum, beans, pea nuts, Soya beans, simsim, tomatoes, cabbages, pineapples, and the traditional cash crops coffee and cotton. A typical farm of a peasant in the district comprises some perennial crops like coffee and bananas with fruit trees (orange, mangoes, avocados, jack fruit and papaya) and shade trees like *Ficus natalensis* and *Albizia sp.* grown adjoining the homestead area constituting about 25 percent of the total land holding. The rest of the land is usually under annual crops and fallow. Families keep small numbers of cattle, goats, sheep, pigs and chicken mostly under free-range for cattle and chicken and tethering for goats, pigs and sheep with supplementary feeding on household crop remains.

The excellent description above of the project setting provided by the NGO allows for a picture to be formed of the **Ecological Context.** Population growth rates and overall densities are high; soils are highly weathered,

productivity is low, and an undulating terrain portends erosion problems. Traditional farming systems still hold an impressive variety of crops, although the genetic diversity per crop is unknown. Such a picture is hardly unique and could represent a wide swath of territory across Africa and beyond. Problems and solutions found here might well have application elsewhere.

Trends and Constraints

The natural and agro-ecosystems of the district have been degrading since the 1970s as the population rapidly grew and the economic situation deteriorated. As the population grew, natural forest and woodlands were cleared for agriculture use, fuelwood, timber and human settlement. Wetlands were massively converted to agriculture use for rice, sugarcane, yam, millet, sweet potato and vegetable production. By 1997, 591 km of wetlands had been converted. Over time, the numbers of wild species have been declining and swamp soils have been drying up, shrinking and becoming sterile due to oxidation, and acid or salt precipitation. The mass clearing of forests, woodlands and wetlands has also resulted in the *increasing scarcity of fuelwood, timber and drinking water from natural wells and springs* which are increasingly drying up at a much faster rate during the dry season. The *increasing soil erosion that is deposited into Lake Victoria is destroying the fish breeding grounds at the lakeshores.*

By 1997, many farmers in the Iganga district were faced with a problem of increasing vulnerability characterized by high poverty levels (above the national average of 45 percent living below the poverty level of one dollar

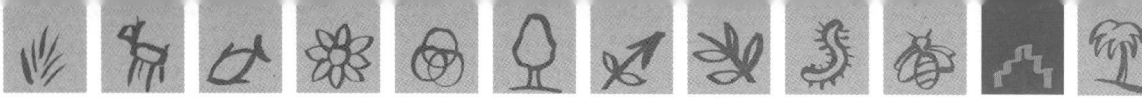

per day) and food insecurity. The causes are many and include:

- The rapidly growing rural population, expanding the frontiers of agriculture into natural forest and wetland ecosystems.
- Degraded natural resource base due to inappropriate farming practices resulting in low crop yields and degraded pastures.
- Fuelwood scarcity. Ninety-five percent of the total quantity of energy consumed in the district is provided by woody biomass, mostly firewood and charcoal production for cooking, lighting in households and fish processing. Energy has become a very critical issue in the community as sources of woody biomass are fast getting depleted and agricultural biomass, which would have been used to replenish soil fertility, is also being used for fuel.
- Weak farmers' organizations.
- Expensive and often inappropriate agriculture inputs.
- Inaccessible and/or inappropriate extension services.
- Poor marketing systems.
- Lack of access to credit.
- Pronounced gender inequality

Operational Guideline #1: Focus on the functional relationships and processes within ecosystems

- Principle #3: Consider the effects on adjacent and other ecosystems
- Principle #5: Conserve structure and function
- Principle #6: Manage within ecosystem limits

The trends and characteristics of the system presented in the case study present an all-too-familiar list of environmental, social and economic weaknesses. The list, however, leaves us to sort out for ourselves what might be the causes and what might be the consequences. Keep in mind that in complex systems cause and effect are often difficult to tease apart because they are often mutually interdependent. Although Decision V/5, as quoted above, laments the lack of widely accepted indicators of agricultural biodiversity, we know enough *from general principles* that several elements indicated on the list, specifically the loss of woody cover, wetlands and riparian zones, are *key causes* of resource-base degradation and a downward spiral in productivity and ecosystem health.

Riparian zones, as narrow borders between aquatic and terrestrial systems, are considered "Critical Transition Zones" (CTZs), characteristically housing some of the highest levels of biodiversity as well as providing key ecosystem services in the form of filtering, processing and otherwise intercepting the flow of soil erosion, toxins and excess nutrients from terrestrial to aquatic systems (hence, the loss of fish breeding grounds) (Ewel *et al.*, 2001). Because of their relatively small size and difficulty in being exploited for various purposes, riparian zones tend to be overlooked for the important ecosystem functions they play. Wetlands themselves are a key habitat for all manner of species that often link back into surrounding terrestrial habitats to play critical roles in terrestrial (natural as well as agricultural) habitats.

Also a familiar trend is loss of woodlands leading to increased pressures for fuel and structural timber. In semi-arid rangeland environments, loss of shrubby vegetation leads to leakage of nutrients in the system (Shachak *et al.*, 1998). In general, biodiversity mediates proximate flows of water and nutrients on and in the soil (Bardgett *et al.*, 2001). All-in-all, the picture is one of mutual interdependency and interaction of soil, water and biological organisms. The description provided in this case study points to a set of positive feedbacks (i.e. amplifying rather than dampening effects) leading to a characteristic "downward spiral" of environmental degradation and increasing pressures on the environment. The task of an ecosystem approach to agriculture (and to environmental management in general) is to arrest this downward spiral, and then to provide the conditions by which the spiral can be reversed and conditions gradually improved.

The social and economic factors listed are *symptomatic of systems* in a decline, but are not necessarily the "drivers" of such a spiral of declining ecosystem health and productivity.

Project Goals and Activities

The three-year project aimed to improve the sustainability of farming in the area through:

(i) improving environmental knowledge and management skills *(Operational Guide #3: use adaptive management practices)*:

- The project employs meetings, workshops and exchange visits to raise general awareness of the linkages between natural resources and local livelihoods.
- Participative appraisal and training methods help farmers plan for desired goals with a 5-10 year horizon.
- Farmers then set out, with the assistance of the project, to implement small projects with their goals in mind.
- The project helps with training on specific focal areas including a variety of organic soil management methods like compost production, mulching and fallow techniques. Other topics include crop and animal husbandry, IPM and agro-forestry;

(ii) improving fuelwood production and efficiency and diets, *(Operational Guide #1: focus on the functional relationships and processes within ecosystems)* Improved woodstoves were introduced, capable of reducing fuelwood consumption by 50-75 percent as well as improving safety in the kitchen;

(iii) to promote a more balanced contribution by both men and women, *(Operational Guide #2: enhance benefit-sharing)*;

(iv) to ensure that successful project activities are scaled up and unsuccessful ones avoided: the project sought technical advice through collaborative relations with several international research agencies, including ICRAF, CIAT, TSBF, as well as Ugandan government and university agencies. *(Operational Guide #5: ensure intersectoral cooperation)*.

Management training workshops aim to improve farmers' ability to network and to interact effectively with external agencies.

- A revolving credit scheme was set up to facilitate resource poor farmers to acquire productive assets including legume seeds for improved fodder production and soil fertility management.

Operational Guidelines #1, #2, and #4 functional relationships, benefit-sharing and appropriate scales for action. The project seems to have a clear vision of the necessary operational goals, and activities required to achieve these goals. Improved wood stove technology is a proven method for fuel-wood conservation, leading to significant environmental, economic and health benefits for the community.

Operational Guidelines #3, #2, and #5: adaptive management, benefit sharing, and inter-sectoral cooperation. The stated goal of improving environmental knowledge and management skills requires activities

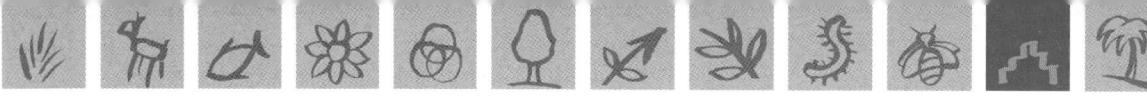

centred on an appropriately constructed and participatory-based training programme. Participatory training we know to be most effective when addressing both male and female sectors of the community. Finally, the project's attention to the issues of "scaling up" and attention to collaborative relationships with other organizations shows taking responsibility to make sure that projects have a realistic chance of having significant impacts at a regional level.

Project Accomplishments

Since 1997 more than 20 000 farmers have benefited from the project. Promotion of an increased diversity of indigenous crop varieties has contributed both to maintaining local agro-biodiversity and to increased food security. Evaluation at the end of the first three-year phase showed 99 percent of the participating farmers reporting increased food supplies, 89 percent of farmers reporting an increased income and 61 percent of households reporting increased savings and effort in accessing fuelwood as a result of improved cook stoves.

Constraints

Constraints include inappropriate national policies that promote, through government extension, monocropping with synthetic fertilizers and pesticides. Problems arose with inexistent or weak farmers' organizations in not being able to effectively market production surpluses. Low literacy rates among female farmers constrained participation in training on farm record keeping. Polygamous family settings were generally not receptive to gender messages promoting increased control of income by women. Finally, regardless of gains, population pressure from a 3.5 percent growth rate will necessarily limit hopes

for long-term sustainability. The project is currently beginning to address both the literacy problems and family-planning issues.

Problems in estimating accomplishments and impacts

Although it is not specifically outlined in the CBD documents, one of the areas with keenest need for attention is impact assessment. Our inability to adequately and consistently evaluate likely benefits of project activities leaves us guessing in the dark. Evaluating impacts is a complex and in many ways uncertain task. This project, as with most, is limited to asserting that simple counts of numbers of participants and amounts of increased income or decreased expenditures is in fact both significant and justifies the effort and expenditure. While this is logical and relatively unambiguous, it misses several different categories of possible benefits.

Consider a project whose immediate benefit might be considered quantitatively small (e.g. the development of a handful of good farmer trainers), but for which the influence is likely to *multiply over time,* carrying forward to have untold and disproportionate impacts on future communities. In such a case, substantial positive impacts may be incremental, cumulative and involve substantial *lag times.* Another such example is the case with improved inputs of soil organic materials, or improved vegetation coverage, resulting in a suite of improved soil fertility characteristics that are only noticeable over several years. Other problems in impact evaluation are likely to exist due to our ignorance of *indirect effects,* which might link the obvious action to results that we may not even be thinking of measuring. For

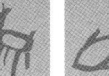

adjustment of crop densities, the management of crop and pruning residues, appropriate use of fertilizers and the introduction of IPM techniques.

During the last eight years, participating farmers have observed a gradual modification of the traditional agro-pastoral system towards a system that has, due to the effect of abandoning slash-and-burn, seen an increase of trees in the pasture-land, and the use of crop harvest residues over the ground of the *milpas* (land cultivated with maize). In this system *jaragua* grass is associated with local tree species such as the aceituna (olive) *(Simaroouba glauca)*, *almendra de agua* (water almond) *(Andrina inermis)* and laurel

(Cordia alliodora). Although it would be ideal to incorporate leguminous trees, the introduction of these has been difficult due to grazing.

In this system, both small and medium livestock farmers also cultivate maize and bean, usually without clearing or burning small areas. When the farms are more than 10 ha, the farmers usually rent part of their land to other farmers for the production of basic grains. Farmers do not use synthetic fertilizers on the grazing land, but have reduced soil degradation and increased conservation of soil moisture by incorporation of crop residues. Many farmers have begun to vaccinate their cattle, and have seen improvements in milk yields and net income.

Summary of activities in view of their relationship to key operational guidelines (see Appendix A)

HONDURAS \ FIVE OPERATIONAL GUIDELINES (CBD V/6)					
ACTIVITY:	FOCUS ON FUNCTIONAL RELATIONSHIPS AND PROCESSES	ENHANCING BENEFIT-SHARING	USING ADAPTIVE MANAGEMENT	MANAGEMENT AT APPROPRIATE SCALES (DECENTRALIZATION)	ENSURING INTER-SECTORAL COOPERATION
To encourage dialogue between producers and resource persons			✓	✓	✓
To encourage farmers to abandon traditional slash-and-burn	✓		✓	✓	✓
To validate improved agronomic methods	✓	✓	✓	✓	✓
To 'rescue' edible indigenous species of plants	✓	✓	✓	✓	✓
To employ participatory training methods to focus on agro-ecosystem analysis by farmer	✓	✓	✓	✓	✓
To build local capacity	✓	✓	✓	✓	✓

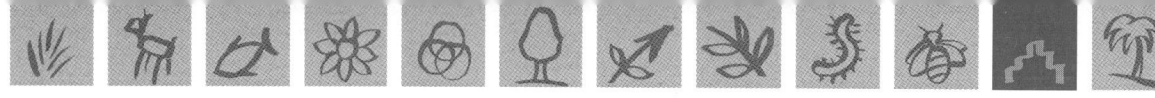

CS-3. Thailand:
organic and fair-trade rice project

The Organic and Fair-Trade Rice project originated more than 10 years ago in northeastern Thailand famous for its Jasmine variety of rice as a cooperative effort between local Thai NGOs in Surin and Yasothorn provinces, and a Swiss-based fair-trade organization, Claro. In 1996 the project expanded its scope to include organic farming as a central objective. The main objective was to establish an organic conversion programme comprising three main components: farmer field school training (FFS), market access, and organic certification.

Project Location and Coverage

Topographically, the northeastern region of Thailand, locally known as Isan, is dominated by the Khorat Plateau. The northern and eastern boarder is marked by the Mekong river. The northen part has some high mountains and various plateaus ranging from 300 to 1 200 meters above sea level. Soil is poor, predominated by entisols, incepticols and ultisols. In various parts of the region, salt deposits can be found in subsoil level. An estimated 2 848 million hectares of land (16.9 percent) are thought to be affected by soil salinity.

An estimated 68.6 percent of agricultural lands in the northeast are cultivated with rice. Rice is predominantly grown in low land during the rainy season, where local rice varieties are cultivated for family consumption while high-yield or high-value rice varieties are for sale. On upland areas, different types of cash crops are grown, including tapioca, sugar cane, maize, bean, jute.

The northeastern region is also the poorest region in the country. The average annual income is around one-third of the national income, i.e. Baht 19 331 (ECU 483 28) compared to an average national annual income of Baht 56 336 (ECU 1 408.40). Almost three quarters of the population in the northeast is involved in the agricultural sector, i.e. 71.9 percent.

Aims and Objectives

The project's main aim is human development through ecological production, with a focus on the farmer as a centre of development, and looks at three inter-woven aspects:

Awareness/Consciousness raising and ethical responsibility;

Appropriate skills and technology; and

Grassroots organization.

Awareness is not sufficient to convince a producer to change his/her production to a more sustainable path. Consciousness raising together with ethical responsibility must be instituted. A dynamic of action-reflection where the producer reflects on his/her practice can better ensure a raised conscience while a fairer return on farm produce offers the producer an incentive to make an ethical commitment and responsibility. The project is convinced that an organic agriculture development project must incorporate practical learning activities where the producer has opportunities to practice and reflect upon his/her own actions and that the project must link to fair-trade market opportunities, possibly to offer fairer prices for the organic produce. This in turn also requires that the organic project must include post-harvest processing and management (e.g. quality assurance and product development) as a key component in its activities.

Practical skills and appropriate technology are key to ecological production. One of the main constraints facing organic farming is the lack of appropriate practical skills and technology that the producer can employ on his/her farm. Indigenous knowledge, rapidly eroded, is often praised but no concrete efforts are made to revitalize it and especially to enhance it so that it can offer practical solutions to tackle the present problems facing local farming communities. The project is committed to participatory technological development and learning approaches and has been using two methodologies the "Farmer Field School" (FFS) and "Participatory Technological Development" (PTD) as main tools for developing appropriate organic farming skills and technology while also incorporating local indigenous knowledge.

The project also believes that producers' organizations play a key role in organic agriculture and that they must be strengthened so that they can become more effective in delivering technological services to their members, reinforcing consciousness and ethical responsibility among their members, and handling organic produce. The project therefore works with producers' organizations committed to organic agriculture and sustainable development.

The Farmer Field School (FFS) approach is a participatory adult learning method. Participating farmers come together on a regular basis (e.g. every 7 10 or 14 days) to study agroecosystem management of a selected study field, known as "field school". In the field school, the plot is divided in half, with one half cultivated with conventional rice and the other with organic rice. The conventional rice field is managed in the same manner as typical local farmers would do to their own conventional field while the organic filed is managed according to organic standards with joint farmers' decision on fertilization plans and pest control strategies. Participating farmers are guided to observe both conventional and organic fields in the actual local conditions comparable to their own fields, comparing crop inhabitants (insects and plants), growth and yield.

This exercise is done in a small group to facilitate active participation and group development. Small groups analyze and present crop conditions, then make organic management recommendations. A trained facilitator (trained extension agent or trained and experienced farmer) synthesizes the group's recommendations. These management recommendations are tested in the school field and farmers observe test results in the next school session. In this process, farmers have an opportunity to share their indigenous knowledge of farming and to learn about different organic management alternatives.

Operational Guide #3: use adaptive management practices
Operational Guide #4: carry out management actions at the scale appropriate for the issue being addressed, with decentralization to lowest level, as appropriate

The core operational goal essential to the success of this project was its focus on education and training. They chose a Farmer Field School (FFS) approach, based on participatory "learning-by-doing" principles, which was first developed by FAO in the late 1980s in Asia, and is now being adopted by an

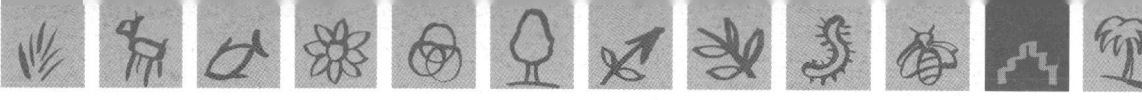

increasing number of NGOs and institutions responsible for farmer training in Asia, Africa and Latin America

> In producer groups that have already undertaken FFS training, but where participating farmers are interested in testing new or additional techniques, a participatory technical development (PTD) activity is organized. Here a specific technique is first discussed and evaluated by farmers (in the spirit of the FFS training). The group sets up field trials on farmer members' fields. The results of these field trials are presented to the group for collective analysis. These trained FFS farmer groups are a natural kernel around which to develop farmers' organizations.

The PTD activity mentioned above is one form of follow-up study. One of the earliest 'lessons learned' in the FFS programme was that a one-shot training of farmers was not, in itself, sufficient to guarantee long-term change. Enthusiastic FFS farmers need some form of future opportunities for continued learning; therefore, follow-up activities were adopted early on in FFS programmes. These activities are devised by farmer-alumni and cover a wide range of possible activities depending on the interests of the farmers' group, including new training in new cropping systems, focused studies on specific soil or pest problems of particular concern, or training for farmers to become trainers.

> Organic certification is taken care of by a local private organic certifier, the Organic Agriculture Certification Thailand (ACT), which was awarded IFOAM accreditation by the International Organic Accreditation Service (IOAS) in November 2001. The farmers'

organizations are assisted to develop internal control systems (ICS) to monitor the compliance of participating farmers to organic standards. Farmer leaders and their staff (who sometimes can also be NGO staff) are trained to run the ICSs, and conduct internal farm monitoring and inspection. This helps to further strengthen producers' organizations and ensure that organic standards and certification requirements are adherred to. Conversion to organic requires 12 months for annual crops, but can be extended if the farm has a history of heavy agro-chemical use. The project provides organic fertilizer credit so that participating farmers can assure that rice yields of converting fields are maintained during the conversion period. This allows more farmers to successfully convert their entire rice fields to organic. From 1998 to 2001 the land under organic production in the project area increased to more than 5 000 Rai (800 hectares), a five-fold increase from year one.

Reported Improvements

- Soil fertility. The most often self-acclaimed benefit expressed by organic farmers is that "soil is darker and softer". Rice straw and stalks, often burnt in the past, are now being ploughed back into the soil. Animal manure and green manure that farmers apply also add organic matter to the soil. All these contribute to the improvement of the soil's physical structure and chemical composition.
- Biodiversity. The cessation of agro-chemical applications contributed to an increase in different animals in rice ecosystem. Farmers noticed an increase in natural enemies, e.g. spiders, birds and fish in rice fields. Fish in the rice fields are from natural sources.
- Health. Organic rice farmers have lower

risk of exposure to pesticides (a serious problem in most farm areas in Thailand). Rice fields are an important food source, where vegetables and animals (mainly fish from natural sources) are caught for family consumption.

● Economic. The Organic and Fair-Trade Rice Project is based on fair-trade principles. The organic rice produce is guaranteed a fair premium price, set at a fixed level taking into account the rice farming cost. At present, the certified organic paddy is purchased from farmers locally at TBT 10 kg (US$0.24), TBT 8 kg (US$ 0.19) for in-conversion paddy, and TBT 7 kg (US$ 0.167) for non-certified organic paddy. This is quite a significant premium as conventional paddy currently costs only TBT 4.7 kg (US$ 0.11). These farm gate prices include both organic and fair-trade premiums. With such a premium price system, organic rice producers have doubled their incomes in the last two years.

Operational Guide #2: Enhance benefit-sharing

Principle #1: Management is a matter of societal choice

Again, societal choice implies adequate knowledge to ensure an informed choice. Organic certification is an expressly defined valuation of what constitutes an adequate quality for agricultural production, and provides a vehicle for society to exercise choice in a manner that is both transparent and fair to both producer and consumer.

Principle #4: manage the ecosystem in an economic context (i) reduce market distortions that affect biological diversity; (ii) align incentives to promote biodiversity conservation and sustainable use; (iii) internalize costs and benefits in the given ecosystem to the extent feasible

More than almost any other country in Southeast Asia, Thailand has a history of heavy pesticide and fertilizer applications, so it should not be surprising that farmers in the Thai organic rice project were initially very wary of "risking" an organic approach. The project, by providing a guaranteed price to the producer, was easing the fears and uncertainties of farmers, allowing them to move in the direction of a more ecologically sustainable farming system. It will be interesting to see if the market-farm links can be established to a sufficiently stable point to allow these farmers to be "weaned" from the price-stabilization mechanisms.

Summary of activities in view of their relations hip to key operational guidelines (see Appendix A):

THAILAND \ **FIVE OPERATIONAL GUIDELINES (CBD V/6)**					
ACTIVITY:	**FOCUS ON FUNCTIONAL RELATIONSHIPS AND PROCESSES**	**ENHANCING BENEFIT-SHARING**	**USING ADAPTIVE MANAGEMENT**	**MANAGEMENT AT APPROPRIATE SCALES (DECENTRALIZATION)**	**ENSURING INTER-SECTORAL COOPERATION**
To raise farmer awareness and sense of ethical responsibility regarding agricultural produciton		✓	✓	✓	✓
To enhance farmers' practical skills in producing crops in an ecologically sound method	✓	✓	✓	✓	✓
To build grass-roots producer organisations committed to organic agriculture and sustainable development		✓	✓	✓	✓

NO. 11

CS-4. Western Iran: sustainable livelihoods project

In 1998 a project in area-based and community-managed sustainable livelihoods was started by the Iranian NGO "Centre for Sustainable Development" (CENESTA) in three regions of the country, including the Houraman Valley, in the lower reaches of which is located the village of Naw. The areas were selected on the basis of being comparatively more deprived and poorer than others. The provinces of Kurdistan, Kohgiluyeh and Hormozgan were on UNICEF's list of those with the greatest disparity in indicators of well-being, especially of women and children, and in each province, consultations were held with those government agencies that deal most closely with rural communities, including health and rural development workers. The programme was designed in three phases: short-medium- and long-term. Phase I (two years) was financially helped by UNICEF, and a number of government departments joined CENESTA in carrying out the action programmes.

The village of Naw in the Houraman Valley of Kurdistan was selected for study for rain-fed agriculture in the arid lands project of FAO. Iran is considered largely arid and semi-arid. The region of western Iran, including Kurdistan, is especially suited, in good years, for rain-fed (dry) farming as its rainfall is adequate for the purpose. The Province of Kurdistan is divided into two climatic zones by the Zagros Mountains. While the western reaches of Kurdistan can have up to 700 mm of rainfall, mostly coming from the Mediterranean region, the eastern reaches of the province are affected more by the warm and dry air coming from the central plateau. Here

the rainfall rarely exceeds 350 mm.

The Houraman valley has a complex and integrated agricultural system, consisting of dry farming, fruit orchards, transhumant pastoralism, and agroforestry and forest products. Rangelands are important factors in the household economy in providing animal feed and access to woodland products. The system of exploitation of natural resources is a contributor to their conservation and sustainable use in the area.

Naw's production system includes, in addition to agriculture and livestock, handicrafts. Women are noticeable in most of the economic activities, in which all participate. The fruits of this work are used equitably. Government institutions have little or no effect on the economic and social activities of this region.

Naw is a part of Houraman Dehestan (village cluster), which is part of a mountainous region in Kurdistan Province in Western Iran in the lower reaches of Houraman valley, some 1 100 metres above sea level. The main access road to Naw is earth and gravel, and the nearest town is 35 km away. The village has about 160 households, with just over 800 people.

Soil resources

The steep slopes of the area (see table 1), the hilly and mountainous terrain, and the occasional hard rains have caused soil erosion, especially in the higher regions. This has resulted in the meagre depth of top soils and the prevalence of gravel and rocks in the surface and depths of the soil, causing many exposed rocky areas, poor soil nutrition and lack of organic matter in the soils. The soils are usually clay to sandy, with poor organic composition. They do not lack

drainage due to natural draining capacity, which is why there is no salinity problems in the area. The agricultural land, which has a surface area of some 900 ha (about 700 ha of farming land and 200 ha of rangelands) is suitable for agriculture but with limitation of frequent high and low terrains as well as high gravel content.

The peasants of Naw classify their soils into three categories: clay, white (clay of Marne origin) and black (volcanic), and consider that most of the soil used for orchard or dry farming needs to receive animal fertilizers. There is no known use of chemical fertilizers or compost in the region. Local peasants use terracing for creating their orchards, a system known as *"Talan."* Top soil is often transported to cover these terraces using animal power, usually mules. They establish stone walls to retain the terraces. Trees are then planted and water is absorbed and stored in the terraces. Even in the rangeland areas they use a system of micro-catchment basins in the form of semicircular stone retainers around oak trees that helps in catching and storing water and humidity.

Water resources

Water resources include surface and ground water. In the rainy seasons the water from rains or the melting snow above moves through a natural flood channel which acts as a drainage mechanism to reach the Sirwan river south of the area. *Local farmers construct traditional stone bunds and earth or stone ditches and canals which take these seasonal floods into the orchards of vines, pomegranates and figs. The height of these stone canals sometimes reaches over two metres. Rain-fed vineyards have usually been constructed along these channels or in snow catching highlands.* Ground water manifests itself in the form of springs inside the village with a yield of 5-40 litres per second. The water needed by the orchards comes from two sets of springs. Some six springs in the higher terrain provide the water needed for the animals. This village has no *qanats* or tube wells. The seasonal and ground water flows have good quality water. The drinking water is taken by goat skins from springs located in the northern highs of the village. Small water retention barriers near the springs help provide the water needed by livestock.

This is a fascinating description indicating a huge effort in terms of initial outlays of labour to capture and control water. Although not mentioned in the case study, it is likely that these stone ditches and canals also serve to capture silt run-off, which might then be redistributed.

Traditional knowledge in this system is clearly highly attentive to arresting soil loss through erosion, yet there is no history of recycling organic inputs, most probably because any such crop residues would go to feed animals.

Biodiversity and natural resource management

The Zagros Mountain Range has its own climatic system with many mountain and piedmont habitats that have given rise to considerable biological diversity in ecosystems, species and genetic resources. Steep slopes, woodlands and forests, which are part of the natural Zagros biome, are found here. Plant communities of oak, wild pistachio and almond are among the prevalent trees and bushes, which create both impressive landscapes and suitable niches and habitats for wild fauna including many species of mammals, reptiles, birds and insects. Some dominant species of wildlife include squirrels, snakes and lizards, mountain sheep and goats, foxes, wolves, jackals, rabbits, boars, bears, leopards and birds such as quails, water birds and raptors including eagles, and insects such as grasshoppers, lepidopterans, beetles, Sunn insects, lady beetles and others. Some agro-forestry also happens on a scattered basis in these woodlands.

Among the plants of biodiversity interest in the area one can name the tree and bush species *Quercus persica* (Persian oak), *Pistacia mutica* (wild pistachio), *Acer persicum*, *Crataegus sp.* (thornapple tree), *Amygdalus orientatus* (wild almonds), *Amygdalus scoparia* (mountain almonds), the *Rhus* sp. (sumac), *Pyrus glabra* (wild pears) and a plant with the Kurdish name "homr".

The leaves and acorns of the oak are used for animal feed, and the wood as well as the tannin extracted from the acorns has industrial and medicinal uses. Acorns, after the extraction of tannin by osmosis in water, are also ground into a meal from which a local type of bread is made for human consumption, especially at times of food shortage. The wild pistachio yields a gum which is used locally as well as processed and exported. The gum extraction work provides income for the inhabitants.

The woodlands, trees and shrubs are owned by the clans and sometimes households in the village, the rights to which are inherited. Sometimes up to 10 households may have use rights for just one or two trees. The *Amygdalus* shrubs have pharmaceutical, industrial and sometimes food value. They are harvested around April or May. The wild pistachio and many of the other species have endemic value. There are also many annual wild plants of medicinal value, including *Echium amoenum*, *Glycyrrhiza glabra* (liquorice), *Fumaria parviflora* (leaved fumitory), *Althaea officinalis*, anis, *Astralagus acutus* (milkvetch) and other plants with such Kurdish names as "azbu", "gole kabud", and "panareh", are used for medicinal and industrial purposes. These plants reach 30-100 cms, and usually grow in the spring and last until the end of autumn.

Environmental management

Most villages in Houraman valley have a vertical migratory pattern, where they spend about a third of the year in higher lands called "hawar". Each "hawar" has a name of its own, and acts like a second village with its own production, habitat and farming system.

In "hawar" each household has a given terrain for its use, and the rangelands are kept and managed as common property. Part of the rangelands is usually kept in a state of "qoroq" or "gia-jar" i.e. no one is allowed to use these resources in the spring until sometime in May. During the remaining months of spring and summer, in addition to extensive grazing, the grass in the "giajar" ranges will be harvested three times. The hay thus harvested is then dried and, if necessary, stamped by oxen or mules and then stored away for use as winter feed for the animals of the village.

In "hawar", the villagers have their traditional ways of protecting and

enhancing the species most appreciated in rangelands. Women and men collect the seeds of such plants, as they dry, while grazing the animals during their steady climb (similar to alpage in Europe). At the time of migration back to the base village, at the end of summer, the seeds collected are spread on these rangelands. The livestock are kept in these lands for some time in order to fertilize the lands. The movement of the animals in the areas causes a "ploughing" of the soil and the burying of the seeds. The coming of the next rainy season will cause the rangelands to be renewed with these plants.

The energy for cooking and processing dairy products is traditionally obtained in the "hawar" by cutting the head branches of the oak trees. Until the recent past the villagers had to sometimes cut trees in order to provide firewood for the cold of the winter. This was one of the contributing factors to losing the plant cover and the occurrence of flooding and damage to the fields. In recent years, government agencies have managed to provide fossil fuel products like paraffin (kerosene), which has resulted in reducing the need to cut trees and increased willingness on the part of the local population to protect the woodlands and forests. As a result local folks can perceive a definite improvement in the plant cover around the community. Local people recall frequent floods in times past, but they have not had them in the past decade or so, since they are no longer having to cut trees for fuelwood. This is especially noticeable in the flood paths and gulleys, which are now rich in shrubs and trees, which helps slow down running water and prevent floods.

(Principle #1: the objectives of management of land, water and living resources are a matter of societal choice)

The apparent success in managing common pasture-lands through collaboration and agreed-upon social norms is surely a sign of a social system still integral, and not yet subject to pressure to the extent that would drive individuals to dispense with social norms.

The replacement by appropriate concentrated alternative fuel sources for fuel-wood gathering, especially in typically fragile semi-arid ecosystems, is an absolutely critical first step towards creating and maintaining sustainable systems.

The complementary and inter-dependent nature of livelihood systems in Naw help protect the biological diversity of the landscape. Wheat and barley are harvested and the straw from them provides a major source of feed for animals. In the higher ranges the pasture plants are harvested and used for winter feed for animals. Oak leaves from the forests are also used, sparingly enough, for animal feed. The partial cutting of some of the bows has a pruning effect on the trees. The period of time that livestock spend in the "hawar" is not too long to cause damage to the trees or the rangelands, since they are regularly taken for pasturing higher up. The acorns are sometimes used for grinding, after soaking in water overnight to eliminate the tannin, acorn bread is made for human use, particularly at times when other sources are lacking.

The Naweans don't use steep slopes for pasturing animals. Furthermore, they preserve the trees in these slopes. The orchards are rich in medicinal plants, which are picked for local use, but not for commercial sale. When bleeding *Amygdalus* and wild pistachio for gum,

they are also careful not to damage the trees and shrubs. Even wildlife has become more abundant in recent years.

Farming Systems in Naw

We have seen that the farming system in Naw is an interactive set of relationships between farming and horticulture, range and forestry (mostly harvesting) activities. Agriculture takes up about 700 ha of which some 40 are partly irrigated orchards. Dry-farming consists of vineyards, grains and pulses. Fig and pomegranate make up most of the orchard horticulture. Grains are planted in autumn and harvested around June.

The agricultural rotation system depends on soil quality, rainfall prognostics and the perception of situation in the market place and availability of labour and agricultural tools. The most prevalent rotations are as follows: Wheat-chickpea-wheat; Wheat-barley-wheat, and Wheat-chickpea-barley-wheat. The rotation system in agriculture is driven partly by the demand in the market place and partly by the needs of the subsistence economy. A major factor determining the dominant role of wheat and barley in the rotation system is the needs of the livestock. In the summer animals feed on the stubble from grains in addition to the pastures, and in winter straw, hay and barley is fed to the animals in addition to the stored range cuttings. Summer horticultural crops are planted on a limited scale, mostly for meeting the needs of the household, amounting to some 0.4 ha per household "hawar". Most of the animal dung in the village is used for the orchards, around 6-7 kg per tree in September. No chemical fertilizers are used. The use of insecticides is conspicuous by its absence. Seeds used for agriculture are local indigenous varieties and the trees used in orchards are also from local genetic stock. There is no felt need in the region for inputs such as seeds and seedlings, chemical fertilizers and pesticides, or agricultural machinery.

The main pest of grains is the Sunn pest, which, in this area, has no major outbreaks, and the farmers often deal with it by generating smoke to chase them away. Farmers say locusts in the area are hunted by birds, and do not cause much damage. In conclusion, in the farming system of this area, it is natural and cultural factors such as the topography, rainfall, land tenure and self-sufficiency that play the major role in the production of rain-fed crops like grains and pulses.

Analysis of agricultural, ecological and economic systems

We can conclude that despite the lack of access of the community to modern "technical" advice and agricultural inputs, and despite the relatively adverse environment given the steep slopes, lack of water for extended irrigated agriculture and relatively poor soils, it has managed to keep a reasonably productive system, mainly for subsistence purposes, through the application of traditional knowledge and wisdom, extensive dry-farming, and the integration of various types of agriculture with forestry, gathering of non-timber forest products, and range management for livestock production. The mutual benefits of agriculture and animal husbandry help maintain higher productivity for both livestock and crops,

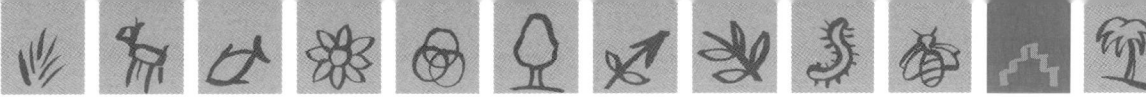

and the maintenance of rangelands and soils in reasonable condition.

The average area of dry farmed land for each household is about 6 hectares, some 10 times the land under annual irrigated crops. Due to the overall low productivity of the agro-ecosystems, and the essentially subsistence character of agriculture, this system does not contribute to savings-based investment. Therefore, many of the households are in a relative state of poverty. The subdivision of land rights due to inheritance may become another problem in the future.

The Naw society has evolved its particular system of subsistence based on its agro-biodiversity, topographic and climatic conditions, the predominance of the tribal traditions, and reliance on seasonal transhumant animal husbandry and agriculture. It has done so in such a way as to make sustainable use of its environment and natural resources. In the spring, with the growth of range species, transhumant migration to higher areas starts. This also keeps them at bay from harmful insects like ticks, flies and mosquitoes. Animals are fed with fresh pasture, and the clean water of the higher range is used for animal and human consumption. New crops are planted at this time as old crops begin to get picked. Pest control takes its natural course with endemic natural biological agents, and livelihood is complimented with small-scale marketing of products from the orchards, forests and handicrafts as well as with occasional seasonal migrant labour.

Sustainable Livelihoods Project

The main purpose of the project is:
- Improvement of the conditions of life of local communities in the areas selected, with special attention to the situation of women and children, through community solidarity and control over decisions concerning planning, implementation and evaluation in their own development, with the effort to overcome the disparity that affects their areas;
- Elaborating, designing and applying models of development that would be environmentally, socially and economically sustainable;
- Establishing endogenous mechanisms for the adaptive replication of the project.

Preliminary results

The project depends heavily on participatory planning and appraisal methods to explore the priorities of the community. The priorities arrived at by the community of Naw include:
- the promotion of shoe-making from local materials;
- protecting and improving organic production of agricultural and horticultural crops;
- processing of horticultural products including pomegranate sauce;
- solar cookers for further improvement of the forest cover;
- marketing agricultural products, and a videoteque project;
- Eco-tourism was also discussed as a priority in the longer term, given the unique beauty and way of life of the Valley of Houraman, including especially the village of Naw.

The project in western Iran introduced above differs from the first three projects in several ways-most importantly because, through their isolation, the Nomads of Naw

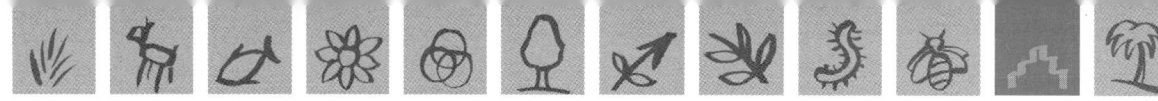

have maintained an intact traditional culture with a long history of adaptation and survival in an extremely arid and fragile climate. However, most all peoples seek increased economic opportunity and well-being, and judging by the results of their participatory workshops, these people are no exception. The advantage held here by the project is that they are able to address issues of community development **before** the environmental and economic factors have collapsed into a downward spiral.

IRAN \ FIVE OPERATIONAL GUIDELINES (CBD V/6)

ACTIVITY:	FOCUS ON FUNCTIONAL RELATIONSHIPS AND PROCESSES	ENHANCING BENEFIT-SHARING	USING ADAPTIVE MANAGEMENT	MANAGEMENT AT APPROPRIATE SCALES (DECENTRALIZATION)	ENSURING INTER-SECTORAL COOPERATION
The promotion of shoe-making from localmaterials,		✓	✓	✓	
Organic production of agricultural and horticultural crops,	✓	✓	✓	✓	✓
Processing of horticultural products including pome-granate sauce		✓	✓	✓	✓
Introduction of solar cookers	✓	✓	✓	✓	✓
Marketing agricultural products, and a videoteque project		✓	✓	✓	✓
Eco-tourism	✓	✓	✓	✓	✓

Summary of activities in view of their relationship to key operational guidelines (see Appendix A)

CS-5. India:
Tribal farmers and hunter gatherers in the tropical deciduous areas of Tamil Nadu

The Keystone project started in the Tamil Nadu region of India in June 1999, with financial assistance from Inter-Cooperation (IC), an organization created for the purpose of dealing with Natural Resource Management by the Swiss Development Cooperation (SDC). The duration of the project is 3 years, until June 2002.

Location

The project is located in South India, in the northwestern part of Tamil Nadu, not far from the city of Ooty, the Nilgiris District's capital, on the border of the States of Kerala and Karnataka. The villages covered under the project are part of Talukas (administrative subdivisions of a district) – Kotagiri and Coonoor. Land holdings are very close to the forested areas at middle elevations of 800-1 000 metres. The entire Nilgiris range rises up to a maximum of 2 600 m. The area is in the humid/semi humid tropics.

Environment

The Nilgiris consist of *one of the most ecologically fragile areas in India.* The hills are steep. Traditional forests have been depleted and are under further threat, because of the increase in large tea plantations and substantial destruction of natural vegetation by the Forest Department, through introduction of exotic commercial tree plantations. Consequently, soil erosion is rampant. Tea and coffee plantations have replaced large parts of its original vegetation and marshes have been converted into agricultural fields. Fifty percent (30 000 ha) of all cultivated area consist of tea plantations. Although no hard figures are available, it is common knowledge that conventional tea plantations make heavy use of chemical fertilizers and pesticides and reduce the water retention capacity of the soil.

It is for the above-mentioned reasons that the area has become part of the Nilgiris Biosphere Reserve, as declared under the Man and Biosphere Programme of UNESCO. During the past 19th and 20th centuries, deforestation, illegal and select cutting of valuable species was carried out. Vast areas of grasslands were replaced with plantations of Wattle, Eucalyptus and Cinchona. Thus large parts of pasturelands, belonging to the indigenous pastoral communities were taken over by the Forest Department. The negative effects of slash-and-burn practices, over-grazing, fire and the development of large plantations in the lower areas have been considerable.

However there are still good tracts of forests, representing the original Nilgiris' vegetation. Here, people live in harmony with the forest and collect Non Timber Forest Produce (NTFP) like wild nutmeg, cinnamon, sugarcane, pepper, honey and herbal plants. These deciduous forests and thorny thickets are found at elevations between 800-1 200 m. Rosewood is the dominant species in the wet areas, teak and sandalwood in the drier zones. But, the forests are much more diverse: *Erythrina, Dendrocalamus, Cedrella toona, Terminalia, Anogiessus latifolia, Pterocarpus marsupium* grow in the wet zones. *Zizyphus* and *Vitis*, many grass varieties and herbs in the drier areas.

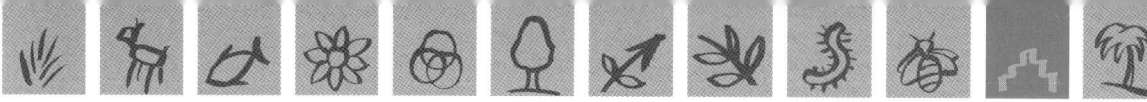

The area is rich in fauna too. Elephants, bisons, spotted and barking deer, Bears, leopards and numerous smaller animals have their habitat in the area.

Guideline #1. "Focus on the functional relationships and processes within ecosystems."

The environmental description presents what is clearly a common pattern: an ecologically fragile environment due to a combination of factors, including conversion of wetlands and forests, use of pesticides and excessive amounts of synthetic fertilizers, over-grazing, soil erosion and lack of organic material inputs. While the CBD laments the lack of environmental indicators in agricultural systems (referring to specific characteristics of soils, or animal or plant species assemblages) it certainly doesn't take much imagination to look to certain suites of agricultural practices, like those mentioned, as indicators of a system that is most likely moving in a downward spiral towards ecological, economic and social decay or collapse.

Crop history

The history of change from traditional cropping to the newer commercial cash crops of tea, coffee and vegetables is interesting. In 1818, when the British entered the District, they found a "primitive population" practising slash-and-burn agriculture. In the 1820s, the British first introduced vegetables. The Badagas, in the plateau area of the Nilgiris took to the cultivation of beans, cauliflower, cabbage and carrots on a large scale. In 1897, 1600 ha of tea were planted. In 1949 the tea area increased to 8 900 ha. Today, tea occupies 50 percent of Nilgiris' total cropped area. It has changed the land-use significantly, destroying grasslands and marshes and replaced a mixed cultivation with a monocrop cash crop. Coffee was introduced on the slopes of the hills in 1838. This was the zone where the hunters/ gatherers lived - Kurumbas, Irulas and Jenu Kurumbas were soon introduced to this crop, which spread within the forested lower zones. The main coffee plantations were in the Gudalur-Wynaad region but also on the slopes of Coonoor and Kotagiri Talukas. Coffee soon became an integral part of the homesteads of tribals and a popular beverage amongst them. Nowadays, coffee is facing a threat from the more lucrative and hardy crop, tea.

People

The Tribal population, which occupies the hills amounted to 25 000 (Census of India, 1991), but may have gone up slightly since. The main hunting and gathering communities consist of Alu Kurumbas (5 000), Irulas (6 000), Jenu Kurumbas (1 000), Betta Kurumbas (3 000) and Kasavas (1 000). They are Dravidian-speaking and belong to the autochthonous Indian population. They are predominantly-forest dwellers-hunters and gatherers, but have been gradually involved in agriculture as small-scale cultivators. They use shifting cultivation and slash-and-burn techniques. Primarily it is a subsistence economy, in which hunting, fishing and collecting are combined with subsistence agriculture and some daily wage labour on the plantations. A study done by Keystone in 1997, among the tribal hamlets revealed that 39 percent are landless; 14 percent have less than one acre; 35 percent between 1-2 acres and 12 percent between 2-3 acres.

With the increase of tea plantations, all communities lost their usufruct rights. For most of the tribal communities, it is living on the edge. Most tribals depend for survival on daily wages, earned for their work on the plantations. Interestingly, the number of women as regular workers is much higher than the number of men. The maximum earnings per week, including NTFP are Rs 200-250 (US$ 4-5). However the type of work depends on the remuneration available and the season of the year. The study reveals that NTFP collection starts in January/February and ends with the honey harvest in May/June. Between July/November, people in the upper plateau have no option than to search for wage labour. In other zones, where it is difficult to get work, people supplement their meagre meals with collected forest roots. Still, to survive is difficult. And if they do, there are little or no reserves. In times of illness, at festivals or funerals, they depend on moneylenders who provide loans at exorbitant rates of interest, up to 120 percent per year.

From the above, it can be concluded that there are two main problems: A highly fragile environment and a marginalized community of tribals.

It has been Keystone's conviction that environment and people are inter-related and improvement of the environment is impossible without a strong involvement of the communities. Keystone mentions as its mission: "A conscious goal to enhance the quality of life and the environment. It means: breaking new paths that are innovative, yet relevant and dealing with diverse problems/issues in an integrated manner".

Keystone's pre-project

In 1995, Keystone started to work with the Irulas and Kurumbas tribal communities on apiculture. The main objective was to improve their techniques of honey gathering and processing. At that time, the idea to develop alternatives for growing tea plantations emerged. The reasons were the following:

Tea plantations, with the exception of the organic ones, make heavy use of chemical fertilizers and pesticides. Tea is a mono-crop, endangering especially the lower slopes of the Nilgiris. Water retention capacity is reduced and vast tracts of primary forested land are destroyed for tea cultivation, limiting food security for the tribals. Moreover, the need for fuelwood, which is commonly used to process tea, increases. Some of these aspects are also true for coffee, except that it is not a monoculture and in fact has a variety of shade trees: jack, silk cotton, pepper, orange, guava, and a number of forest trees. The water retention capacity therefore is much better.

Traditional crops and coffee were introduced, to increase biodiversity and to decrease the dependence of the Tribals on wages. seven kg of millet seeds were bought from various villages in 1996 and together with pumpkin, chillies and mustard seeds, distributed in one village, as an experiment. This yielded sufficient produce to start a seed bank in the same year. In 1997, these traditional seeds were sown on 1,5 acres of land, owned by the Kurumbas in the village of Semmanarai. Many tribal families became interested as they realized the importance of food crops and

No. 11

vegetables, especially for babies and children. In 1998, the experiment got a boost when three other villages expressed interest in clearing their land for planting millet and vegetables. Thus 44 villagers from four villages became involved. The division was as follows:

The story of the village Vagapanai is illustrating. After an initial failure of Keystone to increase beekeeping, new meetings were held in which the earlier problems were analyzed and new approaches agreed upon. In 1998, 18 families agreed and started to clear the land. This increased to the present 27, which virtually means the total settlement. They took care of the land, built small huts on the cultivated land and moved in with their children, goats, dogs and chicken, only to go back once in a while to check their houses and buy provisions. Thus they could more easily weed and protect their fields against monkeys, wild boars, buffaloes and elephants. The first crop harvested was maize and as a festivity served to all guests. Other grains and vegetables followed.

Another initiative by an individual Kurumba to set up a coffee nursery was taken in 1997 in Semmanarai village: 4800 saplings were grown; 350 were distributed throughout his village for free and the remaining 4 050 sold. Keystone monitored the achievements and decided in 1998 to set up two more coffee nurseries in other villages. These village nurseries were used for planting saplings in tribal lands. In total 9 500 saplings were raised, according to organic agricultural standards. Ecologically, coffee is a much better crop than tea, as coffee grows together with many other crops.

Gradually the idea emerged to change the livelihood of the tribals by growing food crops as well as cash crops (coffee) organically and to improve the agro-ecological environment.

The project; its aims and objectives

Keystone's programme looks at land use as a whole. It aims to achieve the following objectives:

- Promote traditional crops cultivation to ensure food security for tribal families.
- Encourage cultivation of millet and vegetables to improve the nutritional intake.
- Discourage the spread of tea on the lower hill slopes.
- Promote Tribals to start and develop their own coffee nurseries. The saplings to be planted in village lands.
- Create a seed bank of rare millet and coarse grain seeds for use by other villages and for the conservation of local food grain varieties.

Overall objective

To promote a form of land use which preserves biodiversity, enhances a mix of cash and food crops and is ecologically sustainable.

Specific objectives for the project

The specific objectives for the project are:

- Initiate soil and water conservation methods for present land holdings.
- Provide options of growing different crops/plants/trees in the land.
- Provide food supplements to tribal families through reviving traditional agriculture
- Integrate off-farm activities of floriculture and beekeeping.
- Promote a mix of commercial crops/oilseeds along with useful traditional varieties.
- Document traditional indigenous practices in hill agriculture systems (for example Darsi seed banks, barter systems, rodent control,

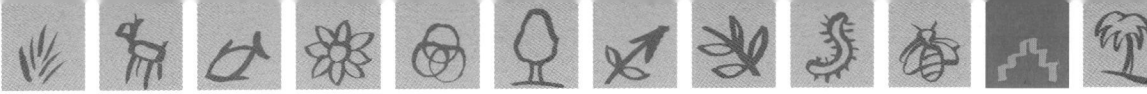

social and cultural mountain rituals).

- Establish a seed bank and nurseries for useful crops for the hills.
- Enable greater biodiversity patches within hill agriculture areas through introduction of multi-tier shade coffee bushes.
- Initiate greater awareness of the importance of sustainable land development against traditional shifting, subsistence agriculture.
- Design a package of benefits combining forest gathering and agricultural produce .

The project has 3 distinct themes:
- development of traditional/organic agriculture, which involves millet cultivation and vegetables, to supplement the diet;
- the development of cash crops, including coffee and vegetables;
- the integration of off-farm activities as floriculture and beekeeping.

Activities included:

Village seed banks: To conserve and increase local varieties of millet (finger-, little- proso and fox-tail millet); and further *Amaranthus* spp; Lablab, maize, mustard and beans. This to increase food security and biodiversity.

Development of nurseries: As a basis for growing coffee and other useful cash crop-species. Species identified by the villagers included silver oak, pepper, silk cotton, tamarind and lime. Individual villagers maintain the nursery in a common place in the village.

Soil and moisture conservation: These most important activities could only take place after much debate and an utmost reluctance of the villagers. They argued that agricultural work on steep mountain land was never done before. Moreover, deep-rooted practices of slash-and-burn were preferred, as hardly any effort is needed. Keystone had to convince them that slash-and-

burn creates serious problems for their food security. Keystone also elaborated on the advantage of using many elements of their traditional knowledge system, favouring an organic approach. A few villagers accepted the ideas and started to work. Keystone initially had to put in ample financial and supervisory inputs. Following a watershed approach, trenches were dug, which were filled up with the first rains. Crop yields increased. Stone bunds were constructed on steep slopes with loose soil and vegetative bunding in less fragile hill areas. Non-perennial water sources were provided with gully plugs.

PRA land use pattern analysis: This participatory exercise is undertaken in the villages where land development work is going on. The lands of the village are marked; their status and land use established. Furthermore stories about village history, old land use, water sources, habitat details and trade practices are collected. The water table movement was followed and key features in settlements established, such as the extent of land in different settlements, acreage owned by individuals, crops grown, output per units of land, perennial and non-perennial sources, usage, whether for human consumption only, user profiles. This research helps to better define problems, and find solutions and approaches for the future.

Documentation of traditional practices: The project documents the traditions and beliefs followed by the Tribals in millet cultivation and the relation of the forest with agriculture. The reason for this documentation is that existing practices, with the exception of slash-and-burn, form the best entrance point for further improvement, as they are often related to organic agricultural approaches and

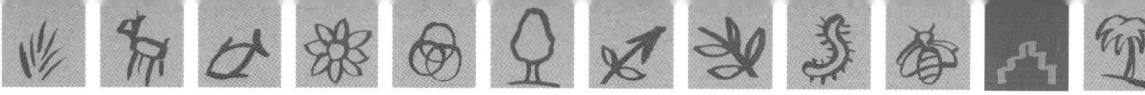

techniques. They include mixed cropping, inter-cropping with vegetables, techniques of seed preservation/selection and rotational cropping.

Construction of a Village Resource Centre: In one of the villages, a resource centre has been completed. Built with local material and using the skills available in the village, it serves as a meeting place and training-centre for farmers from the villages in the project area. Moreover, a village seed bank and an apiary are situated here. One more centre has just been completed in another village. It will serve as a central meeting place, drying-yard for coffee and pepper and value addition activities like making pillows, cushions, bees wax, candles, etc.

Beekeeping and floriculture: The improvement of beekeeping had been the mainstay in Keystone's long-term earlier interventions and has been integrated in the project.

Buy-back mechanisms: Although buy-back arrangements for cash crops are not specifically mentioned under the project, Keystone took this up in addition, as it was regarded as an important component. The buy-back facility is provided to support income options. It encourages the farmers and acts as an incentive. At the same time, it leaves farmers free to market their products elsewhere, if they can fetch a better price.

One of the key elements to this project that looks to be important to its success is the involvement of the tribals from the earliest stages, and through the vehicle of participatory analysis and planning (as the first step and in transition to each subsequent step).

Results

Documentation of traditional knowledge, mapping of areas and biodiversity transects of the villages. The project has just entered its fifth semester and has one more to go. Documentation of traditional knowledge, mapping of areas and biodiversity transects of the villages have been completed. However, it is still to be worked out in a readable document.

Participation of farmers

Meetings and participation from farmers have gone up considerably as interest in working on the land has increased, to the detriment of working as wage labourer on the plantations. In 1999, the community contributed 20 percent to the costs of raising nurseries. This percentage increased to 60 percent in 2000. In the

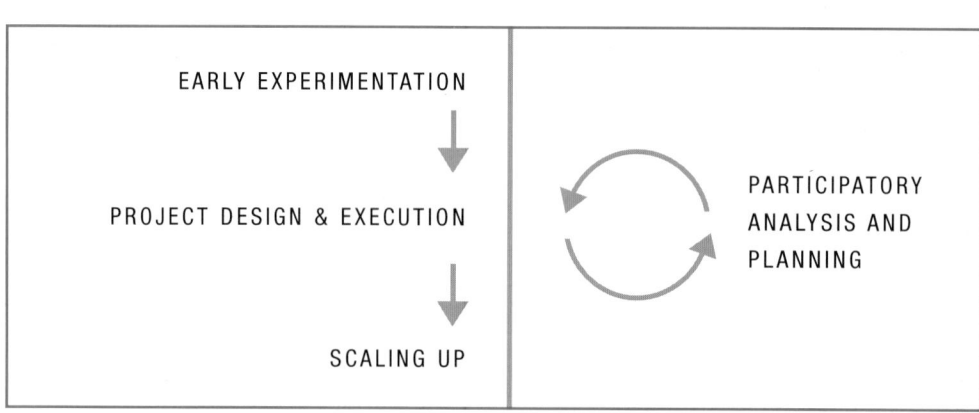

same year the contribution to the costs of land clearing amounted to 25 percent and increased to 50 percent by 2001.

Seed banks and nurseries

Seed banks have been installed in three villages. Twenty-five kg of seeds were distributed in 1999. This has increased to the present 150 kg per season. Nurseries were installed in four villages and produced over 75 000 saplings in 1999-2000. In 2000-2001, when fewer saplings were needed, three nurseries in three villages produced over 40 000 saplings. Seed banks and nurseries are not necessarily interlinked. They may be in different villages.

Soil and water conservation

Training on soil and water conservation has worked. In 1999, minor irrigation was established in four villages; live fencing and earthen check dams in one village; 644 trenches were dug and 583 m. of stone fencing constructed. This drastically increased to 2 800 m of stone bunding, 6 300 cubic feet of gully plugs, 30 000 trenches and 14 000 running feet for staggered trenches.

Buy-back mechanism and marketing

The buy-back mechanism has proved to be an important incentive for the farmers to remain "organic". A buy-back is guaranteed for any quantities of honey and bees wax, collected. The present quantity is about 4-5 tons per annum.

The purchase price is about 50 percent higher than the local rates. The price is set at the beginning of the season, as per the custom. Honey is marketed at rates which are slightly lower than the market rates of companies like Dabur (one of the largest honey companies in India). This indicates the large margin for traders. Buy-back quantities for coffee and pepper are more limited so far, as Keystone has not fully explored the markets in order to absorb full production. The quantities are coffee: one ton and pepper: 700-800 kg. For these products a 10 percent premium is paid over the prevailing wholesale rates

Summary of activities in view of their relationship to key operational guidelines (see Appendix A)

INDIA \ FIVE OPERATIONAL GUIDELINES (CBD V/6)

ACTIVITY:	FOCUS ON FUNCTIONAL RELATIONSHIPS AND PROCESSES	ENHANCING BENEFIT-SHARING	USING ADAPTIVE MANAGEMENT	MANAGEMENT AT APPROPRIATE SCALES (DECENTRALIZATION)	ENSURING INTER-SECTORAL COOPERATION
Village Seed Banks for Local Varieties	✓	✓	✓	✓	✓
Nursuries for Cash crops	✓	✓		✓	✓
Soil and Moisture Conservation	✓			✓	
PRA Land Use Analysis	✓	✓	✓	✓	✓
Documentation of Traditional Practices	✓	✓	✓	✓	
Construction of Village Resource Center	✓	✓		✓	✓
Beekeeping and Floriculture	✓	✓	✓	✓	
Buyback Mechanism		✓	✓	✓	✓

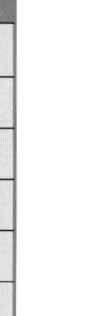

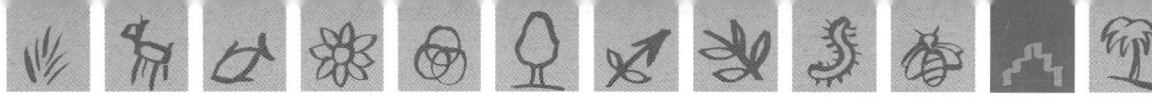

in the nearest market. As it is farm-gate price, no transportation costs have to be incurred by the farmers and thus the actual premium is higher. These products are sold in local markets at slightly higher prices than conventional products. As a matter of fact, more than 60 percent of all cash products are sold in the Nilgiris.

The Tamil Nadu case study is an excellent example of employment of the full range of ecosystem management principles as outlined in the CBD document-applying participatory methods to address issues related to soil fertility and water management, pollination, improving the diversity and quality of crop genetics, and attention to economic sustainability over the longer term.

THE ECOSYSTEM APPROACH AND LESSONS LEARNED IN THE CASE STUDIES

Refer to the Appendix for a full description of the 12 principles set forth in CBD Decision V/6

In applying the 12 Principles of the ecosystem approach, five points are proposed as Operational Guidance. We will use the five Operational Guidelines as reference points for discussion of "lessons learned" in the case studies abstracted above.

Focus on the functional relationships and processes within ecosystems

Principle #3. Consider the effects on adjacent and other ecosystems.

Principle #5. Conserve structure and function.

Principle #6. Manage within ecosystem limits.

Soil fertility management issues are at the centre of all of these case studies.

Farmers managing soils are managing ecosystem processes. Soil ecology is a discipline only just now coming into focus, due to inherent problems in dealing with the size, diversity and complexity of soil systems. Nevertheless, farmers can manage soils, under most conditions, in a sustainable manner employing only a minimum understanding of functional relationships related to biodiversity, nutrient cycling and fertility management.

From the growing number of case studies being accumulated by many concerned organizations, a common pattern emerges of an ecologically fragile environment resulting from a combination of factors, including conversion of wetlands and forests, use of pesticides and excessive amounts of synthetic fertilizers, over-grazing, soil erosion and lack of organic material inputs. While the CBD laments the lack of environmental indicators in agricultural systems (referring to specific characteristics of soils, or animal or plant species assemblages) it certainly doesn't take much imagination to look to certain suites of agricultural practices, like those mentioned, as indicators of a system that is most likely moving in a downward spiral towards ecological, economic and social decay or collapse.

As with other aspects of ecosystem management, optimal soil fertility management depends on many local-specific characteristics. This suggests that a management approach based on the conventional approach of research-station recommendations will inevitably be off the mark and incomplete, as no research and extension agency in any country has the resources to observe the multitude of local specific scenarios and then to then make recommendations for best management. Some suggestions have been made to set up soil testing laboratories in sufficient density that farmers can bring in samples to have analyzed.

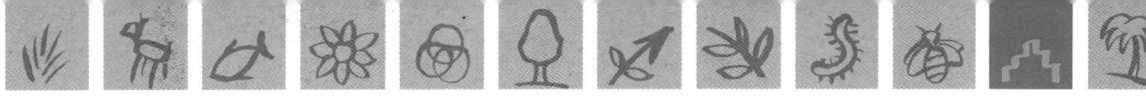

The case studies presented in this document demonstrate that this suggestion misses an important point: that the majority of problems faced by farmers in developing countries with regard to soil fertility management result from mismanagement of a relatively small set of key factors, the solution of which requires not fine-tuned knowledge of nutrient deficiencies but basic fundamental knowledge of the nature and importance of soil erosion, nutrient flows, the critical importance of recycled organic materials and the roles played by below-ground and above-ground biodiversity.

Soil health is linked in a direct relationship to the health of the plants, and indirectly to other important services in the system, including pest and disease suppression in the soil, on the soil, and above the soil surface. As indicated in the Thai organic rice example, farmers who begin to employ improved soil management and based on soil organic amendments frequently report a lowering of pest and disease problems.

Soil health is integrally tied to recycling crop residues, but this often leads into conflicting needs, leading farmers to have to make tough decisions. Residues in subsistence-level communities are often used for cooking fuel, for building construction, or for animal fodder. What is critical is that farmers understand the trade-offs in order that they are able to make informed decisions. It seems the case that farmers in developing countries, while appreciating that organic residues have a role as fertilizers, most often do not appreciate the extensive relationships that soil organic matter has with all aspects of the farming system. This is where NGOs and farmer training programmes, such as those presented in this document, have a critical role to play.

Enhance benefit-sharing

Principle #1: Management is a matter of societal choice

Here "choice" must mean informed choice; knowledge and the means of acquisition of knowledge is both a means to increased benefits, and inherently a benefit.

Principle #4: Manage the ecosystem in an economic context: (i) reduce market distortions that affect biological diversity; (ii) align incentives to promote biodiversity conservation and sustainable use; (iii) internalize costs and benefits in the given ecosystem to the extent feasible)

In the Thai case studies, the project guaranteed a premium price for certified organic produce, which helped initially in buffering the 'pioneer' farmers from initial increased costs in equipment and in the transition from conventional to organic. Similarly, the Indian case study demonstrated that a project-derived "buy-back" scheme was important in helping farmers make the transition. While certain segments of society that see organic farming as a threat would undoubtedly decry the use of credits and buy-back schemes as 'inefficiencies' in a market-based system, they should remember that huge government subsidies for both pesticides and fertilizers-much larger (by orders of magnitude) than what can be found in support of organic agriculture-have been a part of the green revolution formula. One has to examine the size and time frame for these incentives, as well as how they support or detract from progress towards sustainable production systems. In the past the huge subsidies for pesticides and fertilizers moved systems away from both ecological and economic sustainability, often pushing the systems to the breaking point at which new sets of serious problems emerged, such as pest and disease outbreaks and soil-plant problems due to depleted soil fertility.

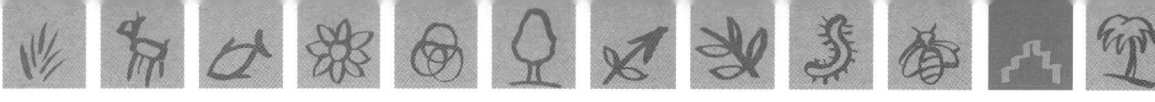

Use adaptive management practices

Principle #1: Management is a matter of societal choice.

Principle #9: Change is inevitable.

Principle #10: Seek a balance between conservation and use of biological diversity.

Principle #11: Consider all forms of relevant information, including scientific, indigenous and local knowledge.

Principle #12: involve all relevant sectors of society and scientific disciplines.

Adaptive management was first developed as a resource management tool in the 1970s (Holling, 1978). Various definitions of adaptive management are available in the literature (Walters, 1986,; Callicott *et al.*, 1999), but the basic concepts are simple and appealing. Adaptive management tries to incorporate the views and knowledge of all interested parties. It accepts the fact that management must proceed even if we do not have all the information we would like, or we are not sure what all the effects of management might be. It views management and policy, not just as a means to achieve objectives, but also as a type of "experiment" or a process for exploring the ecosystem being managed. Thus, learning is an inherent objective of adaptive management. As we learn more, we can adapt our policies to improve management success and to be more responsive to future conditions (Johnson, 1999).

In all the case studies some form of participatory interaction was used in order to help farmers explore the nature of their ecological and economic systems, to help analyze problems and to plan future activities. In the Thai case study we saw an example of 'learning by doing' in the form of the Farmer Field School (FFS). The FFS and other participatory-based systems have emerged as the dominant educational and operational device among development agencies in the past 10 years. Many outstanding examples from the field exist to demonstrate that knowledge and, more importantly, the ability to acquire and to effectively communicate knowledge is a powerful tool in the empowerment of both rural and urban poor.

Carry out management actions at the scale appropriate for the issue being addressed, with decentralization to the lowest level, as appropriate

Principle #2: Management should be decentralized to the lowest appropriate level.

Principle #7: Ecosystem approach should be undertaken at the appropriate spatial and temporal scales.

Principle #8: Management should be set for the long-term.

The most appropriate scale for management of small farming systems is at the level of the farmer. Although seemingly a truism, this fact is overlooked and undermined by past and often current administrative structures for national agricultural research and extension agencies. High heterogeneity, or the dominance of highly localized factors, is simply an unavoidable fact of life for ecological systems, and no more so than in developing countries. Rather than approaching this heterogeneity as a problem to be solved with uniform approaches to agricultural management, we have good reason to believe this high heterogeneity is a source of strength and *resilience* for agro-ecological systems. This realization should be cause for reflecting on the appropriateness of the conventional research-station approach to agricultural development. The latter promotes general

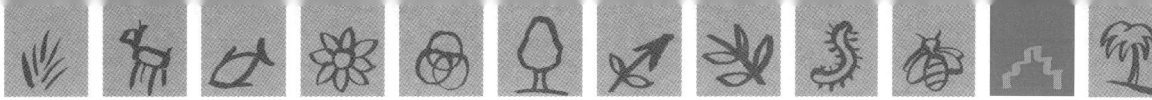

recommendations to be superimposed on a heterogeneous landscape as if conditions were uniform (or should be uniform) from farm to farm. The result is highly inefficient application of resources. Heterogeneity in both systems has been the biggest casualty in the application of industrialized farming methods, leading to increased fragility in these systems. One of the key points to be reiterated by an examination of these case studies is that *farmers are in the best position to observe, and have the greatest motivation to understand and to act to conserve and rejuvenate their own fields.*

Ensure inter-sectoral cooperation

Principle #11: Consider all forms of relevant information, including scientific, indigenous and local knowledge.

Principle #12: Involve all relevant sectors of society and scientific disciplines.

The task of developing a sustainable, productive and equitable future for farming societies is a mighty challenge, and all relevant sectors must contribute to the best of their ability. As we can see-not just from these few case studies, but from the multitude of good work going on world-wide-NGOs play a critically important role in being able to operationalize small-scale programmes geared to specific communities. In many cases NGOs lack the technical resources to be able to address specific problems, or the resources to be able to scale-up successful results. To some extent this problem can be mitigated by developing networks of NGOs who share resources and information. The FAO has also a long history of close collaboration to help address these needs. Universities and international research organizations are showing increasing signs of becoming less-and-less 'public domain' institutions, and more-and-more in line with industry. This is an unsettling prospect as industry tends to have a too narrow perspective on priorities and methodologies. The time is past when we can dream that the problems of agricultural societies will be solved by supposedly "objective" research passed on to extension specialists who will transfer technologies to be executed by farmers. In the domain of agriculture and ecosystem management science needs to be seen in a much broader context as a self-critical, learning process engaged in not just by researchers, but also by policy makers and especially the vast numbers of small farmers in developing countries.

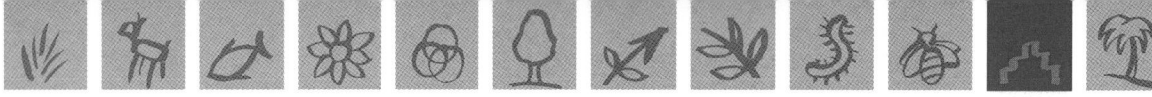

REFERENCES

Bardgett, R. D., J. M. Anderson, V. Behan-Pelletier, L. Brussaard, D. C. Coleman, C. Ettema, A. Moldenke, J. P. Schimel, and D. H. Wall., 2001. The Influence of Soil Biodiversity on Hydrological Pathways and the Transfer of Materials between Terrestrial and Aquatic Ecosystems. Ecosystems 4: 421–429.

Callicott, J. B., L. B. Crowder, and K. Mumford., 1999. Current normative concepts in conservation. Conservation Biology 13: 22-35.

Clark, T. W., and S. C. Minta., 1994. Greater Yellowstone's Future: Prospects for Ecosystem Science. Homestead Publishing, Moose, WY.

Ewel, K. C., C. Cressa, R. T. Kneib, P. S. Lake, L. A. Levin, M. A. Palmer, P. Snelgrove, and D. H. Wall., 2001. Managing critical transition zones. Ecosystems 4: 4552-460.

Funtowicz, S. O., J. Martinez-Alier, M. J., and J. R. Ravetz., 1999. Information tools for environmental policy under conditions of complexity, pp. 34. European Environment Agency, Luxembourg.

Holling, C. S., 1978. Adaptive environmental assessment and management. John Wiley, New York, New York, USA.

Holling, C. S., 1995. What barriers? What bridges?, pp. 593, Barriers and bridges to the renewal of ecosystems and institutions. Columbia University Press, New York.

Johnson, B. L., 1999. The role of adaptive management as an operational approach for resource management agencies. Conservation Ecology 3: 8.

Shachak, M., M. Sachs, and I. Moshe., 1998. Ecosystem management of desertified shrublands in Israel. Ecosystems 1: 475-483.

Walters, C., 1986. Adaptive management of renewable resources. Collier MacMillan, New York, London.

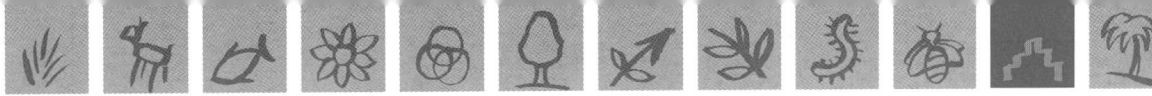

ANNEX 1

CBD COP Decision V/6
Ecosystem Approach

The Conference of the Parties,

1. Endorses the description of the ecosystem approach and operational guidance contained in sections A and C of the annex to the present decision, recommends the application of the principles contained in section B of the annex, as reflecting the present level of common understanding, and encourages further conceptual elaboration, and practical verification;

2. Calls upon Parties, other Governments, and international organizations to apply, as appropriate, the ecosystem approach, giving consideration to the principles and guidance contained in the annex to the present decision, and to develop practical expressions of the approach for national policies and legislation and for appropriate implementation activities, with adaptation to local, national, and, as appropriate, regional conditions, in particular in the context of activities developed within the thematic areas of the Convention;

3. Invites Parties, other Governments and relevant bodies to identify case-studies and implement pilot projects, and to organize, as appropriate, regional, national and local workshops, and consultations aiming to enhance awareness, share experiences, including through the clearing-house mechanism, and strengthen regional, national and local capacities on the ecosystem approach;

4. Requests the Executive Secretary to collect, analyse and compare the case-studies referred to in paragraph 3 above, and prepare a synthesis of case-studies and lessons learned for presentation to the Subsidiary Body on Scientific, Technical and Technological Advice prior to the seventh meeting of the Conference of the Parties;

5. Requests the Subsidiary Body on Scientific, Technical and Technological Advice, at a meeting prior to the seventh meeting of the Conference of the Parties, to review the principles and guidelines of the ecosystem approach, to prepare guidelines for its implementation, on the basis of case-studies and lessons learned, and to review the incorporation of the ecosystem approach into various programmes of work of the Convention;

6. Recognizes the need for support for capacity-building to implement the ecosystem approach, and invites Parties, Governments and relevant organizations to provide technical and financial support for this purpose;

7. Encourages Parties and Governments to promote regional cooperation, for example through the establishment of joint declarations or memoranda of understanding in applying the ecosystem approach across national borders.

Description of the ecosystem approach

1. The ecosystem approach is a strategy for the integrated management of land, water and living resources that promotes conservation and sustainable use in an equitable way. Thus, the application of the ecosystem approach will help to reach a balance of the three objectives of the Convention: conservation; sustainable use; and the fair and equitable sharing of the benefits arising out of the utilization of genetic resources.

2. An ecosystem approach is based on the

application of appropriate scientific methodologies focused on levels of biological organization, which encompass the essential structure, processes, functions and interactions among organisms and their environment. It recognizes that humans, with their cultural diversity, are an integral component of many ecosystems.

3. This focus on structure, processes, functions and interactions is consistent with the definition of "ecosystem" provided in Article 2 of the Convention on Biological Diversity: "'Ecosystem' means a dynamic complex of plant, animal and micro-organism communities and their non-living environment interacting as a functional unit." This definition does not specify any particular spatial unit or scale, in contrast to the Convention definition of "habitat". Thus, the term "ecosystem" does not, necessarily, correspond to the terms "biome" or "ecological zone", but can refer to any functioning unit at any scale. Indeed, the scale of analysis and action should be determined by the problem being addressed. It could, for example, be a grain of soil, a pond, a forest, a biome or the entire biosphere.

4. The ecosystem approach requires adaptive management to deal with the complex and dynamic nature of ecosystems and the absence of complete knowledge or understanding of their functioning. Ecosystem processes are often non-linear, and the outcome of such processes often shows time-lags. The result is discontinuities, leading to surprise and uncertainty. Management must be adaptive in order to be able to respond to such uncertainties and contain elements of "learning-by-doing" or research feedback. Measures may need to be taken even when some cause-and-

effect relationships are not yet fully established scientifically.

5. The ecosystem approach does not preclude other management and conservation approaches, such as biosphere reserves, protected areas, and single-species conservation programmes, as well as other approaches carried out under existing national policy and legislative frameworks, but could, rather, integrate all these approaches and other methodologies to deal with complex situations. There is no single way to implement the ecosystem approach, as it depends on local, provincial, national, regional or global conditions. Indeed, there are many ways in which ecosystem approaches may be used as the framework for delivering the objectives of the Convention in practice.

Principles of the ecosystem approach

6. The following 12 principles are complementary and interlinked: Principle 1: The objectives of management of land, water and living resources are a matter of societal choice. Rationale: Different sectors of society view ecosystems in terms of their own economic, cultural and societal needs. Indigenous peoples and other local communities living on the land are important stakeholders and their rights and interests should be recognized. Both cultural and biological diversity are central components of the ecosystem approach, and management should take this into account. Societal choices should be expressed as clearly as possible. Ecosystems should be managed for their intrinsic values and for the tangible or intangible benefits for humans, in a fair and equitable way.

Principle 2: Management should be decentralized to the lowest appropriate level.
Rationale: Decentralized systems may

lead to greater efficiency, effectiveness and equity. Management should involve all stakeholders and balance local interests with the wider public interest. The closer management is to the ecosystem, the greater the responsibility, ownership, accountability, participation, and use of local knowledge.

Principle 3: Ecosystem managers should consider the effects (actual or potential) of their activities on adjacent and other ecosystems.

Rationale: Management interventions in ecosystems often have unknown or unpredictable effects on other ecosystems; therefore, possible impacts need careful consideration and analysis. This may require new arrangements or ways of organization for institutions involved in decision-making to make, if necessary, appropriate compromises.

Principle 4: Recognizing potential gains from management, there is usually a need to understand and manage the ecosystem in an economic context. Any such ecosystem-management programme should:

(a) Reduce those market distortions that adversely affect biological diversity;

(b) Align incentives to promote biodiversity conservation and sustainable use;

(c) Internalize costs and benefits in the given ecosystem to the extent feasible.

Rationale: The greatest threat to biological diversity lies in its replacement by alternative systems of land use. This often arises through market distortions, which undervalue natural systems and populations and provide perverse incentives and subsidies to favour the conversion of land to less diverse systems.

Often those who benefit from conservation do not pay the costs associated with conservation and, similarly, those who generate environmental costs (e.g. pollution) escape responsibility. Alignment of incentives allows those who control the resource to benefit and ensures that those who generate environmental costs will pay.

Principle 5: Conservation of ecosystem structure and functioning, in order to maintain ecosystem services, should be a priority target of the ecosystem approach.

Rationale: Ecosystem functioning and resilience depends on a dynamic relationship within species, among species and between species and their abiotic environment, as well as the physical and chemical interactions within the environment. The conservation and, where appropriate, restoration of these interactions and processes is of greater significance for the long-term maintenance of biological diversity than simply protection of species.

Principle 6: Ecosystems must be managed within the limits of their functioning.

Rationale: In considering the likelihood or ease of attaining the management objectives, attention should be given to the environmental conditions that limit natural productivity, ecosystem structure, functioning and diversity. The limits to ecosystem functioning may be affected to different degrees by temporary, unpredictable or artificially maintained conditions and, accordingly, management should be appropriately cautious.

Principle 7: The ecosystem approach should be undertaken at the appropriate spatial and temporal scales.

Rationale: The approach should be bounded by spatial and temporal scales that are appropriate to the objectives. Boundaries for management will be defined operationally by users, managers, scientists and indigenous and local peoples. Connectivity between areas should be promoted where necessary. The ecosystem approach is based upon the hierarchical nature of biological diversity characterized by the interaction and integration of genes, species and ecosystems.

No. 11

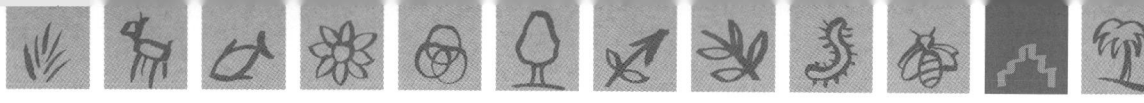

Principle 8: Recognizing the varying temporal scales and lag-effects that characterize ecosystem processes, objectives for ecosystem management should be set for the long term.

Rationale: Ecosystem processes are characterized by varying temporal scales and lag-effects. This inherently conflicts with the tendency of humans to favour short-term gains and immediate benefits over future ones.

Principle 9: Management must recognize that change is inevitable.

Rationale: Ecosystems change, including species composition and population abundance. Hence, management should adapt to the changes. Apart from their inherent dynamics of change, ecosystems are beset by a complex of uncertainties and potential "surprises" in the human, biological and environmental realms. Traditional disturbance regimes may be important for ecosystem structure and functioning, and may need to be maintained or restored. The ecosystem approach must utilize adaptive management in order to anticipate and cater for such changes and events and should be cautious in making any decision that may foreclose options, but, at the same time, consider mitigating actions to cope with long-term changes such as climate change

Principle 10: The ecosystem approach should seek the appropriate balance between, and integration of, conservation and use of biological diversity.

Rationale: Biological diversity is critical both for its intrinsic value and because of the key role it plays in providing the ecosystem and other services upon which we all ultimately depend. There has been a tendency in the past to manage components of biological diversity either as protected or non-protected. There is a need for a shift to more flexible situations, where conservation and use are seen in context and the full range of measures is applied in a continuum from strictly protected to human-made ecosystems.

Principle 11: The ecosystem approach should consider all forms of relevant information, including scientific and indigenous and local knowledge, innovations and practices.

Rationale: Information from all sources is critical to arriving at effective ecosystem management strategies. A much better knowledge of ecosystem functions and the impact of human use is desirable. All relevant information from any concerned area should be shared with all stakeholders and actors, taking into account, inter alia, any decision to be taken under Article 8(j) of the Convention on Biological Diversity. Assumptions behind proposed management decisions should be made explicit and checked against available knowledge and views of stakeholders.

Principle 12: The ecosystem approach should involve all relevant sectors of society and scientific disciplines.

Rationale: Most problems of biological-diversity management are complex, with many interactions, side-effects and implications, and therefore should involve the necessary expertise and stakeholders at the local, national, regional and international level, as appropriate.

C. Operational guidance for application of the ecosystem approach

7. In applying the 12 principles of the ecosystem approach, the following five points are proposed as operational guidance.

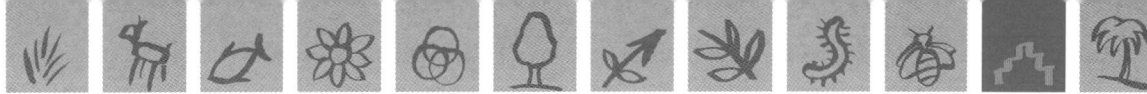

Focus on the functional relationships and processes within ecosystems

8. The many components of biodiversity control the stores and flows of energy, water and nutrients within ecosystems, and provide resistance to major perturbations. A much better knowledge of ecosystem functions and structure, and the roles of the components of biological diversity in ecosystems, is required, especially to understand: (i) ecosystem resilience and the effects of biodiversity loss (species and genetic levels) and habitat fragmentation; (ii) underlying causes of biodiversity loss; and (iii) determinants of local biological diversity in management decisions. Functional biodiversity in ecosystems provides many goods and services of economic and social importance. While there is a need to accelerate efforts to gain new knowledge about functional biodiversity, ecosystem management has to be carried out even in the absence of such knowledge. The ecosystem approach can facilitate practical management by ecosystem managers (whether local communities or national policy makers).

Enhance benefit-sharing

9. Benefits that flow from the array of functions provided by biological diversity at the ecosystem level provide the basis of human environmental security and sustainability. The ecosystem approach seeks that the benefits derived from these functions are maintained or restored. In particular, these functions should benefit the stakeholders responsible for their production and management. This requires, inter alia: capacity-building, especially at the level of local communities managing biological diversity in ecosystems; the proper valuation of ecosystem goods and services; the removal of perverse incentives that devalue ecosystem goods and services; and, consistent with the provisions of the Convention on Biological Diversity, where appropriate, their replacement with local incentives for good management practices.

Use adaptive management practices

10. Ecosystem processes and functions are complex and variable. Their level of uncertainty is increased by the interaction with social constructs, which need to be better understood. Therefore, ecosystem management must involve a learning process, which helps to adapt methodologies and practices to the ways in which these systems are being managed and monitored. Implementation programmes should be designed to adjust to the unexpected, rather than to act on the basis of a belief in certainties. Ecosystem management needs to recognize the diversity of social and cultural factors affecting natural-resource use. Similarly, there is a need for flexibility in policy-making and implementation. Long-term, inflexible decisions are likely to be inadequate or even destructive. Ecosystem management should be envisaged as a long-term experiment that builds on its results as it progresses. This "learning-by-doing" will also serve as an important source of information to gain knowledge of how best to monitor the results of management and evaluate whether established goals are being attained. In this respect, it would be desirable to establish or strengthen capacities of Parties for monitoring.

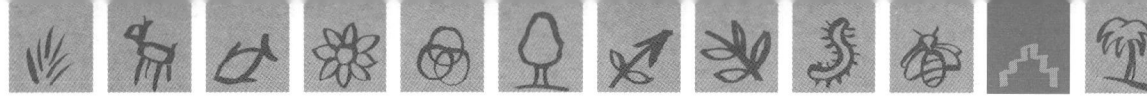

Carry out management actions at the scale appropriate for the issue being addressed, with decentralization to lowest level, as appropriate

11. As noted in section A above, an ecosystem is a functioning unit that can operate at any scale, depending upon the problem or issue being addressed. This understanding should define the appropriate level for management decisions and actions. Often, this approach will imply decentralization to the level of local communities. Effective decentralization requires proper empowerment, which implies that the stakeholder both has the opportunity to assume responsibility and the capacity to carry out the appropriate action, and needs to be supported by enabling policy and legislative frameworks. Where common property resources are involved, the most appropriate scale for management decisions and actions would necessarily be large enough to encompass the effects of practices by all the relevant stakeholders. Appropriate institutions would be required for such decision-making and, where necessary, for conflict resolution. Some problems and issues may require action at still higher levels, through, for example, transboundary cooperation, or even cooperation at global levels.

Ensure intersectoral cooperation

12. As the primary framework of action to be taken under the Convention, the ecosystem approach should be fully taken into account in developing and reviewing national biodiversity strategies and action plans. There is also a need to integrate the ecosystem approach into agriculture, fisheries, forestry and other production systems that have an effect on biodiversity. Management of natural resources, according to the ecosystem approach, calls for increased intersectoral communication and cooperation at a range of levels (government ministries, management agencies, etc.). This might be promoted through, for example, the formation of inter-ministerial bodies within the Government or the creation of networks for sharing information and experience.

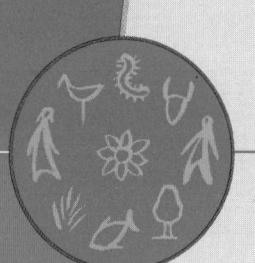

RESPONSIBLE
TECHNICAL DIVISION

Land and Water Development Division
Land and Plant Nutrition Management Service

Parviz Koohafkan

GLOBALLY IMPORTANT

INGENIOUS AGRICULTURAL HERITAGE SYSTEMS

12

GLOBALLY IMPORTANT
INGENIOUS AGRICULTURAL HERITAGE SYSTEMS

ACKNOWLEDGEMENTS

This paper has been prepared based on a concept and drawing on material developed by Parviz Koohafkan, Chief of the Land and Plant Nutrition Management Service (AGLL). Ms. Sally Bunning (AGLL), Mr. Jean Bedel (consultant) and Mr. David Boerma (AGLL) and the stakeholders and experts who participated in the Stakeholder Workshop of the GIAHS project held in Rome, August 2002, further contributed to the elaboration of this concept. The case studies were provided by Mr. Jose Furtado (Rice-Fish), Ms. Aude Verwilghen (Oases and Wayana) and David Boerma (Maasai). The material was summarized and assembled into the present paper by David Boerma with the assistance of Aude Verwilghen (consultant).

INTRODUCTION

Agricultural genetic resources are the result of farmers' careful selection of outstanding varieties of plants and animals, as well as co-adaptation among plants, animals and humans, under specific agro-ecological conditions. The conservation *in situ* of genetic resources for food and agriculture cannot be achieved outside dynamic farming systems and local human cultures in which these resources were developed. Following years of international consultations, with a view to protecting some of the most relevant farming systems that hold important genetic resources, including some that are particularly at risk, FAO in 2002 launched a FAO/UNDP-GEF project to support Globally Important Ingenious Agricultural Heritage Systems (GIAHS). The project seeks to promote the international recognition, conservation and sustainable management of these systems-including where necessary their revitalisation-and the support of the outstanding role these systems play in household food security and in the maintenance of agricultural biodiversity, as well as their contribution to natural, landscape and cultural heritage and indigenous knowledge systems.

Heritage for the future

GIAHS, and their associated agro-ecosystems and landscapes, have been created, shaped, maintained and passed between generations of farmers, herders, forest dwellers and fisher-folk. Based on diverse species and their interactions, and the use of locally adapted, distinctive and often ingenious combinations of management practices and techniques, they have contributed, and continue to contribute, to sustaining and enriching globally significant agricultural biodiversity, resilient ecosystems, and valuable cultural heritage. Moreover, such systems ensure the sustained provision of multiple goods and services, food and livelihood security, and quality of life for people.

Globally Important Ingenious Agricultural Heritage Systems are defined as:

Remarkable Land Use Systems and landscapes which are rich in biological diversity evolving from the ingenious and dynamic adaptation of a community/population to its environment and the needs and aspirations for sustainable development.

GIAHS throughout the world testify to the inventiveness and ingenuity of people in their use and management of biodiversity, inter-species dynamics, and the physical attributes of the landscape, codified in traditional but evolving knowledge, practices and technologies. This ingenuity has resulted in well-balanced agro-ecological systems in marginal, extreme or very specific ecologies, which could not otherwise have sustainably supported human life and agro-biodiversity at its present high level. These systems are organized and managed through highly adapted social and cultural practices and institutions.

Such systems, however, often face great threats and challenges in evolving and adapting to economic change and new and sometimes inappropriate policy environments, particularly in the contexts of land tenure, environmental change and globalization. To survive, they must also adapt their productive capacity to meet the rising expectations of their members, in terms of food security and quality of life.

At all scales, from household to global, diversity is a survival factor in the face of uncertainties, economic or environmental changes, hazards, shocks or disasters. The

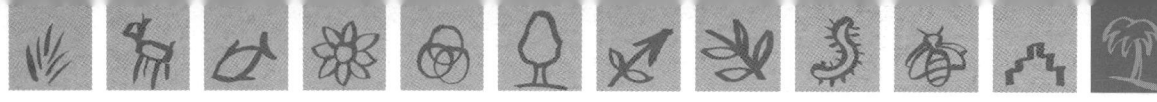

several kinds of diversity cannot be safeguarded or preserved in isolation, as in an archive, gene bank or museum, but only within living, evolving livelihood systems. Examples of GIAHS might include multi-storied home gardens, oases, certain rice-fish systems, qanat[1]-based orchards and gardens in arid areas, agro-forestry or transhumance livestock systems.

The GIAHS programme: concept and goals

The GIAHS concept recognizes and is centred on the profound inter-relatedness of biodiversity, agriculture, ecology, culture and social organization and institutions, ethics, local livelihoods and food security. The programme aims to safeguard the continued co-evolution of these elements. This integrated ecosystem approach builds on existing indigenous knowledge, practices, customs and institutions for the management of agricultural systems, in ways that are socially, economically and culturally appropriate to the identity, needs and aspirations of farming communities.

The underlying strategy of the programme is to avoid or reverse the loss or degradation of the resilience and the essential features and attributes of these systems-especially their biodiversity-while allowing their necessary evolution and at the same time enhancing the socio-economic development of resource users, as well as national and global benefits. The programme firstly attempts to mitigate threats to the resilience of GIAHS, by supporting farmers' and their communities' capacities to continue to sustainably manage these systems, with the involvement of national governments, scientists and other stakeholders. It also seeks to support these communities and their

governments in developing appropriate legal and policy environments and instruments, conducive to their continued existence, and which allow for their evolution and development. The programme offers an opportunity to build, in a step-by-step way, cooperation amongst communities that effectively manage their rich *in situ* heritages, in a sustainable development context, including through the exchange of experience, knowledge and technologies.

Economic viability

GIAHS have an array of value elements or benefits, both local and national or global, which is much wider than the immediate economic return, including an array of social, cultural, environmental and food security and risk management benefits. The aim of GIAHS is, in todays' local and global context, to identify ways to support their continued biodiversity conservation, sustainability and productivity. Promoting knowledge and understanding of GIAHS and wide recognition of their benefits, particularly positive externalities, may be enough to help some of these systems survive.

Some GIAHS may need more specific support, for example through brand creation and promotion, and the development of niche markets for certain produce, or through the creation of institutions that enable returns to communities for environmental services that are by-products of their land-use system. Other GIAHS may need enabling legal and policy environments that allow for their maintenance and socio-economic (self-) sustainability. There may even be some that will be served by more classical sustainable development initiatives that lift barriers and address root causes of the threats they face.

---

1   Horizontal, tunnel-type well

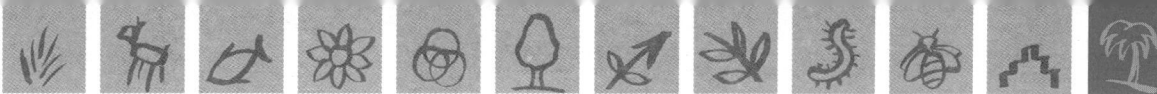

GIAHS CASE STUDIES

The case studies presented here are examples of GIAHS, which may be included in a group of 10 initial systems[2] on which a further programme methodology will be developed. The ingenuity of GIAHS, which are almost invariably based on high levels of agricultural biodiversity, often lies in the adaptive management of bio-physical, economic or socio-cultural resources that have evolved under specific ecological and socio-economic constraints and opportunities. In these contexts, the ingenious management of such systems has developed on the levels (social organization, soil and water management, biodiversity, knowledge transfer, etc.) which are most appropriate and efficient. These case studies demonstrate the different levels at which the ingenious management of such systems can be found, as well as the various services that these systems provide.

Traditional oasis in south Tunisia. An efficient agro-ecosystem generating biodiversity

An oasis can be defined as an agrarian system; an "area where irrigation is necessary and the farming system is highly productive, with the omnipresence of date palm". Their ingenuity is due to the human interactions that shape and influence the agro-ecosystem, and in turn provide many ecological and social services. In effect, an oasis is more than a site of agricultural production. Historically, it is a crossing point and a place of life, of rest, of leisure, of conquest. Thus, it is a complex system with agronomic, ecological, economic, social, political and strategic dimensions. Oases are the inheritance of ancient agricultural civilizations. The Gafsa oasis, for instance, dates back to the Capsian civilization, a Mesolithic culture from 8400 years BC.

The ingenious management of the system through high levels of biodiversity

The management of biodiversity in oases is inherited from adapted, rich and diversified indigenous knowledge systems. These show evidence of the efforts that have been made by successive generations in order to maintain a fragile balance in an environment with severe constraints. Species and varieties are carefully chosen as to be adapted to local environmental constraints. For instance, there is a prevalence of the olive tree in the periphery of oases because of its drought resistance, and the Degla date palms are preferentially planted in southwest Tunisia where climatic conditions are favourable for fructification, whereas common date palm varieties are more frequent in coastal areas. There is an intensive occupation of space for the optimum use of water resources and their functions in regulation of the oasis microclimate, for the maximization of harvest security by producing plants that provide multiple products and through carefully diversified production spacing and timing (cropping pattern and rotation). The latter is done with a three-level system, which includes date palm, arboriculture, annual and pluriannual plants and crops (vegetables, fodder, ornamental plants, etc.) and a high density of species and varieties. The management practices and techniques reveal ingenuity of local population in using biodiversity, for instance in terms of crop management (plantation, pollen transfer and thinning techniques, biological control of pests and diseases, etc.) and irrigation techniques (plant resilience to dought and water reserve in soil, management of and adaptation to salt, sand and wind).

---

2 The GIAHS (FAO/UNDP-GEF) project will focus mainly on systems in low-income, food-deficit countries, where GIAHS are largely subsistence oriented.

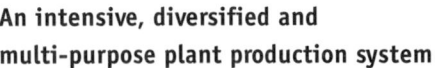

An intensive, diversified and multi-purpose plant production system

The constraining environment and the opportunity and climatological requirement of irrigation leads to a necessary intensification and diversification. The growing of different crops in space and time allows oasis communities to meet the essential needs for human consumption: food, energetic, domestic (building, crafts, etc.), and medicinal requirements. The surplus production is sold in the market and there is a trend to increasing cultivation of cash crops in order to generate income.

In terms of agricultural biodiversity, a large diversity of species and cultivars is planted in oasis agro-ecosystems, with numerous adapted local varieties resulting from meticulous breeding and transmitted from one generation to another. Tunisia has a high qualitative richness for date palm, with numerous very rare varieties and an important percentage of endemism, especially in coastal and mountain oases. There is a very high varietal diversity; for instance, in El Hamma oasis, 55 varieties have been listed. In Tunisia, there are 260 cultivars - named common varieties- of date palm and nameless varieties coming from seedlings, which represent an important potential for future selection. A large number of other species of fruits, vegetables, condiments, fodder, ornamental plants, etc. can also be found in oases, with numerous local varieties. For instance, the local varieties of fig tree: Assal Boutchich, apricot tree: Mechmech Arbi, olive oil tree: Chemchali from El Casba. Furthermore, oasis agro-ecosystems provide habitat and resources for numerous wild species of fauna and flora.

Livestock integration

Livestock raising in the strict oasis area is limited to a few individuals of sheep, goats, donkeys and/or camels. This is functional to the system by providing for food (meat, milk), transport (poeple, agricultural produce, etc.) and manure (soil amendment).

Global changes: threats to the continued viability and sustainability of traditional oasis

The focus on productivity and profitability, the effects of integration in the global market and cash economy (e.g. growing hegemony of a unique date variety, the Degla) and the abuse of modern technology (e.g. expansion of drilling and modernization of irrigation systems, which lead to permanent depletion of underground water sources) contriute to the rapid degradation of oasis systems. They are leading to standardization and specialization, with the consequential depreciation and destabilization of the traditional oasis system. These integrated and complex agro-ecosystems, in terms of the multiplicity and inter-dependence of components, the high ecological interactions, the rich and diversified knowledge and management practices have a highly fragile balance. Oases are havens of agricultural biodiversity in a constraining environment, and their degradation is synonymous with high genetic erosion.

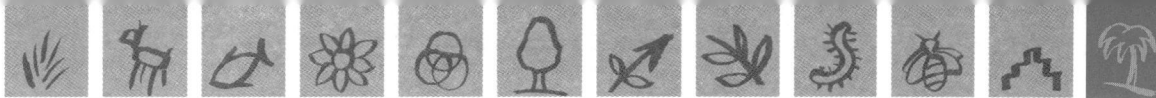

Rice-fish farming systems in Asia. Integrated agro-ecosystems with complex species interactions

Rice-fish farming systems form some of the most striking agricultural landscapes of the world. They have a variety of local designs adapted for cultural, environmental and economic attributes. Those complex and inter-dependent agro-ecosystems are using ecological services, such as biological control and N-fixation as well as landscape integration to ameliorate some persistent failures of elements of the system. Moreover, traditional and low intensity rice-fish systems play an important role in safeguarding the global environment, notably from a biodiversity perspective. Rice-fish systems support and are in turn supported by a large diversity of cultures and their associated institution for the management of these systems.

A large diversity of agro-ecosystems

Rice is the dominant staple crop of tropical Asia. It has a long history of domestication and a rich diversity of cultivated ecotypes based on three varieties of *Oryza sativa: indica, japonica* and *javanica,* which are cultivated in different agro-ecological zones for their differing growth, grain and yield characteristics. There are four basic rice agro-ecosystems each with peculiar edaphic conditions: irrigated ecosystems, upland (terraces) and lowland rainfed ecosystems, and flood-prone (very deep water) ecosystems.

Fish culture can be concurrent (mixed) or rotational with rice, at different intensities. This case considers traditional (capture) and low-intensity culture (no fertilizer, no feed) systems, as they enhance many ecological services. Moreover, those

systems are less risky for the resource-poor farmers than intensive fish farming, because of their efficiency derived from synergisms, their diversity of produce and their environmental soundness.

An integrated system with complex interactions

There is a combined use of habitat and resources for rice and fish. A rich variety of direct and mainly indirect beneficial effects emanate from the interactions between the different elements of the rice-fish agro-ecosystem, enhancing grain and fish production and contributing to the dynamism of the system. For example, rice provides shade for fish, organic matter produced by rice is used by fish, water oxygenation by fish and nutrient recycling benefit rice, biological inter-dependencies provide biological pest control (for example, predation on insects and pests by fish) and N-fixation by *Azolla* spp. for rice. Rice-fish systems are often based on and regulated by complex and highly diverse food webs of microbe, insects and their predators. However, many indirect non-beneficial effects are exacerbated by intensification of rice-fish production.

Global importance in term of food production and environmental issues

The rice-fish systems are globally important in terms of food production. The integration of fish in rice farming systems provides invaluable protein and fatty acids, especially for subsistence farmers managing rain-fed systems. They are also important in terms of the three global environmental issues: climate change (emission of greenhouse gas in rice fields is determined by farming practices, plant metabolism and soil properties; rain-fed systems tend to contributed less emissions than irrigated systems), shared waters (retaining flood

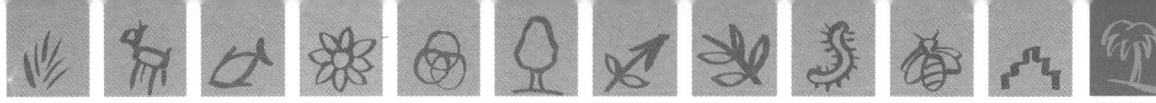

waters in shared catchments and river basins) and biodiversity.

From a biodiversity perspective, rice-fish farming systems contain:

- low to moderate rice genetic diversity due to intense varietal selection primarily for yields and secondarily for system maintenance and economic viability. Higher levels of biodiversity are found in traditional and low-intensity systems (temporal, spatial and genetic diversity resulting from farm-to-farm variations in cropping systems confers at least partial resistance to pest attack). For each agro-ecological, cultural and management system, ecotypes have been selected and developed, optimizing hydro-edaphic, vegetative, reproductive and ripening characteristics and minimizing losses to consumers, to competitors (weeds), as organic wastes and metabolites and to environmental hazards (cool temperatures, soil acidity and salinity, floods, etc.);
- moderate to high fish species' diversity, with low selection of varieties within species. Fish species and aquatic biodiversity appear richer in traditional and low intensity rainfed than in high-intensity irrigated rice-fish systems.

Threats

Many rain-fed rice-fish farming systems are under threat by (excessive) application of chemicals, particularly pesticides, either through modernization of the systems themselves or as the result of negative externalities of upstream agricultural systems. Pesticides mainly destroy the fragile food webs that underpin the rice-fish systems, but they also endanger the fish and human health directly. In some regions rice-fish systems have to cope with increasing

population pressures in systems which are already at the maximum of their productive capacity, leading to unsustainable agricultural practices and migration.

Agrarian system of the Wayana in French Guyana. Cultivated area and surrounding forest as a single agro-ecosystem

The way of life and the production system of Amerindians represent the accumulated experiences of humankind closely interacting with their environment over centuries. The farming system of Wayana society is based on shifting cultivation, with a high agricultural biodiversity. Here agriculture is part of an array of activities taking place within various habitats where Wayana obtain a significant portion of their subsistence requirements through gathering, fishing and hunting. In fact, there is no clear limit between the cultivated and the wild area, which thus can be considered as a single agro-ecosystem.

Cassava: a high varietal diversity for a species of major importance in Amerindian culture

Many crops and multiple varieties of each crop are cultivated on a parcel, supporting both intra-specific and inter-specific diversity. This strategy of minimizing risk by cultivating a diversity of crops and varieties in space and time enhances harvest security and promotes diet diversity.

The central crop of the farming system is cassava *(Manihot esculenta)*, followed by sweet potato *(Ipomea batatas)*. Many other plants are also cultivated, for instance: banana

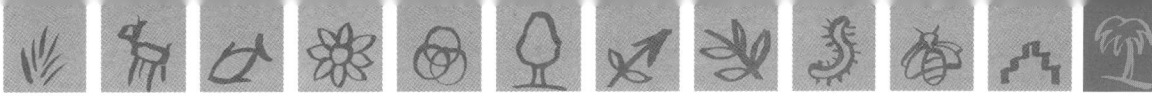

(Musa sapientum), sugar cane *(Saccharum officinarum),* maize *(Zea mays),* yam *(Dioscorea trifida, D. Bulbifera),* water melon *(Citrillus lanatus)* and several *Cucurbitaceae,* cotton *(Gossypium barbadense),* pineapple *(Ananas comosus),* dasheen *(Colocasia antiquorum),* cocoyam *(Xanthosoma sp.),* lima bean *(Phaseolus lunatus)* and cucumber *(Cucumis sativus).*

The Wayana are using and conserving many cultivars of cassava and to a lesser extent of sweet potato. During household surveys, respectively 70 and 13 different cultivars were named. Cassava is at the root of food consumption and products stemmed from processing are numerous, such as *cassava* (a sort of "pancake" used like bread), *couac* (flour), tapioca (to make sauce) or *cachiri* (bier). *Cachiri,* made from cassava mixed with sweet potato, is a drink of fundamental importance in Wayana culture. There are several types of *cachiri* depending on the varieties of cassava and sweet potato used. The diversity of product processing is based on diversity of species and cultivars, and therefore the associated cultural practices and knowledge system in many ways sustain this high varietal diversity.

Inter-relations between the cultivated area and adjacent ecosystem

A number of ecological interactions and ecosystem properties emerge from such diversified spatial and temporal crop arrangements. By enhancing plant diversity in the cultivated area, the system provides alternative habitat and food sources for many organisms that perform various beneficial ecological functions. In the same way, agricultural-natural ecosystem interfaces are of key significance and general ecological services are accrued by natural vegetation growing near the cultivated plot. Many plants

within or around traditional cropping systems are wild or weedy relatives of crops. In fact, farmers often favour certain weeds in or around their fields that have positive effects on soil and crops such as soil improvement and pest repellents, or that serve as food, medicines, ceremonial items, etc.

Many indigenous peoples of the Americas are highly integrated in their surrounding environment of which they feel they are part. There is a very close relationship between humans and nature. In the same way, there is no clear frontier between the domesticated and the wild, between the cultivated area and the surrounding forest. For instance, fallow areas are visited many years after abandoning them, to collect products (for instance fruits, or cassava tubers for more recent fallow) and plant material (*in situ* conservation of cassava genetic resources). They are sometimes also planted with fruit trees to attract game for hunting. On the other hand, wild plants in the forest are sometimes favoured in order to enhance production of non-wood forest products (fruits, bark or leaves for medicinal use, etc.) for further gathering.

Clearly, traditional agriculture commonly encompasses the multiple uses of both natural and artificial ecosystems, thus it appears that crop production plots and adjacent ecosystems are *de facto* integrated into a single ingenious agro-ecosystem.

Threats

As many other forest dwelling indigenous peoples, the Wayana are fragile to external influences. Timber production, the construction of roads and mining activities, lead to destruction or loss of access to resources, as well as to rapid social, cultural and economic changes which are difficult to manage for these communities. In many cases the introduction of new diseases has posed severe threats to these communities. Also endogenic changes

pose challenges to these communities and their production systems, including through the changing preferences and expectations of their members.

Maasai pastoralism: the cultural–ecological strategies of a people in a high-risk environment

The history of Maasai pastoralism is closely intertwined with the evolution of the savannah and highland landscapes of southern Kenya and northern Tanzania. These landscapes are world-renouned for their stunning views and rich wildlife. Tourist revenues from these areas benefit the national economies of the countries involved as well as private tourism companies all over the world. What is often overlooked, when policies are designed and implemented in these areas, is that these landscapes and their wildlife habitats were shaped over centuries by the knowledge, intensive and highly flexible nomadic pastoral strategy of the Maasai community.

The ecological rationale of an opportunistic strategy

The pastoral strategy is highly adaptive to the space and temporal fluctuations of the environment. By moving around herds of cattle, resources (pasture, water, salt) are used where and when they are most available. All habitats are used and there is no functional distinction between wild and agricultural lands. The Maasai have a complex strategy of customary arrangements to commonly manage and use these resources based on their extensive knowledge of the savannah and highland ecosystem. Their settlement patterns and associated social organization are built on the need to spread resource use over a large area to avoid concentration of livestock and consequent over-grazing. Their grazing strategies and burning techniques have turned bushland into pasture and controlled pests, thus also creating a habitat and food source for large wild grazers and their predators. In many ways the abundance of wildlife in these systems is largely due to the pastoral strategy. Over-grazing is sometimes wilfully applied to open up bush-invaded pasture again. The Maasai adjust their herd composition and size to the availability and carrying capacity of certain areas and availability of water (for example: dark cows get warmer in the sun and drink more!).

The human rationale of an opportunistic strategy

The Maasai manage to cope well with the great fluctuations of the environment (seasons, droughts), making the entire system more resilient and sustainable, while providing for their own food and livelihood needs. Their customary institutions for resource access ensure equitable use of resources, with high levels of reciprocity and social security for those who suffer misfortune, whilst being flexible to adjust to environmental circumstances. The many and complex exchanges of cattle taking place provide not only for a rich genetic diversity of cattle in each herd, but serve also as a social strategy to deal with hardship. The genetic heritage of cows is administrated through burning marks on cows, which also have many social, religious and artistic functions. Other users (ethnic groups, including agriculturists and hunter-gatherer groups) are allowed to live and use resources on Maasai territory, which is beneficial for the exchange of goods and services between social groups and livelihood systems, but this is also a potential source of conflict in times of scarcity.

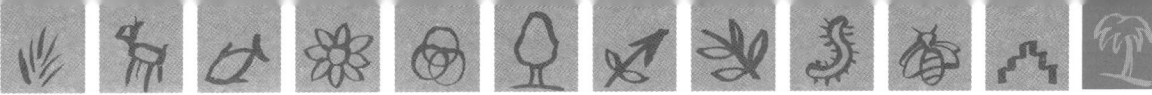

The knowledge base of pastoralism

The Maasai have an intense practical experience and rich knowledge of their environment and the ecological relations between various areas, which is accrued by moving around over large areas and passed on over many generations. They have a vast knowledge of plants and their food and medicinal purposes (human and animal), as well as of animal behaviour. This is borne from the necessity to be able to move their cattle safely through various areas and make use of the resources available in these areas, as they cannot be brought along whilst moving. This knowledge is safeguarded and passed on through many cultural institutions and expressions. One of the them is the considerable freedom of children to move around and discover their environment. Another crucial socio-cultural institution is the stage of warriorhood for young men, now highly in decline. This 3-7 year period combines intensive education by elders in livestock, ecology, social values, justice and leadership, with challenges, rituals and a "military service". The young warriors are expected to take care of themselves and to provide for their needs without the care of their mothers, challenging them to acquire knowledge of animals and plants and their uses, and building social networks with people outside their families. There are also many stories, jokes, sayings, riddles and other cultural expressions that convey knowledge of the environment and social values for the appropriate use thereof.

Threats

When British colonialists first arrived in the Rift Valley they perceived its landscape as a wild habitat. The presence of people and cattle was considered a threat to the landscape and its wildlife. Their background in a sedentary culture made them fail to see the inter-connections and rationale of the nomadic strategy and its role in creating and maintaining the landscape. They also failed to see the resource use efficiency of the pastoral systems when viewed from a larger space-temporal scale than the agricultural zone for a single all-year-around use. Many of these perceptions persist today. Wildlife conservationists and land use planners who are trained in land zoning and planning for a single use, continue to have rigid perceptions of how land and resources should be managed in space and time, with a clear separation of "wild" and "agricultural use" areas. This has consequences for policies, resource access legislation, institutional arrangements for land management and delivery of services, causing great disturbances to the pastoral-ecological dynamics, and the culture and social organization that underpins the system. These perceptions are materialized largely in land tenure legislation by creating restrictions to livestock movement, loss of access to key areas and resources, and subsequent and sometimes deliberate erosion of the culture of the Maasai. This in turn has negative effects on the capacity to deal with ecological risk, causing a decline in food and livelihood security, but also increasingly on wildlife abundance, through invasion of bush and pests on the shared habitats of livestock and wildlife. Many customary institutions for land management and access to resources have been delegitimized and/or replaced. Also, the open system of resource use is not sufficiently safeguarded against agricultural settlers (due to population pressures outside the system)

and land grabbing through corruption, which are both threats of a growing magnitude. HIV/AIDS is also an increasing problem, causing loss of leadership, parental care, labour force and knowledge.

Level of ingenious management

The Maasai pastoral system displays resource management ingenuity in their system of generation and transfer, customary laws for access and use of resources, their social benefit and risk-sharing arrangements and their opportunistic management of ecological space and temporal dynamics.

Global importance

The Maasai pastoral-ecological system provides many food, livelihood, social and ecosystem services. Its efficient use is of local, national and global importance for its use efficiency when measured on the space-temporal scale of entire systems (not necessarily of each separate sub-system), its knowledge of biodiversity and ecological dynamics, its unique and rich livestock genetic resources and its cultural heritage, its contribution to landscapes and associated wildlife that generate tourist revenue and its contribution to food security and risk management.

PROGRAMME DEVELOPMENT

The GIAHS programme was launched in August 2002 at a first stakeholder workshop. This workshop achieved a better and common understanding among partners and a more articulated profile of the GIAHS concept. More importantly, it developed a set of criteria and indicators for the selection of ten pilot sites for the second (PDF-B) phase of the programme. On these pilot sites action programmes will be developed for their support as well as for the development of a further methodology for the programme stages that will follow. Presently, the GIAHS team is developing the PDF-B project proposal and further developing the criteria and a consultative procedure for the selection of the ten pilot sites. FAO and its partners are soliciting proposals for GIAHS candidate systems and are inviting new partners to join the initiative.

Programme management

The GIAHS programme is implemented by FAO in close collaboration and partnership with selected member countries, representatives of local communities and indigenous peoples. Partners include UNDP, UNESCO, ICCROM, WHC, UNEP, CGIARs, IUCN, NGOs and other international institutions, universities, private sector and civil society organizations, as well as interested donors. At the first stakeholder workshop an international Steering Committee was formed that includes project partners, donors, NGOs, other UN Agencies, CGIAR centres, and will in the future include other stakeholders as well, such as governments and indigenous peoples. Technically the programme is supported in FAO by an inter-departmental

task force and internationally by a technical advisory body consisting of a broad range of stakeholders and experts.

System and site selection

Selection of systems and sites will be done on the basis of biophysical, socio-cultural, economic, and programme criteria. Their relative global, national and local importance will be taken into account. The comparative importance of the criteria and indicators cutting across social, ecological and economic aspects are considered the primary ones. The GIAHS programme focuses on the linkages among the socio-cultural and the biophysical-factors rather than viewing them in isolation; it should also include the national policy environment in its considerations. Equitable sharing of the benefits of these systems at different scales

is also considered important. Land tenure gender equity and sensitivity to indigenous and community issues are among the criteria for system and site selection.

To widen the sampling frame of GIAHS candidates and to build on related and field-tested methodologies of existing initiatives and candidate systems sites for the selection process, it was suggested that the GIAHS initiative would link into ongoing initiatives by GEF and other partner projects that address closely related issues, such as: the People Land Management and Environmental Change (PLEC) project, the *in situ* conservation activities of IPGRI and UNESCO's Man and the Biosphere (MAB) and World Heritage programmes.

The selection and valuation of systems will be done by devising a participatory process, to ensure the inclusion of the perspectives, knowledge and values of the different stakeholders. The final selection will be decided by the Steering Committee.

LIST OF PARTICIPANTS[1]

GOVERNMENT DELEGATES

CONGO, DEM. REP.
Modeste Mamingi-Mfudau
Chef de Division
Direction de la production et
protection des végétaux
Ministère de l'agriculture, pêche et élevage
Av. Batetela Kinshasa-Gombe
BP 8722 Kin 1
République démocratique du Congo
Tel: 243-8802519
Fax : 243-8802381
E-mail: minagri@yahoo.fr;
E-mail: modestemalu@yahoo.fr

LIBYA
Nuri Ibrahim Hasan
Ambassador
Permanent Representative of the
Socialist People's Libyan Arab
Jamahiriya to FAO
Via Nomentana, 365
00162 Rome, Italy
Tel: 39-06-8603880
Fax: 39-06-8603880
E-mail: nuribader@maktodo.com

MALAYSIA
Chan Han Hee
Director, Industrial Crops and Flowers
Division
Ministry of Agriculture
Jalan Sultan Salahuddin
Kuala Lumpur - Malaysia
Tel: 60-3-26925674
E-mail: doa09@pop.moa.my

TURKEY
Muzaffer Kiziltan
Deputy General Director
Ministry of Agriculture & Rural Affairs
General Directorate of Agricultural Research
Ist Yolu Bagdat Cd. No. 208
Y. Mahalle, Ankara
Tel: 90-312-3157629
Fax: 90-312-3153448
E-mail: muzaffer.kiziltan@ankara.tagem.gov.tr

INTERNATIONAL AGRICULTURAL RESEARCH CENTRES

International Plant Genetic Resources Institute
Via dei Tre Denari 472/a
00157 Maccarese - Italy
Tel: 39-06-6118212
Fax: 39-06-6197661

Coosje Hoogendoorn
Deputy Director-General
Tel: 99-06-61188200
E-mail: c.hoogendoorn@cgiar.org

Ehsan Dulloo
Genetic Resources Science and
Technology Group
E-mail: c/o Hodgkin t.hodgkin@cgiar.org

Irmgard Hoeschle-Zeledon
Coordinator
Global Facilitation Unit for
Underutilized Species
E-mail: i.zeledon@cgiar.org

Paul Bordoni
Scientific Assistant
Global Facilitation Unit for
Underutilized Species
Tel. 39-06-6118302
Fax: 39-06-61979661
E-mail: p.bordoni@cgiar.org

---

1 This list, reflecting only registered participants, is incomplete.

NON-GOVERNMENTAL ORGANIZATIONS

Agricultural Research Council/Range and Forage Institute

Hoare David
Private Bag X05, Lynn East
0039, South Africa
E-mail: dhoare@lantic.net

Association for the Development of Biodynamic Vegetable Breeding (Kultursaat)

Christina Henatsch
Auguste Victoria Street 4 D-61231
Bad Nauheim, Trantenrother Wg 25
D-58455 Witten
E-mail: Christina-Henatsch@gmx.de

Center of Biotechnology

Kampalappa Ramakrishnappa
Joint Director
Department of Horticulture
Ministry of Agriculture
P.B. No. 7648, Bannerghatta Rd.
Bangalore 560076
Karnataka, India
Tel: 91-80-6582784 - Fax: 91-80-6584906
E-mail: jdhhulimavu@vsnl.net

Centre national de liaison sur les ressources génétiques pour la FAO

Ckeikh Alassane Fall
Responsable du Laboratoire de
Biotechnologies Végétales
Institut Sénégalais de Recherches Agricoles
(ISRA-URCI)
B.P: 7461, Dakar - Senegal
Tel: 221-8324286
Fax: 221-8324286
Tel: 39-06-6118212
Fax: 39-06-6197661
E-mail: sene_grtkf@yahoo.fr

Committee for the Support of the Three United Nations Global Conventions

Biodiversity, Climate Change and
Combating Desertification
CA3C c/o FIDAF
Via Livenza, 6 - 00198 Roma (Italy)
Tel: 39-06-8416 036; 8417305
Fax: 39-06-8845960
E-mail: 3csc@3csc.net
Website: www.3csc.net

Gregory Lazarev
President
E-mail: g.lazarev@agora.stm.it

Emanuele Davia
Board member

Fabio Manzione
Board member

Lelio Bernardi
Board member

Luigi Rossi
Board member

Vittorio Ugga
Board member

ETC

Pat Mooney
ETC group Headquarters
478 River Avenue, Suite 200
Winnipeg MB R3L 0C8 - Canada
Tel: 204-4535259 - Fax: 204-2847871
E-mail: pat@etcgroup.org;
E-mail: etc@etcgroup.org

Indian Consultants Associates Private Ltd.

Subhash Mehta
19 Palace Road
Bangalore 560052
Karnataka - India
Tel: 91-80-2264174
E-mail: icap@vsnl.net

Intermediate Technology Development Group

Mulvany Patrick
Food Security Policy Adviser
Schumacher Centre
Bourton, RUGBY - CV23 9QZ, UK
Tel: 44-1926-634469 - Fax: 44-870-1275420
E-mail: Patrick_Mulvany@CompuServe.com;
E-mail: patrickm@itdg.org.uk
Website: www.ukabc.org; www.itdg.org

International Centre for Integrated Mountain Development (ICIMOD)

Uma Partap
4/80 Jawalakhel, PO Box 3226
Kathmandu, Nepal
Tel: 977-1-525313
Fax: 977-1-524509
E-mail: uma@icimod.org.np
Website : www.icimod.org

International Federation of Organic Agriculture Movements (IFOAM)

Cristina Grandi
IFOAM/FAO Liaison Office
Viale Libia, 22
00199 Roma, Italy
Tel. +39 06 86329403
Fax. +39 06 86385945
E-mail : c.grandi@ifoam.org
Website : www.ifoam.org

League for Pastoral Peoples c/o LIFE

Susanne Gura
Bunghofotr 116 - 53229 Bonn, Germany
Tel: 49-228-9480670
Fax: 49-228-9764777
E-mail: gura@dinse.net

Joyce Njoro
Chief Executive Officer
Community-based Livestock
Initiatives Programme
Box 1249
00606 Sarit Centre
Nairobi, Kenya
Tel/Fax : 254-2-2710083
E-mail : joyce.njoro@itdg.org.ke

Waris M.K. Warsi
Coordinator
LIFE Network
23 Park End - Vikas Marg
New Delhi 110092, India
Tel: 91-11-2024546
E-mail: wmkwarsi@yahoo.com

Rural Development Programme Kampong Thom

Peter Balzer
c/o German Development Service
P.O.Box 628 - Phnom Penh, Cambodia

FAO STAFF

Marcelino Avila
Rural Institutions and Participation Service
Rural Development Division
Sustainable Development Department

Nadine Azzu
Seed and Plant Genetic Resources Service
Plant Protection Division
Agriculture Department

Caterina Batello
Crop and Grassland Service
Plant Production and Protection Division
Agriculture Department

Julia Beckel
Environment and Natural Resource Service
Research, Extension and Training Division
Sustainable Development Department

David Boerma
Land and Plant Nutrition
Management Service
Land and Water Development Division
Agriculture Department

Sally Bunning
Land and Plant Nutrition
Management Service
Land and Water Development Division
Agriculture Department

Ricardo Cardellino
Animal Genetic Resources Group
Animal Production and Health Division
Agriculture Department

Linda Collette
Seed and Plant Genetic Resources Service
Plant Protection Division
Agriculture Department

Fréderic Dévé
Comparative Agricultural
Development Service
Agriculture and Economic Development
Analysis Division
Economic and Social Department

Nadia El-Hage Scialabba
Secretary,
Priority-Area for Interdisciplinary Action on
Biological Diversity for Food and Agriculture
Research, Extention and Trainng Division
Sustainable Development Department

Anton Ellenbroek
Animal Production Service,
Animal Production and Health Division,
Agriculture Department

José Esquinas
Secretariat for the Commission on
Genetic Resources for Food and Agriculture
Agriculture Department

Ariella Glinni
Emergency Operations Service
Field Operations Division
Technical Cooperation Department

Mathias Halwart
Inland Water Resources and
Aquaculture Service
Fishery Resources Division
Fisheries Department

Irene Hoffmann
Animal Production Service
Animal Production and Health Division
Agriculture Department

Samuel Jutzi
Animal Production and Health Division
Agriculture Department

Peter Kenmore
Chairperson,
Priority Area for Interdisciplinary Action on
Biological Diversity for Food and Agriculture
Plant Protection Division
Agriculture Department

Rainer Krell
Sustainable Development Department Group
Regional Office for Europe

Dietrich Leihner
Research, Extention and Trainng Division
Sustainable Development Department

Arturo Martinez
Seed and Plant Genetic Resources Service
Plant Production and Protection Division
Agriculture Department

Patricia Negreros-Castillo
Forest Conservation, Research and
Education Service
Forest Resources Division
Forestry Department

Fernando Patino
Forest Resources Development Service
Forest Resources Division
Forestry Department

Stephen Reynolds
Crop and Grassland Service
Plant Protection Division
Agriculture Department

William Settle
Integrated Pest Management Group
Plant Production and Protection Division
Agriculture Department

Mahmoud Solh
Plant Production and Protection Division
Agriculture Department

Clive Stannard
Secretariat of the Commission on Genetic
Resources for Food and Agriculture
Agriculture Department

Kim-Anh Tempelman
Animal Production Service
Animal Production and Health Division
Agriculture Department

Alvaro Toledo
Secretariat of the Commission on Genetic
Resources for Food and Agriculture
Agriculture Department

Peter Torrekens
Water Resources, Development and
Management Service
Land and Water Development Division
Agriculture Department

TuAnhThan Vu
Plant Protection Service
Plant Production and Protection Division
Agriculture Department

Douglas Williamson
Forest Conservation, Research and
Education Service
Forest Resources Division
Forestry Department

Printed on ecological paper